Mechanics and Thermodynamics of
Biomembranes, Evans and Skalak,
1980

DATE | ISSUED TO

Mechanics and Thermodynamics of Biomembranes

Authors

Evan A. Evans, Ph.D.

Professor
Department of Biomedical Engineering
Duke University
Durham, North Carolina

Richard Skalak, Ph.D.

Professor
Department of Civil Engineering
and Engineering Mechanics
Columbia University
New York, New York

CRC Press, Inc.
Boca Raton, Florida

Library of Congress Cataloging in Publication Data

Evans, Evan A
    Mechanics and thermodynamics of biomembranes.

    Bibliography: p.
    Includes index.
    1. Membranes (Biology)--Mechanical properties.
2. Membranes (Biology)--Thermal properties.
I. Skalak, Richard, joint author. II. Title.
QH601.E82   574.8'75   79-57619
ISBN 0-8493-0127-0

Direct all inquiries to CRC Press, 2000 N.W. 24th Street, Boca Raton, Florida, 33431.

© 1980 by CRC Press, Inc.

International Standard Book Number 0-8493-0127-0

Library of Congress Card Number 79-57619
Printed in the United States

PREFACE

This tutorial provides an introduction to the determination of mechanical properties of biological membranes and methods of analysis useful in their interpretation. These methods are based on fundamentals of continuum mechanics, thermodynamics, and mechanics of thin shells. The displacements and material deformations involved are often quite large; therefore, finite deformation theory is required. The molecular structure of biological membranes is distinctly anisotropic. Consequently, the approach taken here considers a membrane to be a continuous, two-dimensional material made up of a strata of molecular layers. Treatment of the membrane as a two-dimensional, or surface, continuum implicitly integrates over the molecular discontinuity in the membrane thickness dimension. The result is that properties are defined for the membrane as a surface composite material.

We give a brief personal perspective of the field of biomembrane mechanics. Most of the applications that are discussed have been developed only in the last few years. We establish first the more classical aspects of the analysis of surface deformation, rate of deformation, and mechanical equilibrium of a thin membrane material. Then, we derive elastic constitutive equations from thermodynamics of the membrane; these equations include surface elasticity (area compressibility and surface shear) and curvature, or bending, elasticity for multilayered membranes plus chemically induced curvature changes. Next, we demonstrate that thermoelasticity provides the decomposition of reversible mechanical work into internal energies and heats of deformation. We follow the aspects of reversible membrane deformation with a phenomenological treatment of irreversible processes that occur for different regimes of material behavior. These are abstractly represented by viscoelastic solid, semisolid relaxation and creep, and viscoplastic flow relations. Finally, we outline several contemporary experiments on biomembranes and demonstrate the previously developed methodology through the analysis of these experiments.

This article is intended primarily for engineering and physical scientists who are interested in the physical behavior and structure of biological membranes. It is hoped that the presentation will also be of interest to biologists, physiologists, and medical scientists for whom the mathematics and mechanics involved may be less familiar, but who are interested in the physical properties of biological membranes. For this reason, the treatment has been given in a detailed manner, without presupposing a knowledge of tensor theory.

This manuscript grew out of mutual research interests related particularly to the mechanical behavior of red blood cells and amphiphilic layer systems. We feel that a documented account of the modest degree of understanding which has been achieved will be useful to a wide variety of research workers. We hope that it may help to develop further the complementation of biological membrane sciences, as developed through the perspectives of anatomy, biochemistry, physiology, engineering, physics, and mathematics. The ultimate goal is to relate material behavior to biological function and molecular organization of membranes.

The account would not have been possible without the collaboration and encouragement of many teachers, colleagues and collaborators. Professor Y. C. Fung and Dr. Shu Chien have been especially important and helpful in lending guidance and support in many ways. Two particularly close and essential collaborators, Professor R. M. Hochmuth and Dr. P. LaCelle, have contributed significantly to progress in the field of red cell membrane mechanics and to the development of this manuscript. Many students have been instrumental in the research developments, particularly Dr. R. Waugh, Dr. R. P. Zarda, and Dr. A. Tözerin. The editorial assistance of Karen Buxbaum and helpful comments of Dr. S. Horan have been greatly appreciated, along with the participation of students at Duke University who have experienced most of this manuscript in classroom lectures. We thank Ellen Ray for her steady support throughout the typing of the manuscript. We also gratefully acknowledge the research support received over these past years from the National Institutes of Health, the National Science Foundation, and the American Heart Association.

The subject of biological membranes is a frontier of exploration of life processes and this account is necessarily fragmentary at this time. However, we hope that the assembly of the elements of the theoretical foundations and experimental studies presented here on biological membranes will serve to accelerate future developments.

E. A. Evans
R. Skalak
September 8, 1978

THE AUTHORS

Evan A. Evans, Ph.D., is currently a Professor in the Department of Biomedical Engineering, School of Engineering, Duke University, Durham, North Carolina. His teaching activities have included courses in continuum mechanics and thermodynamics at both the graduate and undergraduate levels. His general research interest is to determine the relationship between membrane material properties, membrane structure, and membrane surface-surface interaction associated with the molecular organization of lipids and proteins in membrane systems. His current research includes membrane adhesion and surface affinity, mechano-chemical (thermoelastic) properties of artificial membrane vesicles, lateral diffusion in anisotropic membrane liquids as a function of physical state. These developments are designed to investigate basic mechanisms involved in cell survival, cell surface recognition and removal by phagocytosing cells or agglutination reactions, and membrane material lesions.

Dr. Evans' research publications have been concerned with various aspects of biophysical science, e.g., image holography and super resolution in microscopy of cells, membrane thermodynamics and constitutive behavior, development of sensitive micromechanical techniques, and cell membrane-membrane adhesion. Dr. Evans is currently on the Editorial Board of the *Cell Biophysics* journal.

Dr. Evans was a National Science Foundation, NATO, Fellow and a recipient of a National Institutes of Health Research Career Development Award. He received his doctoral degree in engineering science from the University of California(San Diego).

Richard Skalak, Ph.D., Director of the Bioengineering Institute of Columbia University and James Kip Finch Professor of Engineering Mechanics, has been teaching fluid and solid mechanics for over 30 years. In the last 10 years he has been engaged in various biomechanics research projects including blood rheology, arterial wave propagation, and lung mechanics. He is author of over 80 research publications of which about half are in biomechanics. His current research is in rheology of red and white blood cells, growth phenomena, and lung elasticity.

Professor Skalak is an Associate Editor of the Journal of Biomechanical Engineering and past chairman of both the Applied Mechanics Division of the American Society of Mechanical Engineers and of the Engineering Mechanics Division of the American Society of Civil Engineers. He received his Ph.D. degree in Engineering Mechanics from Columbia University and has been teaching there since. He has studied as an NSF Postdoctoral Fellow at Cambridge University, England, and at the Department of Anatomy of Gothenburg University, Sweden. He is a member of a number of engineering societies and medical organizations and is currently President of Sigma Xi at Columbia University.

TABLE OF CONTENTS

Section I 1
1.1   Introduction 1

Section II 6
2.1   Intensive Membrane Deformation and Rate of Deformation 6
2.2   Deformation 8
2.3   Principal Axes 18
2.4   Invariants of Strain and Independent Deformation Variables 24
2.5   Rate of Deformation 30
2.6   Nonuniform Deformation and Rate of Deformation 41

Section III 46
3.1   Membrane Mechanical Equilibrium 46
3.2   Membrane Force Resultants 46
3.3   Membrane Equations of Mechanical Equilibrium 50
3.4   Mechanical Equilibrium of a Plane Surface Layer 50
3.5   Mechanical Equilibrium of an Axisymmetric Surface Layer 52
3.6   Mechanical Equilibrium of a Flat Membrane With Moment Resultants 62
3.7   Mechanical Equilibrium of an Axisymmetric Membrane With Moment Result-
       ants 65

Section IV 67
4.1   Membrane Thermodynamics and Constitutive Relations 67
4.2   Thermodynamic Outline 70
4.3   Equilibrium Thermodynamics and Membrane Deformation 72
4.4   Isothermal Constitutive Equations for Principal Axes 74
4.5   General Derivation of the Isothermal Constitutive Equations 81
4.6   Surface Pressure and the Tension-Free State 85
4.7   Shear Hyperelasticity: Example of 2-D Elastomer 91
4.8   Hyperelastic Membrane with Small Area Compressibility 97
4.9   Membrane Bending Moments for Coupled Molecular Layers 101
4.10   Chemically Induced Curvature and Moments 112
4.11   Thermoelasticity 117
4.12   Internal Dissipation, Viscosity, Viscoelasticity, Relaxation, and Viscoplastic-
       ity 129

Section V 141
5.1   Biological Membrane Experiments 141
5.2   Elastic Area Dilation Produced by Isotropic Tension 142
5.3   Elastic Extensional Deformation Produced by Membrane Shear 157
5.4   Elastic Curvature Changes: Bending versus Shear Rigidity in Osmotic Swelling
       of Flaccid Cells 169
5.5   Thermoelasticity and Thermodynamics of Cell Membranes 180
5.6   Thermoelasticity and Area Compressibility of Multilamellar Lipid Phases and
       Water 190
5.7   Viscoelastic Recovery of and Response to Membrane Extension 203
5.8   Relaxation and Viscoplastic Flow of Membrane 210

Epilogue 220

General Symbols 221

Appendix 225

References. 238

Index 243

SECTION I

1.1 Introduction

Experiments aimed at the determination of mechanical properties of biological membranes were begun in the 1930s using sea urchin eggs (Cole, 1932)[9] and subsequently, nucleated red blood cells (Norris, 1939).[9] These early experimentalists concluded that the typical cell membrane is a composite material made up of two molecular layers of lipids plus additional materials presumed to be proteins (Seifriz, 1926; Norris, 1939).[64,79] This general picture of the cell membrane ultrastructure has been confirmed and refined in recent years by advances in electron microscopy, biochemistry, and other sciences. Singer and Nicolson (1972)[83] examined the available evidence and presented the conceptual view that the cell membrane is a fluid mosaic in which the mobility of the molecules comprising the membrane is restricted to the plane of the membrane. Such a fluid mosaic would behave as a two-dimensional fluid layer. However, experiments also showed that cell membranes exhibit solid properties to some extent, e.g., elasticity. Since the lipids are in a fluid state, solid characteristics of the membrane must be attributed to connections of proteins and other molecules associated with the membrane (Evans and Hochmuth, 1977).[21] A schematic view of this type of material structure is illustrated for a red blood cell membrane composite in Figure 1.1. This figure idealizes the work of Marchesi et al. (1969, 1970),[57,58] which indicated that a protein called spectrin lies on the cytoplasmic face of the red blood cell membrane. The fluid mosaic model, therefore, was expanded to include the spectrin network (Steck, 1974; Singer, 1974).[82,86] In Figure 1.1, the spectrin network is shown as providing structural rigidity and support for the fluid component of the membrane and is often referred to as a "cytoskeleton." Biochemists are currently studying the association of the spectrin network with the outer lipid and protein mixture (Bennett and Branton, 1977).[2] In cells which are more complex than the red blood cell, a variety of additional structures may exist in association with the cell membrane, e.g., connective tissue, cytoplasmic elements, microtubules, etc. These can also provide structural rigidity and even active deformation. An example of a complicated membrane structure is that of the sea urchin egg. The sea urchin egg has a cortical layer on the order of 3 μm thick adjacent to the plasma membrane (Hiramoto, 1970),[42] and the surface of the membrane is formed into small studded projections (microvilli), which create a carpet-like appearance.

The anisotropic or lamellar configuration of thin membrane structures is peculiar to the preferential assembly of amphiphilic molecules (e.g., lipids, proteins, etc.) into multicomponent mixtures. Strongly anisotropic chemical behavior of an encapsulating membrane surface is evidenced by the very slow rate of exchange of membrane molecules with the adjacent aqueous phases. Consequently, the encapsulating membrane of a biological cell or artificial lipid vesicle behaves as a closed system with fixed mass for short periods of time (less than the order of hours). Thus, mechanical experiments can be used to probe the intact structure of the membrane; these experiments do work on the membrane material for limited time periods. Essentially all membrane mechanical studies related to cells have been done on red blood cells, amphiphilic monolayer and bilayers systems, and sea urchin eggs. Because of the relatively distinct separation of the membrane complex from cytoplasm and other neighboring media, red blood cell membranes and amphiphilic layer systems have been the most popular choices for use in physical experiments where the desire is to relate material structure of the membrane to its composition of lipids and proteins. Most of our examples will involve mechanical experiments on red cell membranes and amphiphilic layer systems, emphasizing the relationship between material properties and molecular structure; however, the general development and methods can be applied to more complicated membranes.

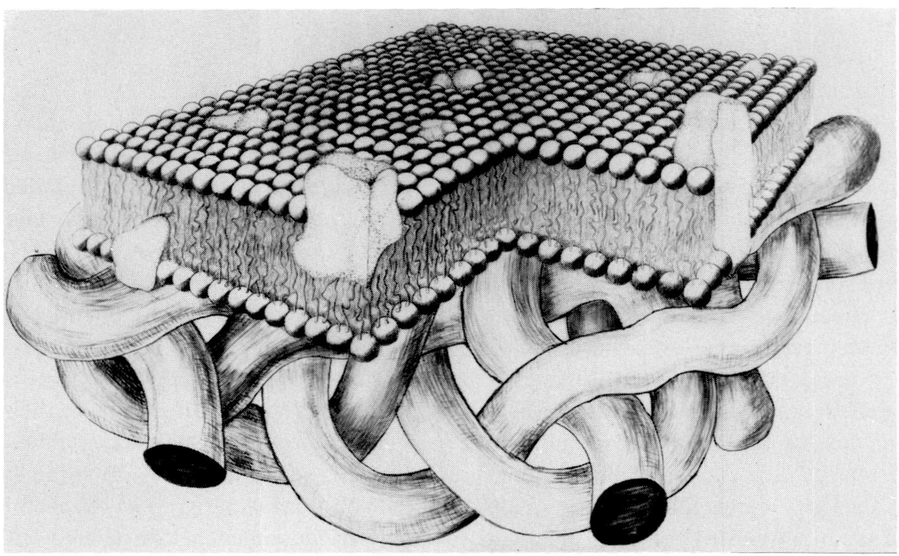

FIGURE 1.1.   An idealized view of the red blood cell membrane composite. (The underneath spectrin network provides structural rigidity and support for the fluid lipid-protein layer of the membrane.)

Mechanical properties characterize the structure of the membrane as a continuum. In other words, the scale in time and space over which these properties are measured must include a sufficient number of molecules such that the fluctuations due to the behavior of individual molecules are small. For a biological membrane, the material can only be considered as a continuum in the two dimensions, which describe the membrane surface. In the third dimension, thickness, the membrane exhibits molecular structure and discontinuity that preclude treatment as a continuous material in this direction. Consequently, membranes are appropriately represented as two-dimensional continua, with possible isotropy in the surface plane.* The fluctuation in a surface property is inversely proportional to the square root of the number of molecules considered within the scale of the continuum. Therefore, if a macromolecule occupies 100 $\mathring{A}^2$ of the membrane surface, then the surface scale would have to be on the order of 0.1 μm or greater to limit the fluctuation to less than 1%. Consequently, surface properties represent the effects integrated over the composite molecular structure in the thickness dimension and over a scale of fractions of a micrometer in the surface.

Although knowledge of the ultrastructure of cell membranes may suggest theoretical models, definitive evidence of membrane mechanical structure is provided only by direct experimental measurement. Conceptually, deformation is produced by applied force; then the deformation is related to the force as a function of time. These results are interpreted in terms of intrinsic properties such as elastic moduli and coefficients of viscosity. In practice, the small scale of the cell membrane makes the measurement of forces and membrane geometry extremely difficult and imprecise. Consequently, progress in membrane mechanics has not come easily. Developments in the field of red cell membrane mechanics provide insight into the difficulties that are encountered. This history shows that the determination of intrinsic mechanical properties is not a simple application of a mathematical formulism to the analysis of an experiment. On the contrary, considerable biophysical insight and creative thought are essential. The

---

*   Surface isotropy means that the material properties do not depend on the orientation of surface coordinates chosen at a specific location; however, properties may be nonuniform, i.e., depend on surface location.

following discussion gives some highlights and nucleating steps in the evolution of red cell membrane mechanics and thermodynamics. In Section V, we will describe and analyze many of the experiments referred to in this discussion.

The first estimates of an elastic modulus for the red blood cell membrane (Katchalsky et al., 1960)[47] were based on osmotic swelling experiments in which the transmembrane pressure had to be derived from the equations of chemical equilibrium. Based on these experiments and later on micropipet suction experiments (Rand and Burton, 1964; Rand, 1964),[70,71] the estimates of the so-called Young's elastic modulus of human red blood cells were on the order of 10^8 dyn/cm^2. These results also involved an estimate of the membrane thickness, which was assumed to be on the order of 10^{-6} cm, but could not be accurately determined. This particular difficulty arose because the investigators (Katchalsky et al; Rand and Burton)[47,71] treated the membrane material as continuous in the thickness dimension. Such approaches give elastic constants in stress units of force per unit cross-sectional area (dyn/cm^2). However, as we have emphasized from the outset, biological membranes are essentially continuous materials only in the two dimensions of the surface. Hence, for a membrane, the surface elastic constants have units of force per unit length (dyn/cm). In these terms, the elastic modulus obtained in the early experiments would be directly equal to the product of the Young's modulus times the thickness. Therefore, the surface elastic modulus based on the experiments of Katchalsky et al. (1960)[47] and Rand and Burton (1964)[71] would be on the order of 10^2 dyn/cm. On the other hand, Rand and Burton (1964)[71] also estimated the red blood cell membrane stiffness to be only on the order of 10^{-2} dyn/cm when the membrane was not forced to assume a spherical shape. The same low order of magnitude was obtained by Hochmuth and Mohandas (1972)[43] from experiments on red blood cells attached to glass and extended by the fluid shear stress of a controlled flow over the cells. The apparent variation of the elastic modulus of the red blood cell membrane by several orders of magnitude in different tests was a result of oversimplification of the analyses. The situation was remedied by recognizing that different material properties were being measured in each experiment, as will be explained below.

Fung (1966)[29] discussed the mechanical equilibrium and deformation of red blood cell membranes as a problem in the theory of thin shells. He pointed out that the effects of bending are likely to be small and that the red blood cell could change shape only to a limited extent without some stretch of the membrane surface. His discussion was based on the extensive developments of the classical theories of thin shells (see, for example, Flugge, 1973)[28] and differential geometry (see Struik, 1961).[87] In shell theory, small deformations and three dimensionally isotropic material properties are usually assumed. Removing the first restriction, Fung and Tong (1968)[31] introduced large deformation theory into an analysis of the osmotic sphering of red blood cells. However, even with the proper equations of large deformation and equilibrium, the results did not correlate with experimental results. The assumption of a three-dimensionally isotropic material was not an appropriate description of red cell membrane material structure.

The resolution of the widely different estimates of the red blood cell membrane elastic moduli was accomplished when the restrictions imposed on membrane behavior by its ultrastructure were recognized. A membrane such as shown in Figure 1.1 is highly anisotropic. It can be isotropic with respect to directions in the plane of the membrane but has a discontinuous, molecular structure in the membrane thickness direction. Further, experiments showed that it is difficult to increase the surface area and that area increases of a few percent resulted in rupture (Rand, 1964; Evans and Fung, 1972; and Chien et al., 1973).[8,18,71] On the other hand, large extensions were readily achieved at constant area (Hochmuth and Mohandas, 1972).[43] These observations and the apparently conflicting moduli were resolved by the introduction of elastic constitutive equations which separated the effects of changes in area from extensional deformations at constant area (Skalak et al., 1973; Evans, 1973).[15,85] These models suggested that the large moduli deduced from sphering experiments were associated with the large resistance to changes in membrane area. The small forces needed to stretch flaccid, unswollen cells were attributable to a small elastic modulus for membrane extension or shear deformation at constant area. Recent experiments and analyses have confirmed this view (Evans, 1973; Hochmuth et al., 1973; Evans and La-Celle, 1975; Evans et al., 1976; Evans and Waugh, 1977; Waugh, 1977; Waugh and Evans, 1978).[16,22a,25,27,44,91,93] The representation is something like a two-dimensional analog of rubber. Rubber greatly resists volume changes as indicated by its large volumetric compressibility modulus, but is capable of very large extensions at nearly constant volume due to its comparatively low shear modulus. However, rubber in the form of a thin sheet is very different from biological membranes because the surface area of rubber sheet can be easily increased with commensurate decrease in thickness (see Green and Adkins, 1970).[35]

Another important aspect of the behavior of biological membranes, which is different from that of macroscopic isotropic materials, is the strong effects of the chemical environment on the shape and properties of cell membranes. In the case of the red blood cell, a range of shapes is observed, including the normal biconcave shape (discocyte), cup shape (stomatocyte), tightly crenated spherical shape (echinocyte), and many other variations which depend on the chemical environment surrounding the cell. Reversible changes of red blood cell shape were first described by Hamburger in 1895.[36] In the 1930s, Ponder (1971)[68] accumulated a vast amount of information on the multiplicity of agents and conditions which induce these transformations. Extensive observations of the specific chemical agents and the shapes that they produce have been assembled by Bessis (1973).[3]

The biconcave disk shape of red blood cells is considered the normal or natural state because it is commonly observed under normal conditions in blood samples from healthy individuals. In addition, micropipet experiments and theoretical considerations indicate that the membrane is unstressed or at very low stress levels in the biconcave state. The pressures inside and outside of the cell cannot be appreciably different in the biconcave state (Fung, 1966).[29] The extraordinary symmetry and smoothness of the normal red cell discocyte have stimulated a great deal of interest, study, and speculation. The normal red cell has a surface area which is about 40% in excess of that required to enclose a sphere of the same volume as the cell. If one imagines the red blood cell being formed by collapsing a larger sphere or by inflating a flat double sheet of membrane material, it is evident that considerable bending or curvature alterations of the membrane must take place. Such ideas led Canham (1970)[5] and Lew (1972)[54] to postulate that the discocyte was a configuration of minimum bending energy. Computations based on this concept have also been carried out by Zarda (1974),[99] using realistic moduli for the membrane properties. The computations show that the biconcave discocyte could in fact be produced by partial deflation of an unstressed sphere and that only a small negative pressure would be required to maintain the shape. However, the process of formation of mature red blood cells in vivo is not a simple deflation of a sphere (Bessis, 1973).[3] The red cell experiences a complicated history of dynamic shear stresses in the circulation. Consequently, the biconcave shape must be regarded only as a reference geometry, not to be confused with the intrinsic material structure.

As Fung (1966)[29] pointed out, bending effects are often negligible compared to the effects of membrane force resultants (tensions) in large deformations of red blood cells. This is indeed the case in micropipet experiments and large deformations produced by fluid shear. However, as Fung also noted, many situations arise in which tensions are small, and bending moments are essential to maintain the stability of the observed shapes. An example is the series of shape changes of the red blood cell during intermediate states of osmotic swelling before the spherical state is reached (Zarda, 1974; Zarda et al., 1977).[99,100] The bending energy and intrinsic rigidity of bilayer membranes have been related to the physical properties of the component monolayers by Evans (1974)[17] and discussed by Evans and Hochmuth (1978)[22] for multilayer membranes. In addition, it was shown that alterations in chemical equilibrium of one or more of the layers could induce bending moments and changes in curvature of the membrane as a whole. Similar ideas led Helfrich (1973, 1974)[37,38] and Dueling et al. (1974)[13a] to postulate the existence of a "curvature elastic energy" which is similar to bending energy in shell theory. Chemically induced changes in curvature correspond to changes in the "spontaneous curvature" introduced by Helfrich (1973).[37] This spontaneous curvature may be regarded as the curvature which specifies the minimum energy configuration of any portion of a membrane.

Quantitative correlation of material behavior with membrane chemistry is provided through thermodynamics. Ideally, this would be carried out in a calorimeter that would measure the heat exchange from the membrane when deformed by external forces. Since it is impossible to construct a calorimeter to measure the heat exchange that occurs during micromechanical experiments on cell membranes, the thermoelastic behavior of red cells is being used to deduce the relative changes in internal energy and heat content that are produced by membrane deformation (Evans and Waugh, 1978).[26] This is possible because the elastic properties of a closed membrane system are associated with reversible thermodynamic changes in the membrane that are produced by deformation. The elastic coefficients are derivatives of the free energy density at constant temperature (work per unit area of the membrane) taken with respect to intensive deformation, i.e., change in surface density and in-plane extension at constant surface density (Evans and Waugh, 1977; Evans and Hochmuth, 1978).[22,24] Temperature-dependent mechanical experiments provide data for decomposition of membrane thermodynamic potential changes into internal energy and configurational entropy contributions (Evans and Waugh, 1977; Evans and Waugh, 1978).[24,26] Comparison of well-defined chemical systems (i.e., vesicles made of specific combinations of lipids, proteins, etc.) with composite, cell membrane systems gives quantitative data on the chemical state of the composite or natural membrane "mixture". For example, we can determine whether the configurational state of molecular complexes is more or less ordered by the deformation and whether the energy of these complexes is appreciably changed by the deformation. For a biological membrane or amphiphilic layer system, thermoelasticity can provide a direct assessment of thermal repulsive forces and natural cohesive forces in the membrane or layer.

The time-dependent deformation of biological membranes under applied forces has also been explored to some extent. This dissipative behavior is represented by viscoelastic and viscoplastic models for membrane rheology. Viscoelastic models for red cell lysis were proposed by Katchalsky et al. (1960)[47] and Rand (1964).[70] However, the unusually large viscosities derived from their osmotic lysis and micropipet experiments (surface viscosities in the range of 10 to 1000 dyn-sec/cm) do not represent a viscous flow process. On the contrary, the results are derived from the temporal dependence of the lytic phenomenon itself, without any perceptible rate of deformation of the surface. Viscoelastic recovery of red blood cells was observed by Hoeber and Hochmuth (1970)[46] in red cells expelled suddenly from micropipets. The time constants involved were on the order of 0.1 sec. This measurement has been combined with known elastic properties and a proposed viscoelastic model to estimate that the surface shear viscosity for the red blood cell membrane is on the order of 10^{-3} dyn-sec/cm (Evans and Hochmuth, 1976; Waugh and Evans, 1976).[19,92] Recently, viscoelastic recovery of red cell extensional deformation has corroborated the earlier estimate (Hochmuth et al., 1978).[45] Irrecoverable deformations of red blood cell membranes have been studied, e.g., plastic behavior produced by membrane shear forces that exceed yield (Evans and Hochmuth, 1976)[20] and the time-dependent relaxation from solid to plastic behavior (Evans and LaCelle, 1975).[22a] However, these phenomena are complex, and the data only represent an initial stage of investigation.

The historical notes above have focused on experiments and theoretical developments connected with red cell membranes. This is because many of the advances and the motivation for the membrane experiments originated in studies of blood rheology. However, the basic mathematical descriptions and the physical concepts employed are applicable to other cell membranes and macroscopic biological membranes as well. With regard to the relation of membrane properties to the gross rheology of blood, there are several recent reviews which summarize the information and theories (Chien et al., 1973 and 1975; Goldsmith and Skalak, 1975).[6,8,34] On the other hand, an entirely separate development has taken place in the study of the membrane-cortex complex of sea urchin eggs. Even though membrane mechanics has its roots in the early experiments of Cole (1932)[9] on sea urchin eggs, the recent work on red cell membranes and amphiphilic layer systems has proceeded independently of the investigations on the sea urchin egg. Nevertheless, extensive mechanical studies have been underway on the sea urchin egg. Several notable early references include the works of Mitchison and Swann (1954),[61] who developed a micropipet aspiration technique, followed by Hiramoto (1963)[41] and Yoneda (1964),[96] both of whom used the compression of the egg between two plates as an experimental method (originally developed by Cole). Even osmotic swelling was used to investigate the mechanical properties (Mela, 1967).[60] Excellent articles have appeared in recent years by Hiramoto (1970)[42] and Yoneda (1972),[97] which give insight into the interests and approaches of investigators in this field. Hiramoto (1970)[42] describes an innovative experiment utilizing a small magnetic particle to test the mechanical properties of different cellular components, including the membrane cortex. The most significant aspect of the membrane cortex of the sea urchin egg (other than its obvious complexity) is the dynamic changes which occur in its properties as a result of the life processes of the cell, i.e., from fertilization through subsequent division. As outsiders to this field, we cannot presume to contribute significantly to the efforts currently in progress on the biomechanics of sea urchin egg membranes; however, we hope to suggest an approach to the development of membrane constitutive relations which can be of use to scientists interested in complicated membranes like that of the sea urchin egg. We will make this suggestion in the form of examples in Section V, where we will analyze a sea urchin egg compression experiment and present a simple model for the magnetic particle experiment of Hiramoto (1970).[42]

SECTION II

2.1 Intensive Membrane Deformation and Rate of Deformation

Experimentally, the determination of material properties of biological membranes (e.g., elastic moduli and viscosity coefficients) involves applying prescribed forces and observing the resulting change in shape of the membrane and the time rate of change of membrane conformation. As we discussed in the introduction, cell membrane materials can only be treated as continuous media in the two dimensions that characterize the surface. Usually, the radii of curvature which describe the cellular envelope are much larger than the thickness of the membrane structure. Consequently, the membrane can be considered to be a thin surface material (as its name implies). Both deformation and rate of deformation are quantitatively analyzed in terms of alterations in

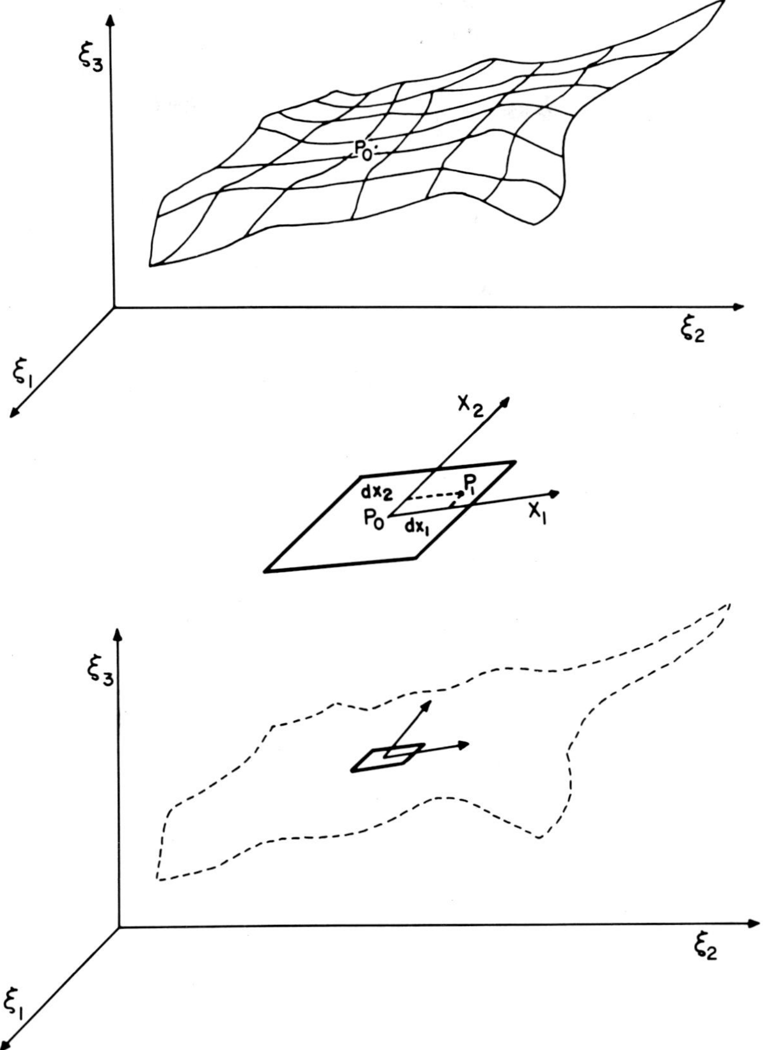

FIGURE 2.1. A conceptual view of the membrane surface as a mosaic of small elements. Although the entire surface may undulate in space, each element of membrane is sufficiently small that we may consider it approximately flat.

the surface geometry of the encapsulating membrane. For this purpose, we conceptualize the surface as a mosaic of differential elements that are sufficiently small to be approximately "flat" when compared to the overall terrain of the surface (see Figure 2.1). The coordinates are considered locally flat, but may traverse ridges and valleys of the surface like roads in the countryside, becoming curvilinear coordinates in general. The relative displacement in time of the material points in each small element specifies the intrinsic deformation of the surface, and the time rate of change specifies the rate of deformation. We will show that the intrinsic deformation of surface material elements can be represented by a sequence of geometric alterations: a dilation or condensation of surface area followed by extension of the surface element at constant area. Similarly, the rate of deformation of the surface can be decomposed into the fractional rate of area change and the fractional rate of extension at constant area.

Most membrane materials are essentially laminated structures. Since the distance between layers are of the order of molecular dimensions (much smaller than the scale of the membrane continuum and radii of curvature), each layer experiences nearly the same surface deformation and rate of deformation. However, deviations from the mean membrane surface deformation are produced within the layers of the surface composite by changes of curvature. The small variations are determined by the inter-layer distances and changes in curvatures. We will consider these effects in Section IV when we investigate curvature elastic phenomena.

2.2 Deformation

The instantaneous position, P_0, of a single element of the surface mosaic is specified by its cartesian coordinates (ξ_1, ξ_2, ξ_3) in space. There are two orthogonal axes, x_1 and x_2, that form a plane tangent to the surface at the point, P_0. Positions in the tangent plane are given by independent coordinates that are functions of the coordinates in space,

$$x_i = x_i(\xi_1, \xi_2, \xi_3) \tag{2.2.1}$$

Here, the index subscript, i, is used to catalog the two tangent plane coordinates (x_1, x_2). Equation 2.2.1 represents two independent functions, one for each plane coordinate. For the simple case of an infinite plane, the tangent plane coordinates may be identified with the spatial coordinates (ξ_1, ξ_2). However, at a specific point, P_0, on a curved surface, only a small region local to the point can be treated as coincident with the tangent plane. In this differential region around P_0, the axes, x_1 and x_2, provide orthogonal bases for determining the locations of neighboring points such as P_1 in Figure 2.1. Therefore, in the elemental region surrounding P_0, any other point, say P_1, is specified by the infinitesimal cartesian coordinates (dx_1, dx_2). The scalar distance between P_1 and P_0 is given by ds,

$$ds = \sqrt{dx_1{}^2 + dx_2{}^2} \tag{2.2.2}$$

Since ds is the measure of distance between local points on the surface, changes in ds will specify the intrinsic deformation of the surface continuum local to P_0 and subsequently over the whole surface. This measure or surface "ruler", Equation 2.2.2, can be written in a quadratic form called a metric,

$$ds^2 = dx_1{}^2 + dx_2{}^2 = dx_k\,dx_k \tag{2.2.3}$$

Here, the repeated subscript is an implicit rule for product summation,

$$dx_k\,dx_k \equiv \sum_{k=1}^{2} dx_k\,dx_k = dx_1\,dx_1 + dx_2\,dx_2$$

(This shorthand summation convention will be employed throughout the manuscript. It is important to recognize that the scalar quantity, ds^2, is independent of the subscript, k, and that such "dummy" indices should not be confused with independent subscripts that label individual quantities as in Equation 2.2.1.)

Because deformation is determined by changes in the distance between points, we will introduce a reference state called the initial state. In the initial state, locations in a plane tangent to the surface at P_0 are defined by coordinates, a_i. These are functions of the spatial coordinates as well,

$$a_i = a_i(\xi_1, \xi_2, \xi_3) \tag{2.2.4}$$

The initial distance between local points, P_0 and P_1, in the differential surface element is obtained in an analogous manner to Equation 2.2.2 by assuming again that the da_i are infinitesimal cartesian coordinates,

$$ds_0 = \sqrt{da_1{}^2 + da_2{}^2}$$

The initial metric is given by the sum,

$$ds_0{}^2 = da_k\, da_k \qquad\qquad (2.2.5)$$

The difference between the two metrics, Equations 2.2.3 and 2.2.5, gives the change in length of the line segment $(P_0\, P_1)$ as the surface deforms from the initial to the instantaneous state. Specifying such changes for all points in the neighborhood of P_0 determines the total deformation in the tangent plane at P_0. This is the local deformation in the domain of a point on the surface. In such a manner, deformation is defined intensively, and the degree of deformation will be quantified by a set of functions that are continuous over the entire surface envelope, i.e., from point to point.

The instantaneous and initial surface coordinates are assumed to be analytically related in the vicinity of the point, P_0, because the surface is considered to be a two-dimensional continuum. This means that a functional relationship exists between the two sets of tangent plane positions:

$$x_i = x_i(a_1, a_2)$$

and,

$$a_i = a_i(x_1, x_2)$$

Consequently, in the elemental region close to P_0, locations (differential coordinates) of neighboring surface points are simply related by the chain rule of differentiation: first, the instantaneous location in the deformed state is related to the initial location by

$$dx_i = \frac{\partial x_i}{\partial a_j}\, da_j \qquad\qquad (2.2.6)$$

or in full,

$$dx_1 = \frac{\partial x_1}{\partial a_1}\, da_1 + \frac{\partial x_1}{\partial a_2}\, da_2$$

$$dx_2 = \frac{\partial x_2}{\partial a_1}\, da_1 + \frac{\partial x_2}{\partial a_2}\, da_2$$

and second, the initial location in the undeformed state is related to the instantaneous location by

$$da_i = \frac{\partial a_i}{\partial x_j}\, dx_j \qquad\qquad (2.2.7)$$

The initial and instantaneous metrics may be expressed using Equations 2.2.6 and 2.2.7 as linear transformations for the differential coordinates,

$$ds^2 = \frac{\partial x_k}{\partial a_i} \frac{\partial x_k}{\partial a_j} \, da_i \, da_j \tag{2.2.8}$$

$$ds_0^{\,2} = \frac{\partial a_k}{\partial x_i} \frac{\partial a_k}{\partial x_j} \, dx_i \, dx_j \tag{2.2.9}$$

Each differential in the original products must be expanded by the chain rule, where the summation of repeated subscripts is over (1,2). (Note: the commutation of algebraic functions has been employed in Equations 2.2.8 and 2.2.9 , e.g.,

$$ds^2 = \left(\frac{\partial x_k}{\partial a_i} \, da_i \right) \left(\frac{\partial x_k}{\partial a_j} \, da_j \right) \equiv \frac{\partial x_k}{\partial a_i} \frac{\partial x_k}{\partial a_j} \, da_i \, da_j$$

The difference between the two metrics may be expressed in terms of the initial and the instantaneous surface coordinates,

$$ds^2 - ds_0^{\,2} = \left(\frac{\partial x_k}{\partial a_i} \frac{\partial x_k}{\partial a_j} - \delta_{ij} \right) da_i \, da_j \tag{2.2.10}$$

$$ds^2 - ds_0^{\,2} = \left(\delta_{ij} - \frac{\partial a_k}{\partial x_i} \frac{\partial a_k}{\partial x_j} \right) dx_i \, dx_j \tag{2.2.11}$$

where the identity matrix or Kronecker delta, δ_{ij}, has been introduced. (It is defined by $\delta_{ij} = 0$ for $i \neq j$ and $\delta_{ij} = 1$ for $i = j$.) Equations 2.2.10 and 2.2.11 suggest the definition of strain components, ε_{ij} and e_{ij},

$$ds^2 - ds_0^{\,2} = 2\epsilon_{ij} \, da_i \, da_j$$

$$ds^2 - ds_0^{\,2} = 2e_{ij} \, dx_i \, dx_j$$

where

$$\epsilon_{ij} \equiv \tfrac{1}{2} \left(\frac{\partial x_k}{\partial a_i} \frac{\partial x_k}{\partial a_j} - \delta_{ij} \right) \tag{2.2.12}$$

$$e_{ij} \equiv \tfrac{1}{2} \left(\delta_{ij} - \frac{\partial a_k}{\partial x_i} \frac{\partial a_k}{\partial x_j} \right) \tag{2.2.13}$$

ε_{ij} is called the Green's strain tensor or material strain. e_{ij} is called the Almansi strain tensor or spatial strain tensor (Prager, 1961).[69] The strain components quantitatively determine the deformation local to P_0. They are four in number for the surface and may be written as a matrix. The strain matrix is always symmetric. The strain components are a way of cataloging the partial derivatives that represent the deformation. The relative changes in element dimensions may be related to the strain components most conveniently from a particular perspective called the ''principal axes system''. This perspective can always be selected by rotation of the axes about P_0. As will be shown from the development of the principal axes system, any surface deformation can be simply decomposed into a change in the area of the element and an extension of the element at constant area.

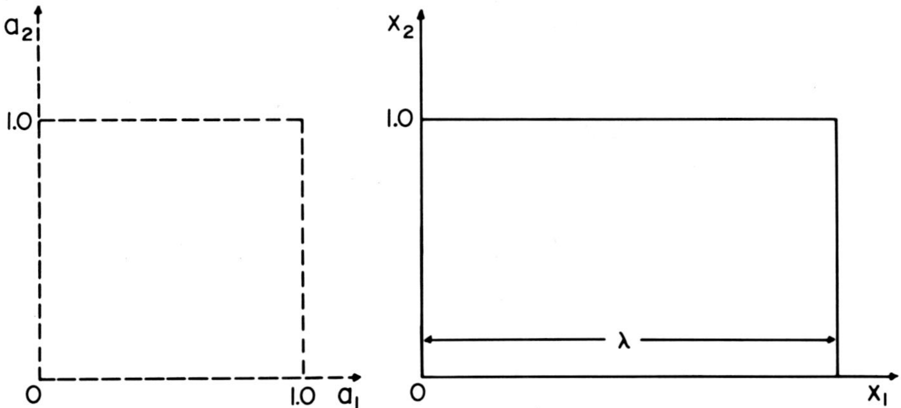

FIGURE 2.2.   Extension of a plane, square element along the a_1 axis with no change in the a_2 direction.

It is apparent from Equations 2.2.10 and 2.2.11 that the changes in point-to-point distances (determined by the metric) may be defined relative to either the initial or the instantaneous state. This distinction disappears when infinitesimal deformations are involved. In much of classical physics, it is common to use the initial state as a reference for the deformation and variables are called material or Lagrangian. On the other hand, in fluid dynamics where the initial state is difficult (if not impossible) to define, the instantaneous state becomes the reference and the corresponding variables are referred to as spatial or Eulerian. Therefore, the strain representations, ε_{ij} and e_{ij}, are called the Lagrangian and Eulerian strain, respectively. If the strain matrices are identically zero, then no deformation has occurred because all local points remain the same distance from P_0.

The following examples demonstrate the analysis of deformation. The first two examples utilize plane geometry to illustrate characteristics of the strain components. The last two examples are analogs of the first two for an axisymmetric, curvilinear surface element (i.e., conical annulus). The latter examples show the application of the deformation analysis to axisymmetric curved surfaces. General surfaces without symmetry may be similarly treated, but the geometry may create algebraic complexity without a commensurate increase in the insight into the analysis of deformation. Therefore, we have chosen to treat the axisymmetric case and direct readers to references related to differential geometry for more general equations (e.g., Struik, 1961, Naghdi, 1972).[63,87]

Example 1

Consider the extension of a plane, square element along the a_1 axis with no change in the a_2 direction as shown in Figure 2.2. The components of the strain matrices are calculated from the relations between (x_1, x_2) and (a_1, a_2) given by

$$x_1 = \lambda_1 \, a_1$$

$$a_1 = \frac{x_1}{\lambda_1}$$

$$x_2 = a_2$$

$$a_2 = x_2$$

Constant λ_1 specifies the extension. The Lagrangian strain components are obtained from Equation 2.2.12,

$$\epsilon_{11} = \frac{1}{2}\left(\frac{\partial x_1}{\partial a_1}\frac{\partial x_1}{\partial a_1} + \frac{\partial x_2}{\partial a_1}\frac{\partial x_2}{\partial a_1} - 1\right) = \frac{1}{2}(\lambda_1^2 - 1)$$

$$\epsilon_{22} = \frac{1}{2}\left(\frac{\partial x_1}{\partial a_2}\frac{\partial x_1}{\partial a_2} + \frac{\partial x_2}{\partial a_2}\frac{\partial x_2}{\partial a_2} - 1\right) = 0$$

$$\epsilon_{12} = \epsilon_{21} = \frac{1}{2}\left(\frac{\partial x_1}{\partial a_1}\frac{\partial x_1}{\partial a_2} + \frac{\partial x_2}{\partial a_1}\frac{\partial x_2}{\partial a_2}\right) = 0$$

and similarly for the Eulerian strain components from Equation 2.2.13,

$$e_{11} = \frac{1}{2}\left(1 - \frac{\partial a_1}{\partial x_1}\frac{\partial a_1}{\partial x_1} - \frac{\partial a_2}{\partial x_1}\frac{\partial a_2}{\partial x_1}\right) = \frac{1}{2}(1 - \lambda_1^{-2})$$

$$e_{22} = \frac{1}{2}\left(1 - \frac{\partial a_1}{\partial x_2}\frac{\partial a_1}{\partial x_2} - \frac{\partial a_2}{\partial x_2}\frac{\partial a_2}{\partial x_2}\right) = 0$$

$$e_{12} = e_{21} = \frac{1}{2}\left(-\frac{\partial a_1}{\partial x_1}\frac{\partial a_1}{\partial x_2} - \frac{\partial a_2}{\partial x_1}\frac{\partial a_2}{\partial x_2}\right) = 0$$

If a simultaneous extension in the a_2 direction is included, then the relation between x_1 and a_1 would remain the same; but for the a_2 direction, we have

$$x_2 = \lambda_2 a_2$$

$$a_2 = \frac{x_2}{\lambda_2}$$

and the strain components are

$$\epsilon_{11} = \frac{1}{2}(\lambda_1^2 - 1)$$

$$\epsilon_{22} = \frac{1}{2}(\lambda_2^2 - 1)$$

$$\epsilon_{12} = 0$$

$$e_{11} = \frac{1}{2}(1 - \lambda_1^{-2})$$

$$e_{22} = \frac{1}{2}(1 - \lambda_2^{-2})$$

$$e_{12} = 0$$

We associate these components with the principal axes system because $e_{12} = \epsilon_{12} = 0$. Elements parallel to either axis remain parallel and are extended or compressed independently of the other coordinate.

If the extension ratio, say λ_1, is close to unity (i.e., small deformation), the nonzero strain components are approximately given by

$$\epsilon_{11} = \frac{(1 + 2\epsilon + \epsilon^2) - 1}{2} \simeq \epsilon$$

$$e_{11} = \frac{1 - (1 - 2\epsilon + \epsilon^2 - \ldots)}{2} \simeq \epsilon$$

where the deviation of λ from unity is ϵ. A Taylor series expansion was used in the Eulerian strain component e_{11}. It is apparent for small deformations that the distinction between Lagrangian and Eulerian variables disappears and that the principal strain is simply the fractional change in length. When ϵ is greater than 10%, the small deformation approximation (the equivalence of Lagrangian and Eulerian strain components) progressively deteriorates.

Example 2

Consider the shear deformation of a square element, as shown in Figure 2.3, with no displacements in the a_2 direction. For this deformation,

$$x_1 = a_1 + a_2 \tan\phi$$

$$a_1 = x_1 - x_2 \tan\phi$$

$$x_2 = a_2$$

$$a_2 = x_2$$

The Lagrangian and Eulerian strain components are again calculated from Equations 2.2.12 and 2.2.13,

$$\epsilon_{11} = 0$$

$$\epsilon_{22} = \frac{\tan^2\phi}{2}$$

$$\epsilon_{12} = \epsilon_{21} = \frac{\tan\phi}{2}$$

$$e_{11} = 0$$

$$e_{22} = -\frac{\tan^2\phi}{2}$$

$$e_{12} = e_{21} = \frac{\tan\phi}{2}$$

For small shear deformation angles, $\tan\phi \simeq \phi$ and $\phi^2 \ll 1$; thus, the above relations reduce to the classical infinitesimal shear components,

$$\epsilon_{11} = e_{11} = 0$$

$$\epsilon_{22} = e_{22} \simeq 0$$

$$\epsilon_{12} = \epsilon_{21} = e_{12} = e_{21} \simeq \frac{\phi}{2}$$

In general, ϵ_{12} and e_{12} are called the shear strain components.

Returning to finite deformations, if we allow for simultaneous extension and

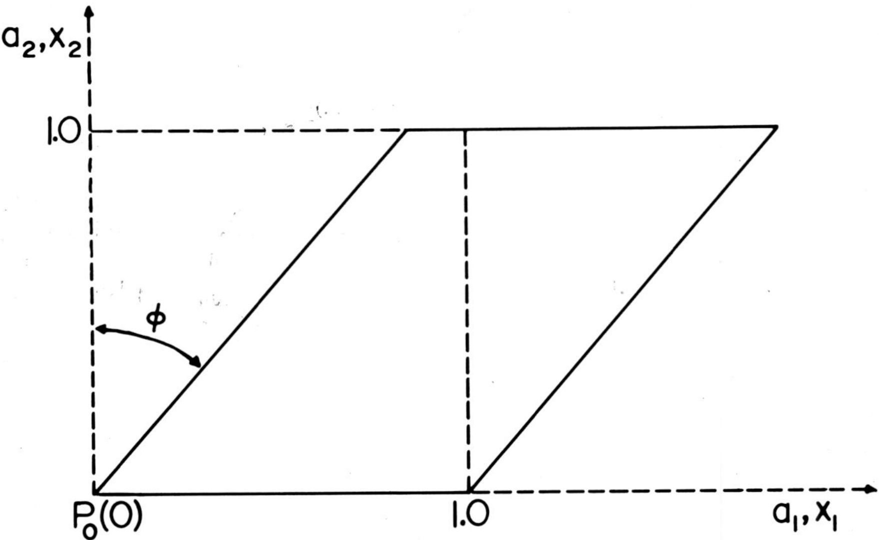

FIGURE 2.3. Simple shear deformation of a plane, square element. The shear deformation angle, ϕ, describes this deformation.

compression along the coordinate axes in Figure 2.3, the coordinate relationships would be given by

$$x_1 = \lambda_x\, a_1 + \lambda_y\, a_2\, \tan\phi$$

$$a_1 = \frac{x_1}{\lambda_x} - x_2\, \frac{\tan\phi}{\lambda_x}$$

$$x_2 = \lambda_y\, a_2$$

$$a_2 = \frac{x_2}{\lambda_y}$$

where (λ_x, λ_y) are the scaling parameters of line elements parallel to the independent axes. The parallelogram in the deformed state is scaled geometrically by the parameters (λ_x, λ_y) for a fixed angle, ϕ. The strain components for extension plus the shear are calculated from Equations 2.2.12 and 2.2.13,

$$\epsilon_{11} = \tfrac{1}{2}\,(\lambda_x^2 - 1)$$

$$e_{11} = \tfrac{1}{2}\,(1 - \lambda_x^{-2})$$

$$\epsilon_{22} = \tfrac{1}{2}\,[\lambda_y^2\,(1 + \tan^2\phi) - 1]$$

$$e_{22} = \tfrac{1}{2}\,[1 - \lambda_y^{-2} - \lambda_x^{-2}\tan^2\phi]$$

$$\epsilon_{12} = \epsilon_{21} = \frac{\lambda_x\,\lambda_y}{2}\,\tan\phi$$

$$e_{12} = e_{21} = \tfrac{1}{2}\,\lambda_x^{-2}\,\tan\phi$$

$$(2.2.14)$$

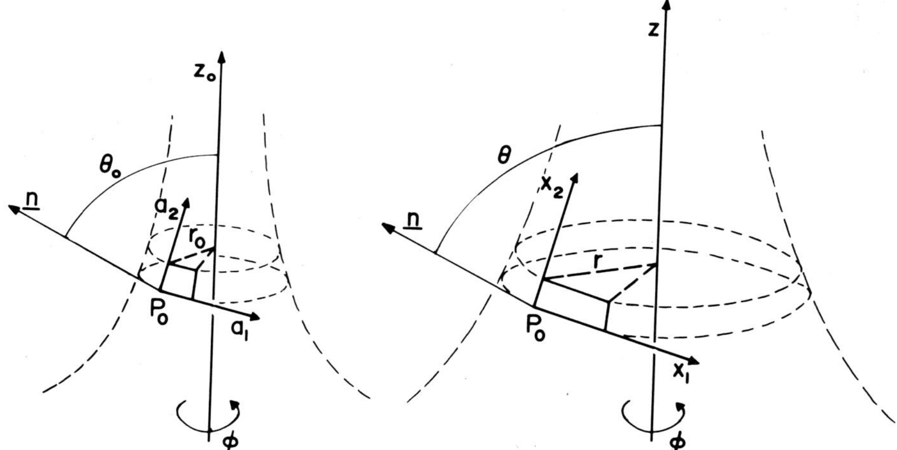

FIGURE 2.4.   Radial expansion of an axially symmetric surface. The curvilinear distance along the meridian is unchanged for corresponding points on the initial and instantaneous surfaces. Elements defined by da_i and dx_i are sufficiently small that we may consider them flat and rectangular.

These expressions reduce the previous set for $\lambda_x = \lambda_y = 1$. The Equations 2.2.14 specify any uniform deformation (i.e., deformations where initially parallel lines remain parallel). For example, consider a deformation which occurs at constant element area. In this case, the area of the parallelogram remains constant and $\lambda_x \, \lambda_y = 1$; Equations 2.2.14 reduce to

$$\epsilon_{11} = \tfrac{1}{2}\,(\lambda_x^2 - 1)$$

$$e_{11} = \tfrac{1}{2}\,(1 - \lambda_x^{-2})$$

$$\epsilon_{22} = \tfrac{1}{2}\left(\frac{\lambda_x^{-2}}{\cos^2 \phi} - 1\right)$$

$$e_{22} = \tfrac{1}{2}\,(1 - \lambda_x^2 - \lambda_x^{-2}\,\tan^2 \phi)$$

$$\epsilon_{12} = \epsilon_{21} = \frac{\tan \phi}{2}$$

$$e_{12} = e_{21} = \frac{\tan \phi}{2\lambda_x^2}$$

Example 3

   Consider the radial expansion of an axially symmetric surface as shown in Figure 2.4. We specify that curvilinear distance along the meridian (the surface generator) is unchanged for corresponding points on the initial and instantaneous surfaces. We now define differential coordinate systems for corresponding surface elements local to a point, P_0, in each state. These are illustrated in Figure 2.4. We assume that the increments, da_i and dx_i, are sufficiently small that the elements can be considered flat and rectangular (the infinitesimal difference between the inner and outer radii of the conical annulus may be neglected). Because of axial symmetry, locations possess the same azimuthal angle, ϕ, in both the initial and final states. Hence, the latitudinal element length is proportional to the radial distance from the symmetry axis. Here, we take da_1 and dx_1 for the differential lengths along a tangent to the latitude circle,

$$da_1 = r_0 \, d\phi$$

$$dx_1 = r \, d\phi$$

The differential distances along the meridian are defined as

$$da_2 = dx_2$$

Now, the strain components can be calculated for the Lagrangian representation by

$$\epsilon_{11} = \tfrac{1}{2}\left(\frac{dx_1}{da_1} \frac{dx_1}{da_1} - 1\right) = \tfrac{1}{2}\left[\left(\frac{r}{r_0}\right)^2 - 1\right]$$

$$\epsilon_{22} = 0$$

$$\epsilon_{12} = \epsilon_{21} = 0$$

and for the Eulerian representation by

$$e_{11} = \tfrac{1}{2}\left(1 - \frac{da_1}{dx_1} \frac{da_1}{dx_1}\right) = \tfrac{1}{2}\left[1 - \left(\frac{r_0}{r}\right)^2\right]$$

$$e_{22} = 0$$

$$e_{12} = e_{21} = 0$$

We see by analogy to Example 1 that the circumferential extension ratio, λ_1, is equal to the ratio of the radial distances (r, r_0) of the instantaneous and initial states, respectively,

$$\lambda_1 = \frac{r}{r_0}$$

If the incremental length, da_2, along the meridian also changes (dx_2), then the second strain component becomes either

$$\epsilon_{22} = \tfrac{1}{2}\left(\frac{dx_2}{da_2} \frac{dx_2}{da_2} - 1\right) = \tfrac{1}{2}\left(\lambda_2^2 - 1\right)$$

or,

$$e_{22} = \tfrac{1}{2}\left(1 - \frac{da_2}{dx_2} \frac{da_2}{dx_2}\right) = \tfrac{1}{2}\left(1 - \lambda_2^{-2}\right)$$

where,

$$\lambda_2 = \frac{dx_2}{da_2} = \frac{ds}{ds_0}$$

Not to be confused with the metrics, the curvilinear distances along the meridian, s and s_0, depend on spatial coordinates (r,z) and (r_0,z_0) for the instantaneous and initial states, respectively. The independent variables may be selected as either the axial coordinates (z,z_0) or the radial coordinates (r,r_0). Parametric representations (s,s_0) may also be used. The deformation embodied in the principal extension ratios (λ_1,λ_2) is obtained from relations of the coordinates r, z, r_0, z_0, according to

$$\lambda_1 = \frac{r}{r_0}, \quad \lambda_2 = \frac{ds(r,z)}{ds_0(r_0,z_0)}$$

The incremental distances along the meridian are given by these spatial representations,

$$ds = \sqrt{1 + \left(\frac{dz}{dr}\right)^2}\, dr = \sqrt{\left(\frac{dr}{dz}\right)^2 + 1}\, dz$$

$$ds_0 = \sqrt{1 + \left(\frac{dz_0}{dr_0}\right)^2}\, dr_0 = \sqrt{\left(\frac{dr_0}{dz_0}\right)^2 + 1}\, dz_0$$

Therefore, the second principal extension ratio, λ_2, is determined by the functions of the spatial derivatives,

$$\lambda_2 = \frac{ds}{ds_0} = \frac{\sqrt{1 + \left(\frac{dz}{dr}\right)^2}}{\sqrt{1 + \left(\frac{dz_0}{dr_0}\right)^2}}\left(\frac{dr}{dr_0}\right) = \frac{\sqrt{\left(\frac{dr}{dz}\right)^2 + 1}}{\sqrt{\left(\frac{dr_0}{dz_0}\right)^2 + 1}}\,\frac{dz}{dz_0}$$

These nonlinear functions can be significant obstacles to finding analytical solutions; however, using a digital computer, these relations are easily evaluated.

Example 4

Consider the shear deformation analogous to Example 2 that is produced by twist of the axisymmetric surface. Local to a point on the surface, this appears as rotation of one circle of the conical annulus, with the other circle fixed, and the incremental distance between the two circles along a meridian is constant as shown in Figure 2.5. The strain components are obtained using the relations between the differential coordinates of the instantaneous state (dx_1, dx_2) and the initial state (da_1, da_2),

$$dx_1 = da_1 + \left(r_0\,\frac{d\phi}{ds_0}\right) da_2$$

$$da_1 = dx_1 - \left(r_0\,\frac{d\phi}{ds_0}\right) dx_2$$

Note that the differential lengths along latitude circles and the differential distance along the meridian are unchanged. Consequently, the strain components are calculated to be

$$\epsilon_{11} = 0$$

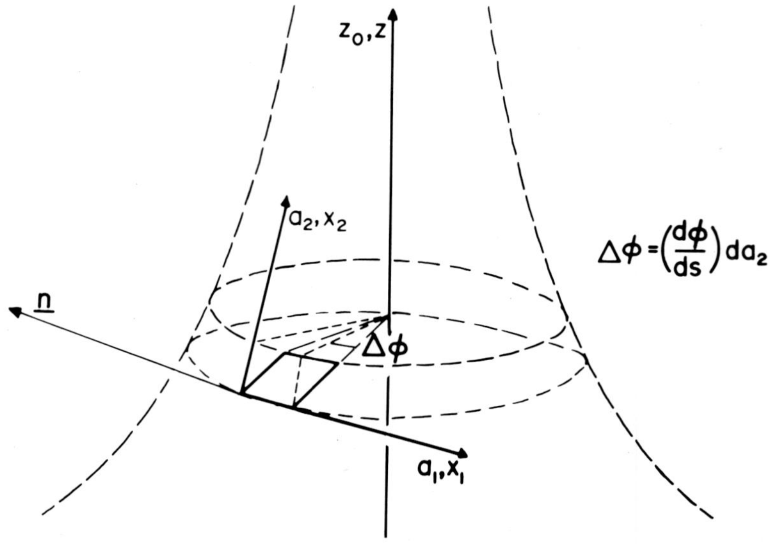

FIGURE 2.5.   Shear deformation caused by twist of an axisymmetric surface. Note that one circle of the conical annulus has been rotated relative to the other.

$$e_{11} = 0$$

$$\epsilon_{22} = \tfrac{1}{2}\left(r_0\,\frac{d\phi}{ds_0}\right)^2$$

$$e_{22} = -\tfrac{1}{2}\left(r_0\,\frac{d\phi}{ds_0}\right)^2$$

$$\epsilon_{12} = \epsilon_{21} = e_{12} = e_{21} = \tfrac{1}{2}\left(r_0\,\frac{d\phi}{ds_0}\right)$$

In terms of spatial coordinates, these components are given by the relation,

$$r_0\,\frac{d\phi}{ds_0} = \frac{r_0}{\sqrt{1 + \left(\dfrac{dz_0}{dr_0}\right)^2}}\,\frac{d\phi}{dr_0} = \frac{r_0}{\sqrt{\left(\dfrac{dr_0}{dz_0}\right)^2 + 1}}\,\frac{d\phi}{dz_0}$$

2.3 Principal Axes

Examination of Equations 2.2.10 and 2.2.11, which determine the deformation in the region local to point P_0, shows that for the four components of the strain matrices, Equations 2.2.12 and 2.2.13, depend on the choice of the directions of the unit base vectors that define the components da_i or dx_i. If the coordinate axes at point P_0 are rotated in the plane tangent to the surface (see Figures 2.6A and B), these components change value. However, we know that merely changing our perspective by choosing different coordinate axes does not change the magnitude of the deformation given by the difference in the quadratic distances between any two points, Equations 2.2.10 and 2.2.11. Therefore, we anticipate that the strain components for any choice of axes will be related to those of any other axes system. We will show that the relation involves a

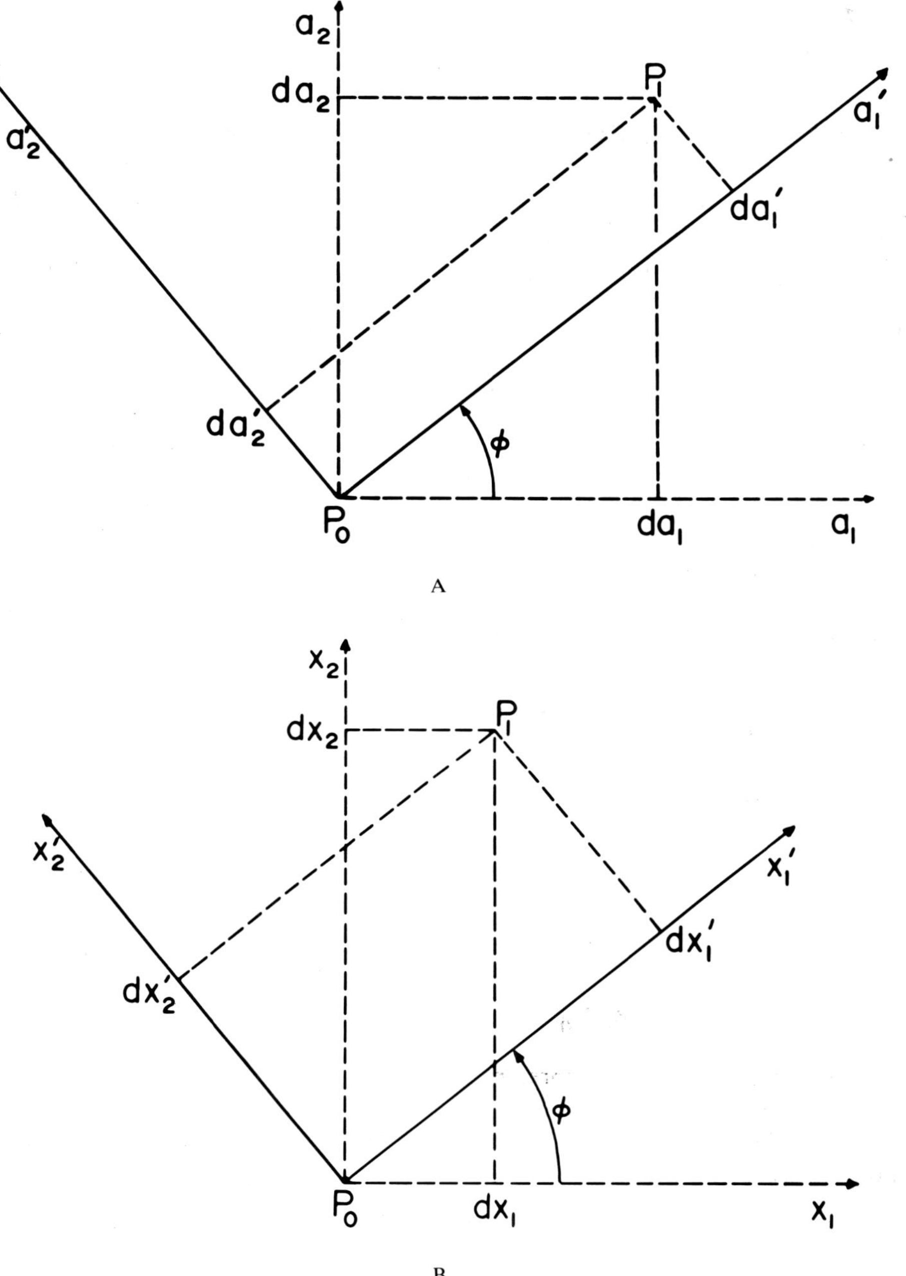

FIGURE 2.6. (A). An illustration of how the differential coordinates may change due to rotation of the axes in the plane tangent to the surface at the point, P_0. ϕ is the rotation angle. (B) A rotation of axes similar to that in (A). In general, a different value for the rotation angle is usually required to obtain the principal axes in the deformed state.

rotation angle, ϕ, and two separate parameters, which are not functions of rotation. These two parameters are called "invariants" of the matrix. They completely characterize the amount of deformation but not orientation.

Mathematically, we begin with the fact that the scalar quantities in Equations 2.2.10 and 2.2.11 are independent of coordinate system. Therefore, using the strain definition in Equation 2.2.10, we write,

$$\epsilon'_{k\ell} \, da'_k \, da'_\ell = \epsilon_{ij} \, da_i \, da_j \qquad (2.3.1)$$

where the primes refer to a coordinate system rotated by ϕ as in Figures 2.6A and B. Equation 2.3.1 expresses the scalar invariance. We use the Lagrangian strain in the development as an example, but the matrix transformation that will be produced may be applied generally to any symmetric tensor. (Note that the subscripts are "dummy" and are completely summed over in Equation 2.3.1, each side being a scalar quantity.)

From Figure 2.6A, we see that the primed coordinates, da_i, can be expressed in terms of the rotation angle, ϕ, and the unprimed coordinates, da_i,

$$da'_1 = (\cos \phi) \, da_1 \; + \; (\sin \phi) \, da_2$$

$$da'_2 = (-\sin \phi) \, da_1 \; + \; (\cos \phi) \, da_2 \qquad (2.3.2)$$

Equations 2.3.2 can be written in indicial notation,

$$da'_\ell = R_{\ell j} \, da_j \qquad (2.3.3)$$

where the rotation matrix $R_{\ell j}$ is identified from Equation 2.3.2,

$$R_{\ell j} = \begin{pmatrix} \cos \phi & \sin \phi \\ -\sin \phi & \cos \phi \end{pmatrix} \qquad (2.3.4)$$

Using Equation 2.3.3 twice in Equation 2.3.1 with caution that summed indices are properly employed, we arrive at the form representing the strain matrix in the unprimed system in terms of the primed system matrix,

$$\epsilon'_{k\ell} \, R_{ki} \, R_{\ell j} \, da_i \, da_j = \epsilon_{ij} \, da_i \, da_j$$

$$\epsilon'_{k\ell} \, R_{ki} \, R_{\ell j} = \epsilon_{ij} \qquad (2.3.5)$$

To find the primed matrix representation in terms of the unprimed matrix, the reverse operation is performed by a similar geometric analysis of the projection of unprimed coordinates onto the primed or rotated system. This is commonly written as an inverse operation or rotation, $R_{j\ell}^{-1}$,

$$da_j = R_{j\ell}^{-1} \, da'_\ell \qquad (2.3.6)$$

From geometry, the simple observation is made that the inverse rotation is the transpose of the rotation (transposition is the interchange of rows and columns of the matrix),

$$R_{j\ell}^{-1} = \begin{pmatrix} \cos \phi & -\sin \phi \\ \sin \phi & \cos \phi \end{pmatrix} \qquad (2.3.7)$$

or,

$$R_{j\ell}^{-1} = R_{\ell j} \qquad (2.3.8)$$

The rotation matrix has the property that $R_{kj} R_{j\ell}^{-1} \equiv \delta_{k\ell}$. Therefore the strain matrix in the rotated system is given by

$$\epsilon'_{k\ell} = R_{ik}^{-1} R_{j\ell}^{-1} \epsilon_{ij} \qquad (2.3.9)$$

or,

$$\epsilon'_{k\ell} = R_{ki} R_{\ell j} \epsilon_{ij}$$

Writing out all of the components of $\epsilon'_{k\ell}$ in terms of the trigonometric functions of the rotation angle gives

$$\epsilon'_{11} = \epsilon_{11} \cos^2 \phi + \epsilon_{12} \sin 2\phi + \epsilon_{22} \sin^2 \phi$$

$$\epsilon'_{22} = \epsilon_{11} \sin^2 \phi - \epsilon_{12} \sin 2\phi + \epsilon_{22} \cos^2 \phi$$

$$\epsilon'_{12} = \epsilon'_{21} = - \frac{(\epsilon_{11} - \epsilon_{22})}{2} \sin 2\phi + \epsilon_{12} (\cos^2 \phi - \sin^2 \phi)$$

$$(2.3.10)$$

where the trigonometric identity $2 \sin\phi \cos\phi \equiv \sin 2\phi$ has been used. Introducing two other trigonometric identities for $\sin^2\phi$ and $\cos^2\phi$, we can transform Equations 2.3.10 into functions of twice the rotation angle, ϕ,

$$\epsilon'_{11} = \frac{(\epsilon_{11} + \epsilon_{22})}{2} + \frac{(\epsilon_{11} - \epsilon_{22})}{2} \cos 2\phi + \epsilon_{12} \sin 2\phi$$

$$\epsilon'_{22} = \frac{(\epsilon_{11} + \epsilon_{22})}{2} - \frac{(\epsilon_{11} - \epsilon_{22})}{2} \cos 2\phi - \epsilon_{12} \sin 2\phi$$

$$\epsilon'_{12} = \epsilon'_{21} = - \frac{(\epsilon_{11} - \epsilon_{22})}{2} \sin 2\phi + \epsilon_{12} \cos 2\phi$$

$$(2.3.11)$$

From the equation for ε'_{12}, the shear strain component in Equations 2.3.11, it is apparent that it is always possible to choose a rotation such that ε'_{12} and ε'_{21} are zero and the matrix is diagonal, i.e.,

$$\epsilon'_{k\ell} = \begin{pmatrix} \epsilon'_1 & 0 \\ 0 & \epsilon'_2 \end{pmatrix}$$

when

$$\tan 2\phi = \frac{2\epsilon_{12}}{(\epsilon_{11} - \epsilon_{22})}$$

[As illustrated in Figure 2.6B, a different value for the rotation angle is usually required to obtain the principal axes which diagonalize the Eulerian strain matrix (or any other matrix based on the deformed geometry). An example at the end of this section will illustrate the approach and show the difference between rotation angles required to obtain principal axes of the deformation in the initial as compared to the instantaneous state.]

If we initially choose the principal axes system, we see that all other strain representations (for coordinate systems rotated relative to the principal axes system) can be calculated from the diagonal matrix,

$$\epsilon_{ij} = \begin{pmatrix} \epsilon_1 & 0 \\ 0 & \epsilon_2 \end{pmatrix}$$

and Equations (2.3.11) as follows:

$$\epsilon'_{11} = \frac{(\epsilon_1 + \epsilon_2)}{2} + \frac{(\epsilon_1 - \epsilon_2)}{2} \cos 2\phi$$

$$\epsilon'_{22} = \frac{(\epsilon_1 + \epsilon_2)}{2} - \frac{(\epsilon_1 - \epsilon_2)}{2} \cos 2\phi$$

$$\epsilon'_{12} = \epsilon'_{21} = - \frac{(\epsilon_1 - \epsilon_2)}{2} \sin 2\phi$$

$$(2.3.12)$$

The components, ε_1 and ε_2, are referred to as the principal values of the matrix.

From Equations 2.3.12 we see that the rotation angle and the principal values of the matrix are sufficient to characterize the components of the strains. The magnitudes of the strains depend on two parameters: $(\varepsilon_1 + \varepsilon_2)$ and $|\varepsilon_1 - \varepsilon_2|$. The results given by Equations (2.3.12) may be graphically illustrated by plotting minus ε'_{12} as ordinate and ε'_{11} or ε'_{22} as abscissa. The values are mapped on a circle with radius $|\varepsilon_1 - \varepsilon_2|/2$, center at $(\varepsilon_1 + \varepsilon_2)/2$, and are located by twice the rotation angle, ϕ, as shown in Figure 2.7. It is assumed that the principal axes are chosen so that $\varepsilon_1 \geqslant \varepsilon_2$. In order to obtain the principal axes system from an arbitrary system, the coordinate system is rotated counterclockwise (in the positive sense of the angle, ϕ) by ϕ radians. On the other hand, a rotation of minus ϕ is needed to generate the primed coordinate system from the principal axes system.

In Figure 2.7, we see that the diagonal components, ε'_{11} and ε'_{22}, range between ε_1 and ε_2. The maximum shear strain occurs at $\phi = \pm 45°$ and is equal to

$$\epsilon_s = |\epsilon_{12}|_{MAX} = \left| \frac{\epsilon_1 - \epsilon_2}{2} \right|$$

From Equations 2.3.12 and Figure 2.7, we see that it is possible to separate the matrix into an isotropic part (independent of rotation, and diagonal for any choice of axes) plus a rotationally dependent part, called the deviator. The isotropic contribution, $\bar{\epsilon}'_{ij}$, is represented by $\bar{\epsilon}'_{ij} = \bar{\epsilon} \, \delta_{ij}$, where

$$\bar{\epsilon}'_{11} = \bar{\epsilon}'_{22} = \bar{\epsilon}, \text{ and } \bar{\epsilon}'_{12} = 0 \tag{2.3.13}$$

Therefore, we define $\bar{\epsilon}$ by $\bar{\epsilon} = \frac{1}{2}(\varepsilon_1 + \varepsilon_2) \equiv \frac{1}{2}\varepsilon'_{kk}$ with the summation rule. The isotropic matrix is proportional to the mean of the principal components.

The deviator, $\tilde{\epsilon}'_{ij}$, is obtained from Equations 2.3.12 as

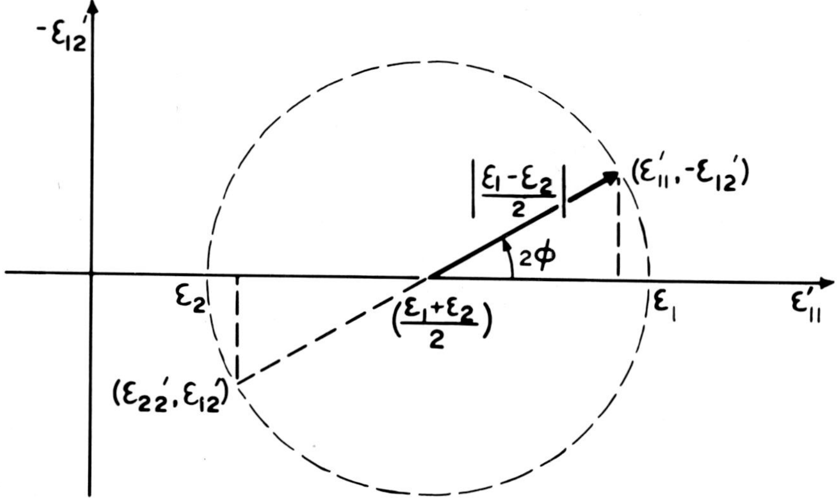

FIGURE 2.7.   A graphic illustration of the diagonal and shear strain components of a matrix as functions of the axes rotation angle. Note, the values are mapped on a circle with radius, $|\varepsilon_1 - \varepsilon_2|/2$, and center located at $(\varepsilon_1 + \varepsilon_2)/2$.

$$\tilde{\epsilon}_{11}' = \frac{(\epsilon_1 - \epsilon_2)}{2}\cos 2\phi$$

$$\tilde{\epsilon}_{22}' = -\frac{(\epsilon_1 - \epsilon_2)}{2}\cos 2\phi$$

$$\tilde{\epsilon}_{21}' = \tilde{\epsilon}_{12}' = -\frac{(\epsilon_1 - \epsilon_2)}{2}\sin 2\phi$$

The deviator is a function only of the maximum shear component and the angle. The deviator matrix is written explicitly as

$$\tilde{\epsilon}_{ij}' = \epsilon_{ij}' - \tfrac{1}{2}\epsilon_{kk}'\,\delta_{ij} \tag{2.3.14}$$

The deviator contains all of the rotationally dependent elements of the matrix and none of the isotropic or directionally independent contribution. This is verified by noting that the trace, $\tilde{\epsilon}'_{kk}$, is identically zero. The isotropic and deviatoric parts represent the mean and deviation from the mean of a matrix with respect to rotation of the local coordinates.

Example 1

Consider the uniform deformation of a square element into a parallelogram with equal sides. What are the angles of rotation for the initial and instantaneous coordinate systems that will give the "principal axes systems"? Equations 2.2.13 give the strain components. Note that $\lambda_y = \lambda_x \cos\phi$ for a parallelogram with equal sides (see Figure 2.3 for the definition of the shear angle, ϕ).

$$\epsilon_{11} = \tfrac{1}{2}(\lambda_x^2 - 1)$$

$$e_{11} = \tfrac{1}{2}(1 - \lambda_x^{-2})$$

$$\epsilon_{22} = \tfrac{1}{2}(\lambda_x^2 - 1)$$

$$e_{22} = \tfrac{1}{2}\left[1 - \lambda_x^{-2}\left(\frac{1 + \sin^2\phi}{\cos^2\phi}\right)\right]$$

$$\epsilon_{12} = \frac{\lambda_x^2}{2}\sin\phi$$

$$e_{12} = \frac{\lambda_x^{-2}}{2}\frac{\sin\phi}{\cos\phi}$$

The rotation angles of the initial and instantaneous coordinate axes that define each principal axes system are ϕ_{R_0} and ϕ_R, respectively. Using the strain components, these angles are given by

$$\tan 2\phi_{R_0} = \frac{2\epsilon_{12}}{\epsilon_{11} - \epsilon_{22}} = \frac{2\lambda_x^2\sin\phi}{0} \rightarrow \infty$$

or

$$\phi_{R_0} = 45°$$

and

$$\tan 2\phi_R = \frac{2e_{12}}{e_{11} - e_{22}} = \frac{\cos\phi}{\sin\phi}$$

$$\phi_R = \tfrac{1}{2}(90° - \phi)$$

Therefore, the deformation of a square into a parallelogram with equal sides simply involves extension and compression of the element diagonals as shown in Figure 2.8 for a specific situation where $\lambda_x = 1.33$ and the shear angle $\phi = 30°$. The coordinates must be rotated by angles of $\phi_{R_0} = 45°$ and $\phi_R = 30°$, respectively, to give the principal axes systems.

2.4 Invariants of Strain and Independent Deformation Variables

For a matrix based on a two-dimensional cartesian space, two parameters specify the matrix components for any given orientation of the coordinates. The parameters are the mean and difference of the principal components of the diagonalized form of the matrix. As we showed in the previous section, these are expressed by the trace of the matrix,

$$2\bar{\epsilon} = \epsilon'_{11} + \epsilon'_{22} = \epsilon_1 + \epsilon_2 = \epsilon'_{kk}$$

and the deviator magnitude,

$$|\epsilon_{12}|^2\,\mathrm{MAX} = \left(\frac{\epsilon_1 - \epsilon_2}{2}\right)^2 = \left(\frac{\epsilon'_{11} - \epsilon'_{22}}{2}\right)^2 + \epsilon'^2_{12}$$

Such parameters are referred to as matrix invariants, i.e., they are invariant to rotation of coordinate axes. We will demonstrate that in addition to rotational invariance, these

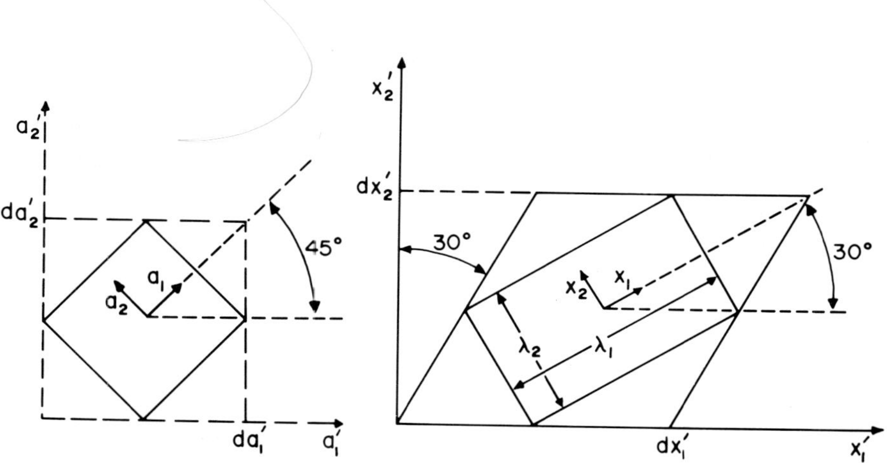

FIGURE 2.8. Deformation of a square into a parallelogram. The deformation simply involves extension and compression of element diagonals.

variables depend only on the fractional change in area of a material element plus the extension of the element at constant final or instantaneous area.

The principal axes system is the most convenient for evaluation of the two invariants that characterize the deformation because the shear strain component is zero in this system, i.e., $\varepsilon_{ij} = 0$, $i \neq j$. The partial derivatives, $\partial x_k / \partial a_i$ or $\partial a_k / \partial x_i$, $k \neq i$, are zero. Here, the geometry is simple; the material increments along each axis are either extended or compressed independent of the other coordinate. Principal axes deformation is demonstrated by the rectangular inserts drawn inside the initial and instantaneous surface elements shown in Figure 2.8. The components of the Lagrangian and Eulerian strain matrices in the principal axes system are related to the principal extension ratios by

$$\epsilon_1 = \tfrac{1}{2}\left(\frac{dx_1}{da_1}\frac{dx_1}{da_1} - 1\right) = \tfrac{1}{2}(\lambda_1^2 - 1)$$

$$\epsilon_2 = \tfrac{1}{2}\left(\frac{dx_2}{da_2}\frac{dx_2}{da_2} - 1\right) = \tfrac{1}{2}(\lambda_2^2 - 1)$$

$$(2.4.1)$$

and

$$e_1 = \tfrac{1}{2}\left(1 - \frac{da_1}{dx_1}\frac{da_1}{dx_1}\right) = \tfrac{1}{2}(1 - \lambda_1^{-2})$$

$$e_2 = \tfrac{1}{2}\left(1 - \frac{da_2}{dx_2}\frac{da_2}{dx_2}\right) = \tfrac{1}{2}(1 - \lambda_2^{-2})$$

$$(2.4.2)$$

The invariants, therefore, are related to the principal extension ratios by

$$\epsilon_1 + \epsilon_2 = \tfrac{1}{2}(\lambda_1^2 + \lambda_2^2 - 2)$$

$$(\epsilon_1 - \epsilon_2)^2 = \tfrac{1}{4}(\lambda_1^4 - 2\lambda_1^2\lambda_2^2 + \lambda_2^4)$$

for the Lagrangian strain matrix. Note that these two invariants are not unique. If we use I_1 to label the first invariant,

$$I_1 \equiv \epsilon_1 + \epsilon_2 = \tfrac{1}{2}(\lambda_1^2 + \lambda_2^2) - 1 \tag{2.4.3}$$

then the equation for $(\varepsilon_1 - \varepsilon_2)^2$ can be written as

$$(\epsilon_1 - \epsilon_2)^2 = I_1^2 - 4\epsilon_1\epsilon_2$$

Therefore, $\varepsilon_1\varepsilon_2$ (when expressed in terms of ε'_{ij}) is also an invariant. As shown in Example 1 below, the second invariant of the strain, I_2, is derived indirectly from matrix algebra and transformation properties,

$$I_2 \equiv -\epsilon_1\epsilon_2 = \tfrac{1}{4}(\lambda_1^2 + \lambda_2^2 - \lambda_1^2\lambda_2^2 - 1) \tag{2.4.4}$$

From the perspective of geometry, a significant feature is observed in Equations 2.4.3 and 2.4.4. The invariants which characterize the deformation are related to the sum of the squares of the extension ratios, $(\lambda_1^2 + \lambda_2^2)$, and the product of the extension ratios, $\lambda_1\lambda_2$. (The extension ratios are always positive numbers, so we can use $\lambda_1\lambda_2$ instead of $\lambda_1^2\lambda_2^2$.) This indicates that it is possible to represent general deformations by independent geometric functions. However, these parameters are not completely independent. They are equal when the extension ratios are same, $\lambda_1 = \lambda_2$. It is possible to make the functions independent by normalizing the sum of squares by the product. The result is a set of functions that are both linearly independent and invariant to rotation of coordinates. Choosing

$$\alpha = \lambda_1\lambda_2 - 1$$

$$\beta = \frac{1}{2\lambda_1\lambda_2}(\lambda_1^2 + \lambda_2^2) - 1 \tag{2.4.5}$$

we have such a representation. The invariants, I_1 and I_2, are given by

$$I_1 = \epsilon_1 + \epsilon_2 = \alpha + \beta + \alpha\beta$$

$$I_2 = -\epsilon_1\epsilon_2 = \tfrac{1}{4}(2\beta + 2\alpha\beta - \alpha^2)$$

The two functions, α and β, completely characterize the geometric features of the deformation: (1) change of area of the element, (2) change of aspect ratio (length to width) or eccentricity of the element. The parameter, α, is the fractional change in the area of the surface element,

$$\alpha = \frac{dA}{dA_0} - 1 = \frac{dx_1}{da_1}\frac{dx_2}{da_2} - 1 \tag{2.4.6}$$

The change in the aspect ratio or eccentricity of the material element is given by

$$\frac{dx_1}{dx_2} - \frac{da_1}{da_2} = \frac{da_1}{da_2}\left(\frac{\lambda_1}{\lambda_2} - 1\right)$$

For an initially square element, the change is $(\lambda_1/\lambda_2 - 1)$. In order to derive a function independent of choice of axes, we take half of the symmetric sum, which is the parameter, β,

$$\beta = \tfrac{1}{2}\left(\frac{\lambda_1}{\lambda_2} - 1\right) + \tfrac{1}{2}\left(\frac{\lambda_2}{\lambda_1} - 1\right) = \frac{1}{2\lambda_1\lambda_2}\,(\lambda_1{}^2 + \lambda_2{}^2) - 1$$

Thus, the deformation parameters, α and β, are determined by linearly independent functions: the product of extension ratios, $\lambda_1\lambda_2$, and the quotient, λ_1/λ_2.

The quotient defines a single extension ratio, $\tilde{\lambda}$, which represents the element elongation at the final area. This is demonstrated by scaling each extension ratio by the square root of the area ratio,

$$\tilde{\lambda} \equiv \frac{\lambda_1}{\sqrt{\lambda_1\lambda_2}} \tag{2.4.7}$$

and

$$\tilde{\lambda}^{-1} = \frac{\lambda_2}{\sqrt{\lambda_1\lambda_2}}$$

From these scaled extension ratios, the deformation parameter, β, is determined to be

$$\beta = \tfrac{1}{2}\,(\tilde{\lambda}^2 + \tilde{\lambda}^{-2}) - 1 \tag{2.4.8}$$

which is invariant to coordinate rotation. Therefore, we see that deformation of any element can be considered simply as a change in area at constant element shape followed by the scaled extension of the element at constant element area (Figure 2.9). The linear independence of α and $\tilde{\lambda}$ is demonstrated by evaluating the Jacobian, J. In a domain where α and $\tilde{\lambda}$ are linearly independent, we require that

$$J \equiv \begin{vmatrix} \dfrac{\partial\tilde{\lambda}}{\partial\lambda_1} & \dfrac{\partial\tilde{\lambda}}{\partial\lambda_2} \\[2ex] \dfrac{\partial\alpha}{\partial\lambda_1} & \dfrac{\partial\alpha}{\partial\lambda_2} \end{vmatrix} \neq 0$$

or,

$$J = \left(\frac{\partial\tilde{\lambda}}{\partial\lambda_1}\right)\left(\frac{\partial\alpha}{\partial\lambda_2}\right) - \left(\frac{\partial\alpha}{\partial\lambda_1}\right)\left(\frac{\partial\tilde{\lambda}}{\partial\lambda_2}\right) \neq 0$$

From Equations 2.4.7 and 2.4.8, the Jacobian is equal to $\sqrt{\lambda_1/\lambda_2}$. Because the extension ratios are positive-definite quantities, the deformation parameters, α and $\tilde{\lambda}$, are always linearly independent.

The independence of α and β is immediately evident if only changes in area occur or extension takes place at constant area. If only area changes of the element occur ($\alpha \neq 0$ and $\lambda_1 \equiv \lambda_2$), it is apparent from Equation 2.4.5 that the parameter, β, remains zero. On the other hand, if the element area remains constant ($\alpha = 0$) and extensional

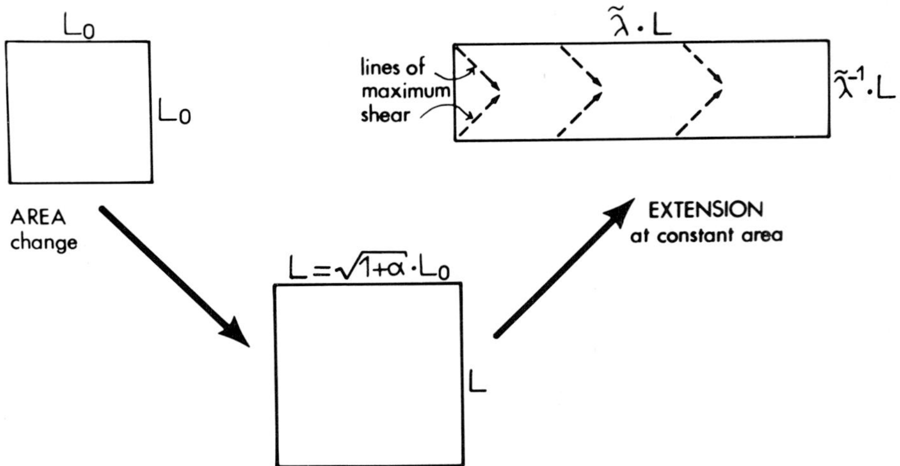

FIGURE 2.9.   An illustration that deformation of any element can be considered simply as a change in area at constant element shape followed by the scaled extension of the element at constant element area.

deformation of the square to a rectangle takes place, ($\lambda_1 = 1/\lambda_2$), the parameter β is always greater than zero. When we consider molecular surfaces that contain a fixed number of molecules, α is the change in specific area per molecule in the surface; β quantifies the extension of the molecular complex at constant area per molecule. Therefore, the derivation of mechanochemical equations of state from the material thermodynamic potentials will depend on these two parameters if we assume that the surface is locally isotropic in the initial state.*

In an analogous manner, the Eulerian invariants can be expressed in terms of parameters, α' and β',

$$\alpha' \equiv \frac{1}{\lambda_1 \lambda_2} - 1 = \frac{-\alpha}{1 + \alpha}$$

$$\beta' \equiv \frac{\lambda_1 \lambda_2}{2} \left(\frac{1}{\lambda_1^2} + \frac{1}{\lambda_2^2} \right) - 1 = \beta$$

These functions reflect the intrinsic reference to the instantaneous state. The invariants of the Eulerian strain matrix can be expressed in terms of α' and β',

$$I_1' = e_1 + e_2 = -(\alpha' + \beta' + \alpha'\beta')$$

$$I_2' = -e_1 e_2 = \tfrac{1}{4}(2\beta' + 2\alpha'\beta' - \alpha'^2)$$

The functional forms are the same for both sets of invariants in terms of (α,β) and (α',β').

The existence of principal axes and two invariants, discussed above for the strain matrices, is a general property of symmetric matrices in two dimensions. The general method of finding principal values and invariants valid for any symmetric matrix is given in the following example.

* Local isotropy means that the molecular structure in the plane of the surface is independent of the rotation angle around point P_0.

Example 1

For a symmetric matrix like the Lagrangian strain matrix, choosing a unit vector, n_j, that lies along a principal axis and contracting it with the matrix will produce a vector projection along the principal axis,

$$\epsilon_{ij} n_j = \epsilon n_i \tag{2.4.9}$$

The projected vector is scaled by ϵ which is called a principal value. This may be easily verified by considering the case where we are already in principal axes.

The components, n_i, describe a unit vector in one principle axis direction. Since there are two principal axes, there will be two such unit vectors and two scalar values for ϵ, called principal values or eigenvalues of the matrix. These are diagonal components, ϵ_1 and ϵ_2, in the principal axes system. Equation 2.4.9 can be written as

$$(\epsilon_{ij} - \epsilon \delta_{ij}) n_j = 0 \tag{2.4.10}$$

The eigenvalue equations, Equations 2.4.10, are a set of two linear, homogenous equations that determine the components, n_j; therefore, Equations 2.4.10 can only be valid if the determinant of the coefficients is identically zero.

$$\left| \epsilon_{ij} - \epsilon \delta_{ij} \right| \equiv 0$$

This determinant yields the result known as a characteristic equation,

$$(\epsilon_{11} - \epsilon)(\epsilon_{22} - \epsilon) - \epsilon_{12}^2 = 0$$

or

$$\epsilon^2 - (\epsilon_{11} + \epsilon_{22})\epsilon - (\epsilon_{12}^2 - \epsilon_{11}\epsilon_{22}) = 0 \tag{2.4.11}$$

which is quadratic in the principal values of ϵ. The two roots of this equation are the principal values, ϵ_1 and ϵ_2.

The determinant is a scalar quantity that is invariant to rotation. Consequently, the coefficients of powers in ϵ in the characteristic Equation 2.4.3 are invariant to rotation,

$$\epsilon^2 - I_1 \epsilon - I_2 = 0$$

$$I_1 \equiv \epsilon_{11} + \epsilon_{22} = \epsilon_{ii}$$

$$I_2 = \epsilon_{12}^2 - \epsilon_{11}\epsilon_{22} = \tfrac{1}{2}(\epsilon_{ij}\epsilon_{ji} - \epsilon_{ii}\epsilon_{jj}) \tag{2.4.12}$$

The forms of the two invariants, I_1 and I_2, and the other forms previously discussed are valid for any symmetric matrix. Other such matrices will be introduced in later sections.

For the deviator of the strain matrix, $\tilde{\epsilon}_{ij}$, the invariants are determined using Equations 2.4.12 and 2.3.15,

$$\tilde{\epsilon}_{ij} = \epsilon_{ij} - \tfrac{1}{2}\epsilon_{kk}\delta_{ij}$$

$$I_1 = \tilde{\epsilon}_{ii} = 0, \ (\delta_{ii} \equiv 2)$$

$$I_2 = \tfrac{1}{2}(\tilde{\epsilon}_{ij}\tilde{\epsilon}_{ji}) = \tfrac{1}{2}(\epsilon_{ij}\epsilon_{ji} - \epsilon_{ii}\epsilon_{jj})$$

$$I_2 = \left(\frac{\epsilon_{11} - \epsilon_{22}}{2}\right)^2 + \epsilon_{12}^2 = \epsilon_s^2 \tag{2.4.13}$$

Here, I_2 is the magnitude of the maximum shear component squared.

2.5 Rate of Deformation

In the previous sections, we have developed an intensive description of the deformation of small elements of the membrane surface that is indepedent of the extrinsic shape of the membrane envelope. We considered initial and instantaneous states of surface element geometry without regard to the time during which the deformation took place. The implication is that the surface geometry may be evolving and changing with time. Any material deformation always takes time because it is impossible to produce immediate relative displacement of molecules. There will always be inertial and frictional resistance that limit the rate of relative displacement or deformation. Therefore, it is important to analyze the rate of deformation of material.

The rate of deformation in the vicinity of P_0 is determined by the time rate of change of distances between P_0 and its neighboring material points, given by the metric, Equation 2.2.3. By definition, we regard the initial state and the differential coordinates (da_i) as fixed in time. Therefore, we can conveniently measure the rate of material deformation using the initial state as the reference. The instantaneous state of deformation is given by Equations 2.2.10 and 2.2.12,

$$ds^2 - ds_0^2 = 2\epsilon_{ij}\, da_i\, da_j$$

The time rate of change of this equation specifies the rate of deformation,

$$(\dot{ds^2}) = 2\dot{\epsilon}_{ij}\, da_i\, da_j \tag{2.5.1}$$

where the operator ( ) signifies the time rate of change as viewed from the material reference frame (i.e., with the observer fixed at the point, P_0, in the material surface). We call $(\dot{ds^2})$ the rate of material deformation relative to the moving point, P_0. The derivatives, ( ), of quantities that depend on the initial coordinates and time simply become partial derivatives with respect to time; for example,

$$\dot{\epsilon}_{ij} \equiv \frac{\partial \epsilon_{ij}}{\partial t}$$

$$(\dot{ds_0^2}) \equiv \frac{\partial (ds_0^2)}{\partial t} = 0$$

$$(\dot{da_i}) \equiv \frac{\partial (da_i)}{\partial t} = 0$$

It is important to emphasize that the differential operator, $d(\ )_i$, represents local (infinitesimal) spatial variation of a property with respect to the value at P_0. For instance, the differential coordinates are given by da_i and dx_i relative to P_0. Changes with respect to small time variations, δt, will be indicated by the differential operator, $\delta(\ )$.

Equation 2.5.1 gives the rate of change of the metric using the initial coordinates as the reference geometry. This is a Lagrangian representation with the rate of deformation being given by the time derivative of the Lagrangian strain matrix as viewed from the material point, P_0. In general, it is often difficult to keep track of the evolution of material points. Therefore, it is also necessary to define rate of deformation based on spatial or instantaneous coordinates. This is represented by the rate of deformation matrix, $V_{k\ell}$, which is an Eulerian variable, with the instantaneous coordinates as the reference geometry,

$$(\dot{ds^2}) \equiv 2V_{k\ell}\, dx_k\, dx_\ell$$

Thus, the rate of deformation for the instantaneous (Eulerian) reference state is obtained from the transformation of initial to instantaneous differential coordinates and Equation 2.5.1,

$$V_{k\ell}\, dx_k\, dx_\ell = \left(\frac{\partial \epsilon_{ij}}{\partial t}\ \frac{\partial a_i}{\partial x_k}\ \frac{\partial a_j}{\partial x_\ell}\right) dx_k\, dx_\ell \tag{2.5.2}$$

The matrix, $\partial a_i / \partial x_k$, obviously represents the deformation of the initial element geometry to the instantaneous state.

Rate of deformation (as viewed from a fixed location in space not traveling with the material) is evidenced by velocity variations from point to point of material. The differential velocities of material points near a specified point, say P_0 (that is, relative to the motion of P_0), are determined by partial derivatives with respect to time of the instantaneous local coordinates. Since

$$dx_i = f(da_i, t)$$

$$dv_i = \frac{\partial}{\partial t}(dx_i) = (\dot{dx_i}) \tag{2.5.3}$$

Relative velocities, at any instant in time, of material points near P_0 are also expressed by the spatial gradients of the instantaneous velocity field times the differential coordinates, dx_i, because of surface continuity,

$$dv_i = \frac{\partial v_i}{\partial x_j}\, dx_j \tag{2.5.4}$$

This linear approximation is valid close to P_0 and is illustrated in Figure 2.10. In Equation 2.5.4, the partial derivative, $\partial(\ )/\partial x_j$ implies that time is held constant, i.e., we are considering only the spatial variation of velocities of material points at a specific time.

The time rate of change of the metric, Equation 2.5.1, may also be obtained by the partial time derivative of $dx_k\, dx_k$,

$$(\dot{ds^2}) = 2\, dx_k\, \frac{\partial(dx_k)}{\partial t}$$

From Equation 2.5.3, the time rate of change is given in terms of instantaneous differential coordinates and velocities relative to P_0 by

$$(\dot{ds^2}) = 2dx_k\, dv_k$$

the spatial derivatives of the instantaneous velocity field reduce the equation to

$$(\dot{ds^2}) = 2\,\frac{\partial v_k}{\partial x_\ell}\, dx_\ell\, dx_k \tag{2.5.5}$$

As shown in Figure 2.11, the relative velocity, dv_i, produces an infinitesimal relative displacement, $\delta(dx_i)$, of P_0 for the infinitesimal time increment, δt. Figure 2.10 also shows that the infinitesimal displacement, $\delta(dx_i)$, is composed of two parts: (1) a small

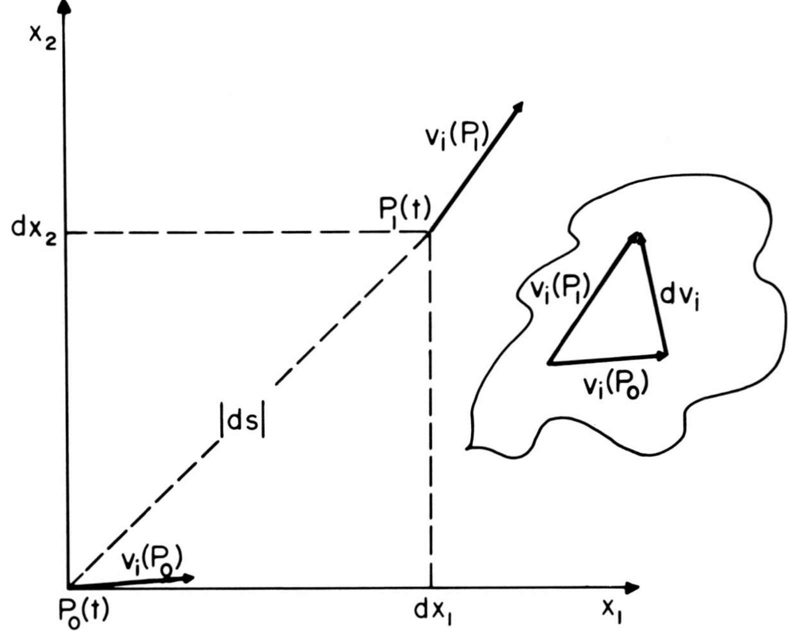

FIGURE 2.10.   An illustration of the spatial distribution of velocities close to P_0. Relative velocities of points near P_0 are approximated by the spatial gradients of the instantaneous velocity field times the differential coordinates.

rotation of P_1 about the origin at P_0, and (2) an extension or reduction in the scalar distance between P_1 and P_0. Obviously, a rigid body rotation does not contribute to deformation and may be extracted from the infinitesimal displacement field. The two components of the displacement are simply given by the antisymmetric and symmetric parts of the matrix, $\partial v_k / \partial x_\ell$, respectively.

A symmetric matrix is defined as one in which $A_{ij} = A_{ji}$ for all i,j. For an antisymmetric matrix, $A_{ij} = -A_{ji}$ for i ≠ j, and equals zero for i = j. Any matrix can be expressed as the sum of symmetric and antisymmetric components. For example,

$$D_{k\ell} \equiv \frac{\partial v_k}{\partial x_\ell} = \frac{1}{2}\left(\frac{\partial v_k}{\partial x_\ell} + \frac{\partial v_\ell}{\partial x_k}\right) + \frac{1}{2}\left(\frac{\partial v_k}{\partial x_\ell} - \frac{\partial v_\ell}{\partial x_k}\right)$$

or

$$D_{k\ell} = D_{k\ell}^S + D_{k\ell}^A$$

which is an identity relation. The superscripts, S and A, indicate symmetric and antisymmetric parts.

If we use the decomposition of the matrix, $D_{k\ell}$, Equation 2.5.5 is reduced to

$$(\dot{ds^2}) = 2(D_{k\ell}^S + D_{k\ell}^A)\, dx_\ell\, dx_k = 2D_{k\ell}^S\, dx_\ell\, dx_k \qquad (2.5.6)$$

The reduction is possible because the product, $D^A{}_{k\ell}\, dx_k\, dx_\ell$, is zero: $D^A{}_{k\ell}$ is antisymmetric and $dx_\ell\, dx_k$ is symmetric. We illustrate the geometric significance of the terms in Equation 2.5.6 in the following discussion.

The infinitesimal displacement, $\delta(dx_i)$, can be expressed as

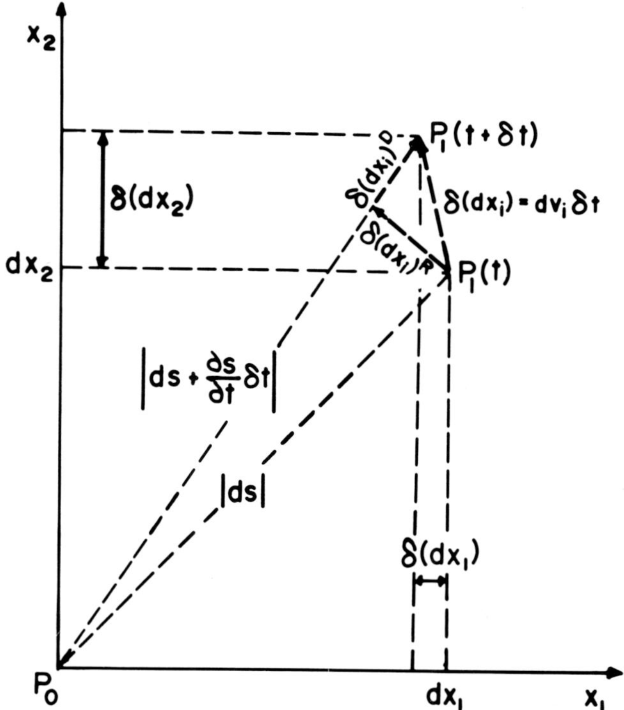

FIGURE 2.11. An infinitesimal displacement, $\delta(dx_i)$, of P_1 produced by the relative velocity, dv_i, for the infinitesimal time increment, δt. Note that $\delta(dx_i)$ is composed of a small rotation of P_1 about P_0 and a change in the scalar distance between P_1 and P_0.

$$\delta(dx_i) = \delta(dx_i)^R + \delta(dx_i)^D$$

illustrated in Figure 2.11. The first term on the right-hand side is a rigid rotation about P_0, and the second term is the remaining displacement of P_1 relative to P_0

$$\delta(dx_i)^R = D^A_{ij} \, dx_j \, \delta t$$

$$\delta(dx_i)^D = D^S_{ij} \, dx_j \, \delta t \qquad (2.5.7)$$

The scalar product of the rotation component, $\delta(dx_i)^R$, with the material vector, dx_i (i.e., the projection onto the (P_0, P_1) line element), is zero,

$$0 \equiv \delta(dx_k)^R \, dx_k = (D^A_{k\ell} \, dx_\ell) \, dx_k \, \delta t$$

Therefore in the rate of deformation, given by Equation 2.5.6, the term, $D^s_{k\ell}$, completely specifies the time rate of change of relative distances between neighboring material points in terms of the instantaneous coordinates (Eulerian variables). By analogy to the definition of strain matrices, the rate of deformation matrix, V_{ij}, is defined as

$$V_{ij} \equiv \tfrac{1}{2}\left(\frac{\partial v_i}{\partial x_j} + \frac{\partial v_j}{\partial x_i}\right) = D^S_{ij} \qquad (2.5.8)$$

Consequently, the time rate of change of the metric becomes

$$(\dot{ds^2}) = 2V_{k\ell}\, dx_k\, dx_\ell \qquad\qquad (2.5.9)$$

As we discussed previously Equations 2.5.1 and 2.5.9 yield a relationship between the rate of deformation expressed in Lagrangian variables and that in Eulerian variables, as given in Equation 2.5.2. Since the differential coordinates are linearly independent functions, the rate of deformation matrix is related to the time rate of change of the Lagrangian strain at a material point by

$$V_{k\ell} = \left(\frac{\partial \epsilon_{ij}}{\partial t}\right) \frac{\partial a_i}{\partial x_k} \frac{\partial a_j}{\partial x_\ell} \qquad\qquad (2.5.10)$$

The equivalence of Equations 2.5.8 and 2.5.10 may be directly verified by first taking the partial derivative of the Lagrangian strain matrix with respect to time,

$$\frac{\partial \epsilon_{ij}}{\partial t} = \tfrac{1}{2}\left(\frac{\partial v_m}{\partial a_i}\frac{\partial x_m}{\partial a_j} + \frac{\partial x_m}{\partial a_i}\frac{\partial v_m}{\partial a_j}\right)$$

and then multiplying by the coordinate deformation matrices, $\partial a_i / \partial x_k$ and $\partial a_j / \partial x_\ell$, to obtain

$$\frac{\partial a_i}{\partial x_k}\frac{\partial a_j}{\partial x_\ell}\left(\frac{\partial \epsilon_{ij}}{\partial t}\right) = \tfrac{1}{2}\left(\frac{\partial v_m}{\partial x_k}\frac{\partial x_m}{\partial x_\ell} + \frac{\partial x_m}{\partial x_k}\frac{\partial v_m}{\partial x_\ell}\right)$$

Since the coordinates, (x_i), are linearly independent (i.e., based on orthogonal axes),

$$\frac{\partial x_m}{\partial x_\ell} \equiv \delta_{m\ell}$$

$$\frac{\partial x_m}{\partial x_k} \equiv \delta_{mk}$$

and hence,

$$\frac{\partial a_i}{\partial x_k}\frac{\partial a_j}{\partial x_\ell}\left(\frac{\partial \epsilon_{ij}}{\partial t}\right) = \tfrac{1}{2}\left(\frac{\partial v_\ell}{\partial x_k} + \frac{\partial v_k}{\partial x_\ell}\right) = V_{k\ell}$$

This is the result derived previously using spatial derivatives of the velocity field.

By analogy to the development of the strain matrices, we can characterize the rate of deformation matrix by its invariants and also by the isotropic and deviatoric parts of the matrix. The invariants are

$$I_1 = V_{kk}$$

$$I_2 = \tfrac{1}{2}\left(V_{ij}V_{ji} - V_{ii}V_{jj}\right) \qquad\qquad (2.5.11)$$

In the principal axes system, the invariants reduce to

$$I_1 = V_1 + V_2$$

$$I_2 = -V_1 V_2 \tag{2.5.12}$$

As we have demonstrated, the isotropic part is proportional to the trace of the matrix or its mean,

$$\overline{V}_{ij} = \tfrac{1}{2} V_{kk} \delta_{ij} = \tfrac{1}{2} I_1 \delta_{ij} \tag{2.5.13}$$

The mean part describes a uniform rate of area expansion or condensation. The deviator, which characterizes the rate of shear deformation, is

$$\widetilde{V}_{ij} = V_{ij} - \overline{V}_{ij} = V_{ij} - \tfrac{1}{2} I_1 \delta_{ij} \tag{2.5.14}$$

The second invariant of the deviator is equal to the square of the maximum rate of shear deformation, V_s,

$$V_s^2 = \widetilde{I}_2 = \left(\frac{V_1 - V_2}{2} \right)^2 \tag{2.5.15}$$

The components of the rate of deformation matrix are given by

$$V_{11} = \tfrac{1}{2} I_1 + V_s \cos 2\phi$$

$$V_{22} = \tfrac{1}{2} I_1 - V_s \cos 2\phi$$

$$V_{21} = V_{12} = -V_s \sin 2\phi \tag{2.5.16}$$

for any choice of coordinates rotated by an angle, $-\phi$, relative to the principal axes system.

Assuming that the directions of the principal axes are not time dependent, we may simply view the surface element as being either extended or compressed at a finite rate parallel to each coordinate axis (Figure 2.12). The principal components of the rate of deformation are

$$V_1 = \frac{dv_1}{dx_1}$$

$$V_2 = \frac{dv_2}{dx_2} \tag{2.5.17}$$

Since the principal axes are time independent, the principal axes for V_{ij}, ε_{ij}, and $\dot{\varepsilon}_{ij}$ are all the same and

$$V_1 = \frac{\partial \epsilon_1}{\partial t} \left(\frac{da_1}{dx_1} \right)^2$$

$$V_2 = \frac{\partial \epsilon_2}{\partial t} \left(\frac{da_2}{dx_2} \right)^2 \tag{2.5.18}$$

From the principal extension ratios, λ_1 and λ_2, for each coordinate and Equations 2.4.1, the principal components of the rate of deformation are given by

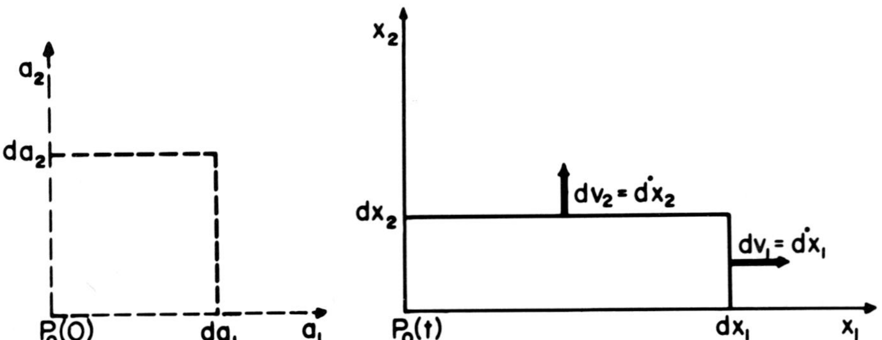

FIGURE 2.12.   Deformation of a plane, square surface element at a finite rate parallel to each coordinate axis. In this case, there is extension in the x_1 direction at a rate dv_1 and extension in the x_2 direction at a rate dv_2.

$$V_1 = \frac{1}{\lambda_1} \frac{\partial \lambda_1}{\partial t} = \frac{\dot{\lambda}_1}{\lambda_1},$$

$$V_2 = \frac{1}{\lambda_2} \frac{\partial \lambda_2}{\partial t} = \frac{\dot{\lambda}_2}{\lambda_2} \tag{2.5.19}$$

which are the dimensionless growth rates parallel to each instantaneous principal coordinate. The first invariant is

$$I_1 = V_1 + V_2 = \frac{\dot{\lambda}_1}{\lambda_1} + \frac{\dot{\lambda}_2}{\lambda_2} = \frac{\dot{\alpha}}{1 + \alpha} \tag{2.5.20}$$

where, again, α is the fractional change in area, Equation 2.4.5. Therefore, the isotropic part of the rate of deformation matrix represents the rate of area expansion or condensation of the material in the surface.

The maximum rate of shear deformation is given by

$$V_s = \tfrac{1}{2} \left| \frac{\dot{\lambda}_1}{\lambda_1} - \frac{\dot{\lambda}_2}{\lambda_2} \right| = \tfrac{1}{2} \left| \frac{dv_1}{dx_1} - \frac{dv_2}{dx_2} \right| \tag{2.5.21}$$

as shown in Figure 2.13, and occurs at $\pm 45°$ to the principal axes of extension in the instantaneous state. With the scaled extension ratio, $\tilde{\lambda}$, Equation 2.4.7, it can be shown that the maximum rate of shear deformation is the growth rate of $\tilde{\lambda}$ referred to the instantaneous element area,

$$V_s = \left| \frac{1}{\tilde{\lambda}} \frac{\partial \tilde{\lambda}}{\partial t} \right| \tag{2.5.22}$$

The condition of incompressibility of surface material is that the time rate of change of element area is zero, $\dot{\alpha} = 0$, or $V_{kk} = 0$. From Equations 2.5.17 and 2.5.20 this is given by the following two equivalent expressions,

$$\frac{\dot{\lambda}_1}{\lambda_1} + \frac{\dot{\lambda}_2}{\lambda_2} = \frac{dv_1}{dx_1} + \frac{dv_2}{dx_2} = 0 \tag{2.5.23}$$

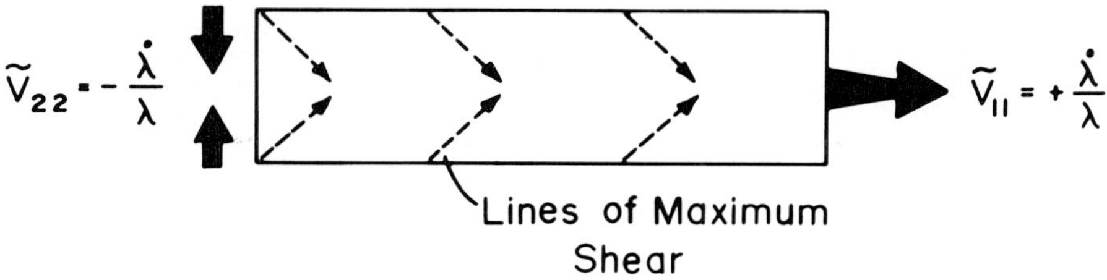

FIGURE 2.13.   Maximum rate of shear deformation occurring at ±45° to the principal axes of extension in the instantaneous state. The maximum rate of shear deformation is the growth rate of $\tilde{\lambda}$ referred to the instantaneous element area.

The second equation is known in fluid mechanics as the two-dimensional continuity equation; and because it is an invariant, it has the same form for any choice of instantaneous coordinates (x_i),

$$\frac{\partial v_k}{\partial x_k} = 0 \tag{2.5.24}$$

This states that the divergence of the surface velocity, v_k, is zero.

For the incompressible flow of surface material, we see that the maximum rate of shear deformation is the magnitude of the growth rate of either principal extension ratio,

$$V_s = \tfrac{1}{2} \left| \frac{\dot{\lambda}_1}{\lambda_1} + \frac{\dot{\lambda}_1}{\lambda_1} \right| = \left| \frac{\dot{\lambda}_1}{\lambda_1} \right| \tag{2.5.25}$$

or the one-dimensional divergence of either velocity component,

$$V_s = \tfrac{1}{2} \left| \frac{dv_1}{dx_1} + \frac{dv_1}{dx_1} \right| = \left| \frac{dv_1}{dx_1} \right| \tag{2.5.26}$$

The second invariant of the rate of deformation matrix in the incompressible case, $(\dot{\alpha} = 0)$, gives a simple form for the maximum rate of shear deformation squared,

$$I_2 = \left(\frac{\dot{\lambda}_1}{\lambda_1} \right)^2 = \left(\frac{dv_1}{dx_1} \right)^2 = V_s^2 \tag{2.5.27}$$

The following set of examples is based on the deformations in Section 2.2 for an incompressible surface in which the area per molecule or surface density is constant. These examples will provide illustrations of the rate of deformation and the relationship between Lagrangian and Eulerian variables.

Example 1

Consider the shear deformation of a square element at constant area presented in Example 2 of Section 2.2. If the tangent of the angle, ϕ, is a function of time, then the rate of deformation components are given by

$$V_{11} = \left(\frac{\partial \epsilon_{11}}{\partial t} \frac{\partial a_1}{\partial x_1} \frac{\partial a_1}{\partial x_1} + 2 \frac{\partial \epsilon_{12}}{\partial t} \frac{\partial a_1}{\partial x_1} \frac{\partial a_2}{\partial x_1} + \frac{\partial \epsilon_{22}}{\partial t} \frac{\partial a_2}{\partial x_1} \frac{\partial a_2}{\partial x_1} \right)$$

$$V_{22} = \left(\frac{\partial \epsilon_{11}}{\partial t} \frac{\partial a_1}{\partial x_2} \frac{\partial a_1}{\partial x_2} + 2 \frac{\partial \epsilon_{12}}{\partial t} \frac{\partial a_1}{\partial x_2} \frac{\partial a_2}{\partial x_2} + \frac{\partial \epsilon_{22}}{\partial t} \frac{\partial a_2}{\partial x_2} \frac{\partial a_2}{\partial x_2} \right)$$

$$V_{21} = V_{12} = \left(\frac{\partial \epsilon_{11}}{\partial t} \frac{\partial a_1}{\partial x_1} \frac{\partial a_1}{\partial x_2} + \frac{\partial \epsilon_{12}}{\partial t} \frac{\partial a_1}{\partial x_1} \frac{\partial a_2}{\partial x_2} + \frac{\partial \epsilon_{12}}{\partial t} \frac{\partial a_2}{\partial x_1} \frac{\partial a_1}{\partial x_2} \right.$$

$$\left. + \frac{\partial \epsilon_{22}}{\partial t} \frac{\partial a_2}{\partial x_1} \frac{\partial a_2}{\partial x_2} \right) \tag{2.5.28}$$

If we use $c = \tan \phi$ in the coordinate transformations of Example 2, Section 2.2, the rate of deformation components are given by

$$V_{11} = (0 + 0 + 0 + 0) = 0$$

$$V_{22} = \left[0 + 2 \left(\tfrac{1}{2} \frac{\partial c}{\partial t} \right) (-c) + c \frac{\partial c}{\partial t} \right] = 0$$

$$V_{21} = V_{12} = \left(0 + \tfrac{1}{2} \frac{\partial c}{\partial t} + 0 + 0 \right) = \tfrac{1}{2} \frac{\partial c}{\partial t}$$

$$V_s = \tfrac{1}{2} \left| \frac{\partial c}{\partial t} \right|$$

For the Eulerian variables, the rate of deformation components are

$$V_{11} = \tfrac{1}{2} \left(\frac{\partial v_1}{\partial x_1} + \frac{\partial v_1}{\partial x_1} \right) = \frac{\partial v_1}{\partial x_1}$$

$$V_{22} = \tfrac{1}{2} \left(\frac{\partial v_2}{\partial x_2} + \frac{\partial v_2}{\partial x_2} \right) = \frac{\partial v_2}{\partial x_2}$$

$$V_{21} = V_{12} = \tfrac{1}{2} \left(\frac{\partial v_1}{\partial x_2} + \frac{\partial v_2}{\partial x_1} \right) \tag{2.5.29}$$

In the instantaneous coordinate system, the relative velocities are given by

$$dv_1 = \frac{\partial c}{\partial t} dx_2$$

$$dv_2 = 0$$

therefore,

$$V_{11} = V_{22} = 0$$

$$V_{21} = V_{12} = \tfrac{1}{2} \, \frac{\partial v_1}{\partial x_2} = \tfrac{1}{2} \, \frac{\partial c}{\partial t}$$

in which $\partial c / \partial t$ is the relative velocity due to the shear between lines parallel to the x_1 coordinate axis and a unit distance apart.

Example 2

Consider the deformation of a conical annulus illustrated in Example 3 of Section 2.2, with the assumption that the area remains constant. We need only specify the radius as a function of time and initial position because the product of principal extension ratios, $\lambda_1 \lambda_2$, is always equal to unity,

$$\lambda_1 \lambda_2 = 1$$

hence, for example,

$$\lambda_1 = \frac{r}{r_0}$$

$$\lambda_2 = \frac{ds}{ds_0} = \lambda_1^{-1} = \frac{r_0}{r}$$

where the differential element lengths are defined in Example 3 of Section 2.2.

This deformation preserves the principal axes directions along the surface. We may use Equations 2.5.19 and 2.5.25 to obtain the principal rate of deformation components,

$$V_1 = \frac{\dot{\lambda_1}}{\lambda_1} = \frac{1}{r} \, \frac{\partial r}{\partial t}$$

$$V_2 = - \, \frac{\dot{\lambda_1}}{\lambda_1} = - \, \frac{1}{r} \, \frac{\partial r}{\partial t}$$

$$V_s = \left| \, \frac{1}{r} \, \frac{\partial r}{\partial t} \, \right| \qquad\qquad (2.5.30)$$

The time derivative of the radius is the radial velocity, v_r, of the point P_0,

$$V_s = \left| \, \frac{v_r}{r} \, \right|$$

To determine the components of the rate of deformation in terms of Eulerian variables, we use Equations 2.5.17. The relative velocities of neighboring points are obtained from the time rate of change of the differential element length along the meridian,

$$dv_2 = \frac{\partial (ds)}{\partial t} = dv_s$$

dv_s is the relative velocity between points along the meridian of the conical surface, taken as the x_2 direction. The differential coordinate for points on a latitude circle relative to P_0 is given by

$$dx_1 = r\, d\phi$$

therefore,

$$dv_1 = \frac{\partial r}{\partial t}\, d\phi$$

which is the velocity (relative to P_0) of local material points on the latitudinal circle of the conical surface. Note that the absolute velocity tangent to a latitude circle is zero because each material point moves only radially or axially and, hence, either diverges away from or converges toward P_0 as the latitudinal circle moves outward or inward, respectively. With the relative velocity components, we obtain

$$V_1 = \frac{1}{r}\frac{\partial r}{\partial t} = \frac{v_r}{r}$$

$$V_2 = \frac{dv_s}{ds}$$

$$V_s = \frac{1}{2}\left| \frac{dv_s}{ds} - \frac{v_r}{r} \right|$$

The equation of continuity is equal to zero because the surface density remains constant (no area change),

$$V_1 + V_2 = \frac{dv_s}{ds} + \frac{v_r}{r} = 0$$

therefore,

$$V_s = \left| \frac{dv_s}{ds} \right| = \left| \frac{v_r}{r} \right|$$

Next, consider constant surface density flow over a fixed axisymmetric surface shape. Flow is along the meridian. In Figure 2.4, we see that the velocity in the radial direction is the negative projection of the velocity along the meridian onto the plane normal to the axis of symmetry,

$$v_r = -v_s \sin\theta$$

The continuity equation (conservation of surface area) can be expressed in the curvilinear coordinate, s, as

$$\frac{dv_s}{ds} + \frac{v_r}{r} = 0$$

or,

$$\frac{dv_s}{ds} = \frac{v_s}{r} \sin \theta$$

The sine of the angle between the surface normal and the axis of symmetry, $\sin \theta$, is given by

$$\sin \theta = - \frac{dr}{ds}$$

Consequently, the continuity equation for surface flow is

$$\frac{dv_s}{ds} = - \frac{v_s}{r} \frac{dr}{ds} \tag{2.5.31}$$

and

$$v_r = v_s \frac{dr}{ds}$$

The maximum rate of shear deformation is either

$$V_s = \left| \frac{dv_s}{ds} \right| \tag{2.5.32}$$

or

$$V_s = \left| \frac{v_s}{r} \frac{dr}{ds} \right| = \left| \frac{v_r}{r} \right|$$

which is the same result as obtained with Lagrangian variables. The maximum rate of shear deformation is the magnitude of the gradient of the flow velocity along the meridian of the surface if the shape of the surface is fixed.

2.6 Nonuniform Deformation and Rate of Deformation

In developing the analyses of deformation and rate of deformation of a surface, we investigated the change in distances between points local to a position on the surface. We chose the local domain to be small enough that we could treat the surface element as flat, described by cartesian differential coordinates. We derived differential relations, which specify strain matrices and rate of deformation matrices, by using the infinitesimal approximation that the deformation is uniform over this small domain. Consequently, deformation and rate of deformation variables are continuous functions of position and are intensive representations of geometry changes, i.e., these variables are point definitions that do not depend on the extrinsic size of the surface envelope. Clearly, no system of units or scale is required to measure these properties. This powerful method, based on differential geometry, may be used to determine nonuniform deformation and rate of deformation of a surface, such as variations of surface extension and density from one location to another. In order to illustrate nonuniform extension and rate of extension in a surface, we will give some simple examples for the condition of constant surface density (i.e., the areas of differential elements in the surface are unchanged by the deformation).

Example 1

Consider an infinite flat surface with a hole of radius, R_0, in the center. Now, with deformation of the surface at constant area, the size of the hole is changed to a new value, given by radius, R. The initial dimensions of a surface element are $r_0 \, d\phi$ and dr_0, where r_0 is the radial position in the initial state and $d\phi$ is a differential portion of azimuthal angle. The instantaneous or deformed dimensions of the same surface element are $r \, d\phi$ and dr, respectively, where r is the new radial position of the surface element ($d\phi$ is the same because of circular symmetry). The material extension ratios are given by

$$\lambda_1 \equiv \frac{r}{r_0}$$

and

$$\lambda_2 = \frac{dr}{dr_0}$$

The assumption of constant surface density or element area is defined by $\lambda_1 \lambda_2 \equiv 1$, or

$$r \, dr = r_0 \, dr_0$$

This simple relation shows that the area of a thin annular ring is equal before and after deformation. By integrating this equation from the common location at the hole radius, we establish the equivalence of surface areas in both states (initial and instantaneous),

$$r^2 - R^2 = r_0^2 - R_0^2$$

this relation gives the nonuniform distribution of extension ratios throughout the surface. For instance, in a spatial reference frame (instantaneous state), the extension of material along the radius (meridian) line is given by

$$\lambda_2 = \frac{dr}{dr_0} = \sqrt{1 + \frac{(R_0^2 - R^2)}{r^2}} = \lambda_1^{-1} = \frac{r_0}{r}$$

The degree of extension is maximum at the edge of the hole and decreases inversely with increasing distances from the hole, approaching the undeformed level ($\lambda = 1$) at large radii. If the new radius is smaller than the initial hole size, we see that the surface elements are elongated along the meridian or radius line and compressed circumferentially. For a larger hole size, the opposite deformation occurs.

Example 2

Next, consider the rate of deformation that occurs in Example 1 when the radius of the hole changes at a finite rate.

$$R = R(t)$$

For the Lagrangian approach, the extension ratios must be defined relative to a coordinate system fixed with the material, i.e., r_0,

$$\lambda_1 = \sqrt{1 + \frac{(R^2 - R_0^2)}{r_0^2}}$$

The rate of deformation is given by the fractional rate of change of the extension ratio,

$$V_s = \left| V_1 \right| = \left| V_2 \right| = \left| \frac{1}{\lambda_1} \frac{\partial \lambda_1}{\partial t} \right| = \left| \frac{1}{\lambda_2} \frac{\partial \lambda_2}{\partial t} \right|$$

Only the rate of shear deformation is nonzero because the surface is assumed to be incompressible. Taking the time derivative of the extension ratio in the material reference frame, we obtain,

$$V_s = \left| V_1 \right| = \left| \frac{1}{\lambda_1} \frac{\partial \lambda_1}{\partial t} \right| = \left| \frac{R}{[r_0^2 + (R^2 - R_0^2)]} \frac{\partial R}{\partial t} \right|$$

This can be written in terms of the instantaneous spatial coordinate, r,

$$V_s = \left| \frac{R}{r^2} \frac{\partial R}{\partial t} \right|$$

The rate of shear deformation falls off inversely with the square of the radial coordinate.

Now, for the Eulerian approach, the radial velocity field, v_r, needs to be established. For the incompressible flow of surface, the continuity equation gives,

$$\frac{dv_r}{dr} + \frac{v_r}{r} = 0$$

which can be integrated to yield a conservation relation,

$$v_r \cdot r = \text{constant}$$

The velocity at the edge of the hole is $v_r^0 = \partial R / \partial t$; therefore, the conservation relation specifies the radial velocity field,

$$v_r \cdot r = v_r^0 \cdot R = R \frac{\partial R}{\partial t}$$

and

$$v_r = \frac{R}{r} \frac{\partial R}{\partial t}$$

The rate of shear deformation is given by the following equivalent relations:

$$V_s = \left| \frac{dv_r}{dr} \right| = \left| \frac{v_r}{r} \right|$$

therefore, we obtain the same result as before the Lagrangian approach,

$$V_s = \frac{R}{r^2} \frac{\partial R}{\partial t}$$

Example 3
   As another example, consider the deformation of an initially flat surface into an

infinite paraboloid envelope at constant surface density. Again, we will only need to use one principal extension ratio, say, along the meridional generator of the paraboloid surface. This extension ratio, λ, will be given by the ratio of initial to instantaneous radial coordinates as demonstrated in Example 3 of Section 2.2.

$$\lambda = \frac{r_0}{r}$$

The paraboloid envelope is described by the quadratic function of spatial coordinates (z, r).

$$z = c\,r^2$$

The constant element area assumption is stated explicity as

$$r\,ds = r_0\,dr_0$$

where ds is the differential length of the element along the meridian. It is given by

$$ds = \sqrt{\left(\frac{dz}{dr}\right)^2 + 1}\;\;dr$$

or in particular by

$$ds = \sqrt{4c^2\,r^2 + 1}\;\;dr$$

Therefore, surface incompressibility is represented by the integral relation,

$$\int_0^r r' \cdot \sqrt{4c^2\,r'^2 + 1}\;\;dr' = \int_0^{r_0} r'\,dr' = \frac{r_0^{\,2}}{2}$$

and corresponding points are given by

$$r_0^2 = \frac{1}{6c^2}\,[(4c^2\,r^2 + 1)^{3/2} - 1]$$

The square of the extension ratio for differential elements along the meridian is determined to be

$$\lambda^2 = \frac{1}{6c^2\,r^2}\,[(4c^2\,r^2 + 1)^{3/2} - 1]$$

For small distances from the symmetry axis, $r \ll c$, the extension ratio is approximated by

$$\lambda \simeq 1 + c^2\,r^2 = 1 + c\,z$$

For large distances from the symmetry axis, $r \gg c$, the extension ratio approaches,

$$\lambda \rightarrow \frac{4}{3}\,cr = \frac{4}{3}\,\sqrt{cz}$$

Therefore, the surface extension increases quadratically near the origin with radius but eventually increases linearly with radius at large distances from the origin.

Example 4

Consider incompressible flow of the surface over a paraboloid shape as defined in Example 3. The flow of magnitude, $\dot{A}$, will be assumed to emanate from a "source" at the origin. This flow is the surface area produced per unit time (flux of surface at the source). The continuity equation for surface flow excluding the origin is given by

$$\frac{\partial v_s}{\partial s} + \frac{v_s}{r} \frac{\partial r}{\partial s} = 0 = \frac{\partial}{\partial s} (r \cdot v_s) = 0$$

where v_s is the velocity of the surface along the meridian of the paraboloid. This equation gives the conservation relation in terms of the source area flux,

$$r \cdot v_s = \frac{\dot{A}}{2\pi}$$

and the distribution of flow velocity along the meridian is

$$v_s = \frac{\dot{A}}{2\pi r}$$

The rate of shear deformation is given by

$$V_s = \left| \frac{v_s}{r} \frac{\partial r}{\partial s} \right|$$

If we take the curvilinear distance, s, along the meridian as positive from the origin of the paraboloid outward, the derivative of this distance with respect to radius is obtained from Example 3,

$$\frac{ds}{dr} = \sqrt{4c^2 r^2 + 1}$$

Therefore, the rate of shear deformation is given as a function of radial or axial position on the paraboloid by

$$V_s = \left| \frac{\dot{A}}{2\pi r^2 \sqrt{4c^2 r^2 + 1}} \right|$$

or

$$V_s = \left| \frac{c\dot{A}}{2\pi z \sqrt{4cz + 1}} \right|$$

SECTION III

3.1 Membrane Mechanical Equilibrium

In Section 2, the deformation and rate of deformation of a membrane surface was quantitatively described in terms of intensive variables (e.g., the principal extension ratios and their time derivatives) defined at each point of the membrane. In this section, we will determine the intensive distribution of forces and moments in the membrane, which result from external forces applied to the membrane. The membrane will be assumed to support the external forces without acceleration, i.e., we assume equilibrium between the forces applied to the membrane surface and the forces and moments within the material itself.

We consider the membrane as a laminated composite of molecular layers. A single layer of the composite can only oppose forces that locally act in the plane of the layer. Such forces are distributed per unit length in the surface, e.g., dyn/cm or erg/cm^2 in CGS units. This is illustrated in Figure 3.1 where we cut the surface layer and expose the force distributed along the edge. The force per unit length along the edge may be decomposed into two parts, normal and tangent to the edge. The surface force resultants are called the tension, T_n, and shear resultant, T_s, respectively. These variables are Eulerian because they are measured in the instantaneous state relative to the deformed surface element coordinates (dx_1, dx_2). Since the force resultants are expressed per unit length of the material, they are intensive membrane functions that can vary from one location to another on the surface.

Thin strata of molecular layers under external loads also respond with in-plane forces. However, if the layers are strongly associated, then couples of force resultants between layers can be produced that create a moment resultant, M, which acts on the exposed edge shown in Figure 3.2. Because these couples are formed by small interlayer distances (e.g., the order of molecular dimensions), the moment resultant usually contributes negligibly to the membrane equilibrium in response to external forces. There are important situations, however, where the moment resultant is the dominant response to the external forces. Moment resultants have units of dyn-cm/cm, dynes, or erg/cm in the CGS system. In addition to the moment resultant, a transverse shear resultant, Q_T, may be produced. The transverse shear resultant acts to cut the membrane normal to its plane and has CGS units of dyn/cm or erg/cm^2. Figure 3.2 illustrates the resultants on an exposed edge of a laminated membrane structure. Uncoupled molecular layers only exhibit in-plane force resultants like individual layers, since they can slide relative to each other. The energy produced by moments in a single molecular layer is proportional to the angular rotation about a molecule. Therefore, it will be negligibly small until the surface radii of curvature approach molecular dimensions. Curvatures of such a large magnitude are essentially incompatible with the concept of a continuous medium and are best treated as discontinuities or defect structures in the material. Continuous curvature effects in coupled and uncoupled membrane strata are developed in Section IV.

3.2 Membrane Force Resultants

As described in Section 3.1, the surface force resultants are made up of tension and shear components along any exposed edge (Figure 3.1). If we excise a small material element (as we did in Section II to study deformation), the element will have a set of tension resultants, T_{11} and T_{22}, and a set of shear resultants, T_{21} and T_{12}, as shown in Figure 3.3. The first index specifies the direction of the force resultant relative to the instantaneous axes (x_1, x_2), e.g., the components T_{1j} act in the x_1 direction. The second index specifies the particular material edge along which the resultant acts, e.g., given

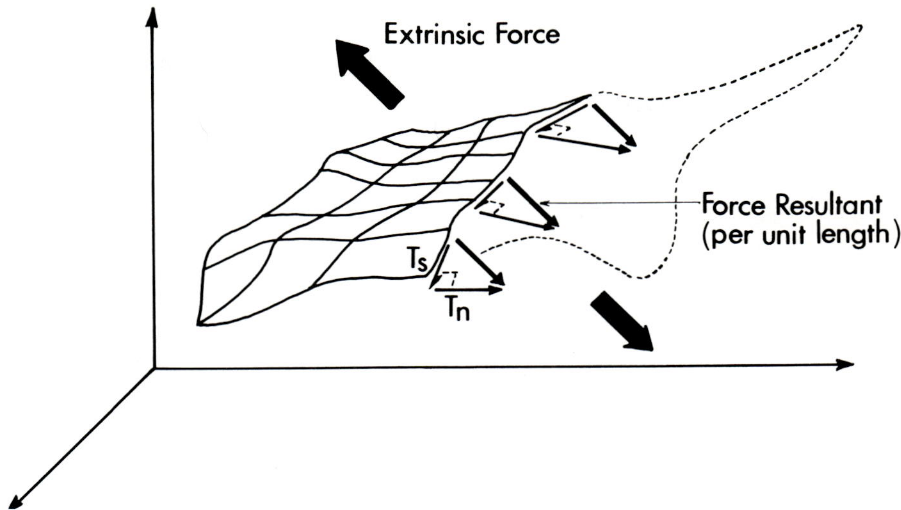

FIGURE 3.1.   Free-body diagram for a surface layer. The total force that is applied to the surface layer is shown by the heavy arrows. The force resultants are distributed per unit length along with imaginary cut made in the layer and are shown by the light arrows. Here, T_s is the shear resultant and T_n is the tension. The CGS units of T_s and T_n are dyn/cm. (From Evans, E. A. and Hochmuth, R. M., *Current Topics in Membranes and Transport,* Vol. 10, Kleinzeller, A. and Bronner, F., Eds., Academic Press, New York, 1978, 1. With permission.)

by either differential length dx_1 or dx_2. The element edges are defined directionally by the normal to the edge. For instance, the edges with length dx_2 have normal directions parallel to the x_1 axis (taken as positive out of the material element). Thus, the force resultants, T_{i1}, act on the element edges having the length, dx_2. Figure 3.3 implies that all forces are distributed uniformly in the region of the material element because the force resultants are balanced identically on opposite sides of the element. For small elements, this is correct to first order. The second order departure from uniformity will lead to the differential equations of equilibrium for the surface. On an infinitesimal scale, the intrinsic force resultants will be uniform as shown in Figure 3.3. In addition to the balance of intrinsic forces in any direction, the net moment about the surface normal at the material element center of mass must be zero for equilibrium unless an intrinsic surface couple exists that acts about the surface normal. Without molecular surface couples, the moment of the intrinsic forces about the surface normal is the sum of the two couples produced by the shear resultant pairs shown in Figure 3.3,

$$(T_{21} \, dx_2) \, dx_1 \; - \; (T_{12} \, dx_1) \, dx_2 \; = \; 0 \qquad\qquad (3.2.1)$$

where the force components are in parentheses. For the particular sense of the shear resultants chosen, it is apparent that the assumption of zero intrinsic moment (no surface couple sources) about the surface normal implies that the components, T_{21} and T_{12}, are equal.

The resultant quantities, T_{11}, T_{22}, T_{21}, and T_{12}, have values that depend on the coordinate axes system chosen at any point, and they form a matrix which transforms according to the previous rule for any coordinate transformation. For example, if we rotate the axes system (x_1, x_2) 90°, the normal components are interchanged, and the shear components are also interchanged with opposite signs as in Equations 2.3.10 or

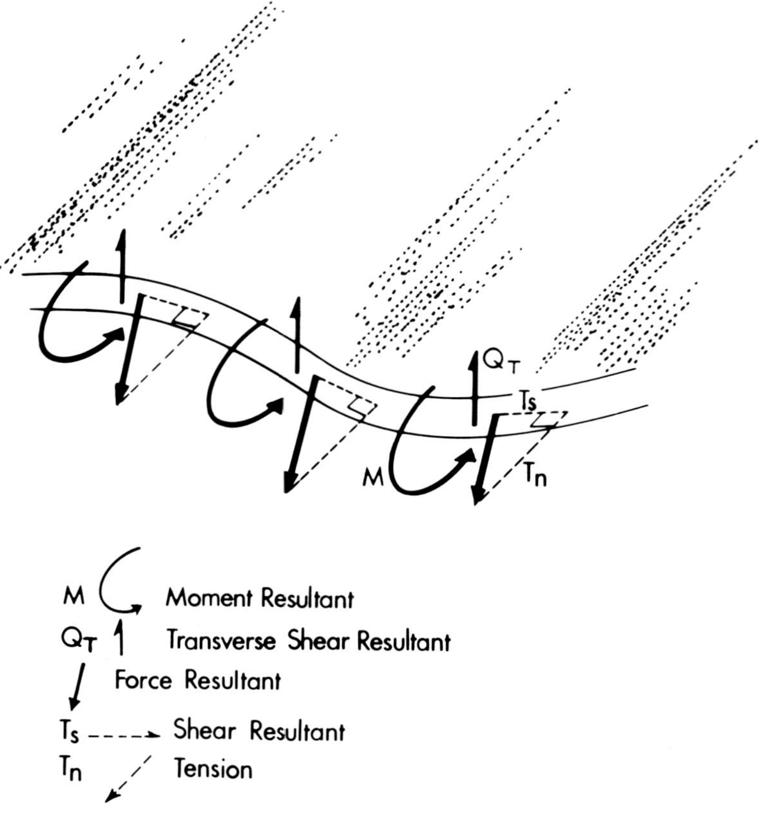

FIGURE 3.2.   Force and moment resultants which are produced in a thin, multi-layer membrane as the result of coupling between adjacent layers. Here, in addition to the shear resultant, T_s, and tension, T_n, shown previously in Figure 3.1, the membrane may have a moment resultant per unit length, M, (CGS units of dyn-cm/cm) and a transverse shear resultant, Q_T (CGS units of dyn/cm). (From Evans, E. A. and Hochmuth, R. M., *Current Topics in Membranes and Transport,* Vol. 10, Kleinzeller, A. and Bronner, F. Eds., Academic Press, New York, 1978, 1. With permission.)

Figure 2.7. As for any symmetric matrix, the force resultants can be expressed in terms of principal components, called principal tensions T_1 and T_2,

$$T_{11} = \frac{(T_1 + T_2)}{2} + \frac{(T_1 - T_2)}{2} \cos 2\phi$$

$$T_{22} = \frac{(T_1 + T_2)}{2} - \frac{(T_1 - T_2)}{2} \cos 2\phi$$

$$T_{21} = T_{12} = - \frac{(T_1 - T_2)}{2} \sin 2\phi \tag{3.2.2}$$

where minus ϕ is the angular rotation from the principal axes system required to arrive at the desired coordinate axes by analogy to Equations 2.3.12. Again, it must be emphasized that the intensive force distribution represented by the resultant matrix is based on the instantaneous coordinates, i.e., measured in the instantaneous material state, and is thus an Eulerian or spatial variable. In the principal axes system, there

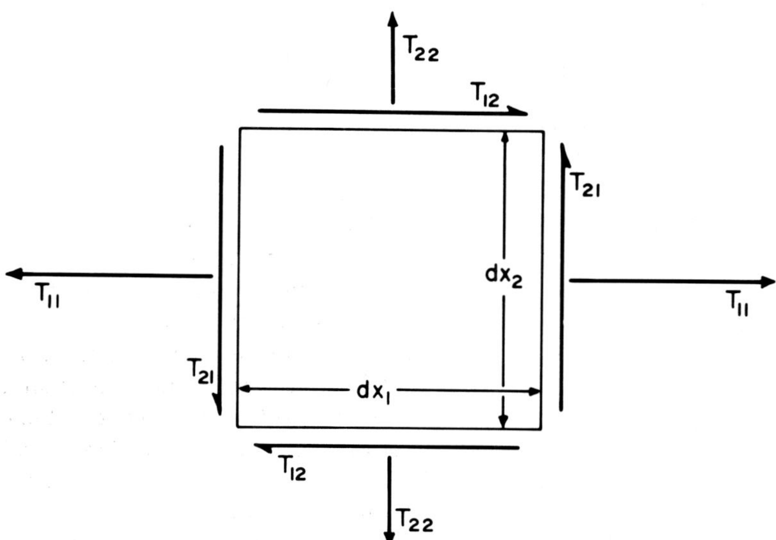

FIGURE 3.3.   A diagram of the surface force resultants supported by a small material element. The element has a set of tension resultants, T_{11} and T_{22}, and a set of shear resultants, T_{21} and T_{12}. Note that the first index specifies the direction of the force resultant relative to the instantaneous axes (x_1, x_2) and that the second index specifies the particular material edge along which the resultant acts. The figure implies that all forces are distributed uniformly in the region of the element.

are no shear resultant components. The material element is simply subjected to tension or compression normal to the material element edges.

Equations 3.2.2 show that the resultant matrix is characterized by two parts: (1) an isotropic or mean, tension, $\overline{T}$; (2) a deviatoric part, $\tilde{T}_{ij}$, representing shear contributions,

$$T_{ij} = \overline{T}\delta_{ij} + \tilde{T}_{ij} \tag{3.2.3}$$

The isotropic tension is

$$\overline{T} \equiv \frac{T_1 + T_2}{2} \tag{3.2.4}$$

The deviatoric part is given simply by the maximum shear resultant, T_s, and the coordinate rotation angle, ϕ,

$$T_s = \left| \frac{T_1 - T_2}{2} \right| \tag{3.2.5}$$

For $T_1 > T_2$, the deviatoric components are

$$\tilde{T}_{11} = T_s \cos 2\phi$$

$$\tilde{T}_{22} = -T_s \cos 2\phi$$

$$\tilde{T}_{12} = \tilde{T}_{21} = -T_s \sin 2\phi = T_{12} = T_{21} \tag{3.2.6}$$

The shear components of the deviator are the same as in the original matrix. Throughout the development of equations of mechanical equilibrium, we will implicitly use the symmetry property of the force resultant matrix, i.e., $T_{12} \equiv T_{21}$.

3.3 Membrane Equations of Mechanical Equilibrium

Equilibrium equations relate the external forces applied to a membrane to the force and moment resultants within the membrane. The equilibrium equations follow from the requirement that for any body in steady motion (constant velocity or at rest), the sum of all forces and the sum of all moments that act on the body must be zero. This applies to any part or element of the body if account is taken of the forces exerted on the element by the remainder of the body. If an element is in unsteady motion, the sum of the forces equals the mass of the element times the acceleration of its centroid. Equations which express this fact are called (Newton's) equations of motion. Similarly, the sum of moments that act on the element may be different from zero due to angular accelerations. The equations of mechanical equilibrium describe forces and moment balances for the body or membrane without any acceleration, i.e., they are equations of motion with the acceleration terms set equal to zero. It is important to recognize that mechanical equilibrium is not the same as thermodynamic equilibrium. In mechanical equilibrium, forces may be dissipative (nonconservative) as well as elastic (conservative) in character. Thermodynamic equilibrium is associated with reversible (conservative) processes and, consequently, only involves a balance of elastic forces. This is sometimes referred to as "static" equilibrium.

The equations of mechanical equilibrium may usually be applied to biological membranes even if they do experience acceleration because the product of the membrane mass times the acceleration is negligible compared to other forces. This is so because the membrane is normally less than a few hundred angstroms thick and, thus, has an extremely small surface mass density. In addition, the applied forces and accelerations must be small enough that living tissues can survive without damage. However, some cases, such as analysis of vibration distal to a stenosis in a massive arterial wall, may require taking the inertial terms into account.

Equations of equilibrium may include body forces proportional to the surface density. For a membrane with uniform surface density, body forces are distributed in proportion to the area of the membrane element. Examples are gravitational forces and forces produced by external electromagnetic fields. Intrinsic surface or body couples are also created by external fields. Surface couples can be important if the membrane surface contains polarized molecular complexes, but they are not expected to be important in most biological membranes and are not included here. (See Naghdi, 1972[63] for an exposition including such effects.)

We will first develop equations of mechanical equilibrium for the specific case of surface layers with negligible moment resultants. We will then consider the more general thin membrane case. Plane and axisymmetric membrane geometries will be used to develop the equilibrium equations as functions of the spatial coordinates. (For more general equations see Naghdi, 1972; Reissner, 1949.)[63,73]

3.4 Mechanical Equilibrium of a Plane Surface Layer

An element of a plane layer is shown in Figure 3.4. We first assume that the x_1, x_2 axes are the principal axes of the force resultant matrix and that moment resultants are negligible. Forces from the adjacent environment produce tractions or stresses on the surface layer that are distributed per unit area of the surface. Shear stress and pressure are forces per unit area (dyn/cm^2) exerted by fluids or solids in contact with the surface. The net normal traction is the difference between the pressures, P and P′,

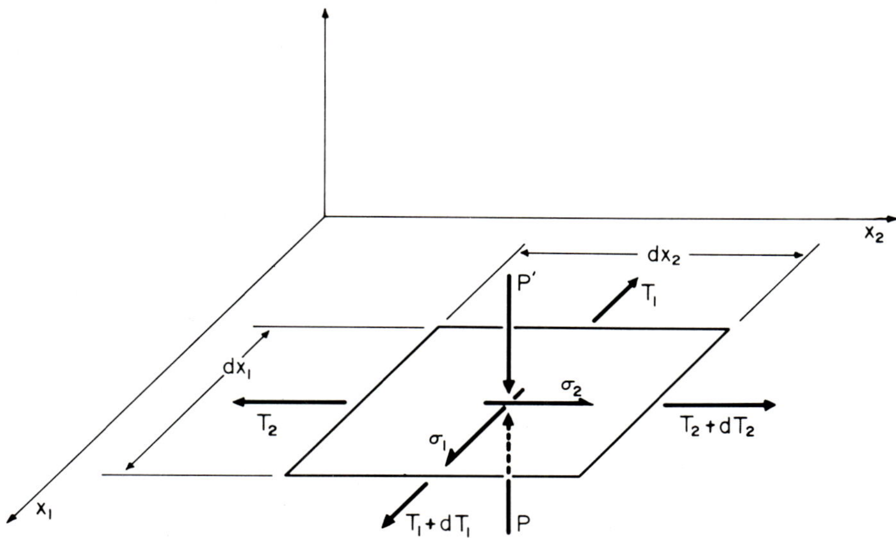

FIGURE 3.4.   An element of a plane layer with the x_1, x_2 axes as the principal axes of the force resultant matrix. Moment resultants are negligible. The net normal traction is the difference between the pressures, P and P′, and the net shear stresses in the x_1 and x_2 directions are the tangential tractions, σ_1 and σ_2. Shear stress and pressure are forces per unit area (dyn/cm²).

acting on the lower and upper sides of the surface, respectively. The tangential tractions, σ_1 and σ_2, are the net shear stresses in the x_1 and x_2 directions as shown. The tractions, σ_1 and σ_2, are the sum of shear forces per unit area that act on both faces of the layer. Added to the forces shown in Figure 3.4 are the body forces per unit mass of material, which act with components f_1, f_2, f_3 in the x_1, x_2, x_3 directions.

The balances of forces in three orthogonal (independent) directions define the equations of mechanical equilibrium,

$$\Sigma F_1 \equiv 0; \ \Sigma F_2 \equiv 0; \ \Sigma F_3 \equiv 0;$$

$$\Sigma F_i \equiv 0 \tag{3.4.1}$$

where ΣF_i stands for the sum of all forces that act on the element in each independent direction. Since we are considering an infinitesimally small element of the surface, the increments of the force resultant components from one side of the element to the other can be approximated by the first order term in a Taylor series expansion. Therefore, we write the incremental changes in force resultants from one side of the element to the other as differential forms, dT_1 and dT_2. The force resultant components and the differential increments are illustrated in Figure 3.4. The balance of forces in the x_1 direction for the small element is given by

$$(T_1 + dT_1) \, dx_2 - T_1 \, dx_2 + \sigma_1 dx_1 dx_2 + \rho f_1 dx_1 dx_2 = 0 \tag{3.4.2}$$

where density, ϱ, is the mass per unit area of surface material. The differential increment in principal tension in the x_1 direction is given by its spatial gradient times the differential coordinate dimension,

$$dT_1 = \frac{\partial T_1}{\partial x_1} \, dx_1 \tag{3.4.3}$$

Equations 3.4.3 and 3.4.2, divided by the element area, $dx_1 \, dx_2$, give the first equilibrium equation,

$$\frac{\partial T_1}{\partial x_1} + \sigma_1 + \rho f_1 = 0 \tag{3.4.4}$$

Similarly, the balance of forces on the element in the x_2 direction gives the equation,

$$\frac{\partial T_2}{\partial x_2} + \sigma_2 + \rho f_2 = 0 \tag{3.4.5}$$

Force equilibrium in the x_3 direction does not involve T_1 and T_2 because the resultants lie in the $x_1 \, x_2$ surface which is flat. The force balance in the direction normal to the element is

$$P - P' + \rho f_3 = 0 \tag{3.4.6}$$

Equations 3.4.4, 3.4.5, and 3.4.6 are the equations of equilibrium for a plane layer in which the principal tensions are assumed to be parallel to x_1 and x_2. Two interesting observations are apparent from these equations:

1. Variations in the force resultants are produced by tangential, shear traction, σ, or body force components parallel to the plane.
2. The pressures, P and P′, are equal if f_3 is zero. We see that a membrane cannot sustain a pressure difference without curvature or moment resultants.

If the axes (x_1, x_2) are not the principal axes of the force resultant matrix, the equations of equilibrium of a plane layer without moment resultants are obtained by replacing T_1, T_2 by T_{11}, T_{22} and adding shear resultants, T_{12}, to each edge of the element. As before, the application of Equations 3.4.1 yields the equilibrium equations,

$$\frac{\partial T_{ij}}{\partial x_j} + \sigma_i + \rho f_i = 0 \tag{3.4.7}$$

Equations 3.4.7 are the general equations of mechanical equilibrium for plane membranes without moments. Because the mass per unit area of biological membranes is so small (the surface density is on the order of 10^{-7} to 10^{-8} g/cm^2), the body force contributions can usually be neglected as were the acceleration forces. Throughout the remainder of the discussion, this assumption will be employed.

3.5 Mechanical Equilibrium of an Axisymmetric Surface Layer

The coordinate systems that we use to describe an axisymmetric membrane are shown in Figure 3.5A. The cylindrical coordinates (r, z, ϕ) are defined with z as the axis of symmetry. It is also convenient to define curvilinear coordinates (s, θ, ϕ) for any point on the surface. "s" is defined as the distance along the meridional curve measured from an arbitrary base location; the coordinate, θ, is defined as the angle that vector, **n**, normal to the surface makes with the z axis; the coordinate, ϕ, in the s, θ, ϕ system is the same azimuthal angle as in cylindrical coordinates. The principal radii of curvature for the surface are labeled R_1 and R_2. For the axisymmetric surface, R_1 is the radius of curvature that describes a local arc of the surface in the plane which contains the surface normal and is orthogonal to the surface meridian. It is the distance from the symmetry axis to the surface along the surface normal. The other principal radius of the curvature, R_2, describes a local arc of the surface meridian.

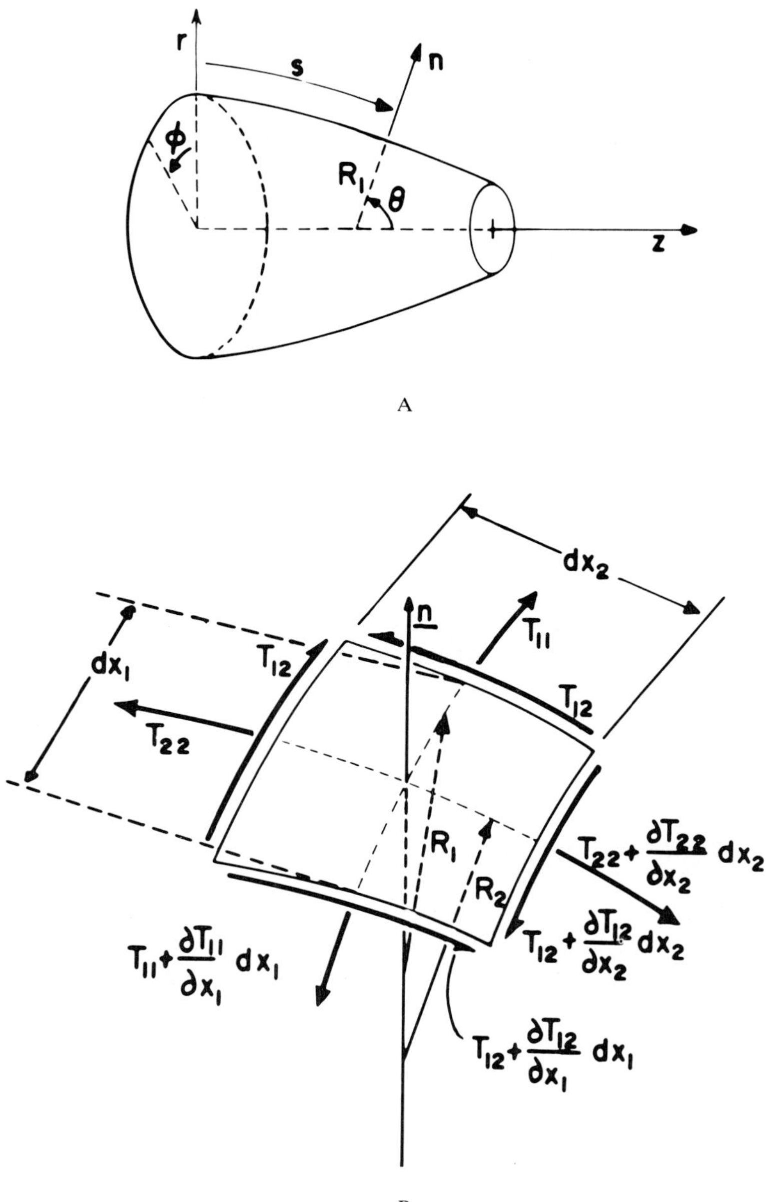

FIGURE 3.5. (A) The coordinate systems used to describe an axisymmetric membrane. (B) A small differential element, bounded by two meridional lines and two latitudinal circles, excised from the axisymmetric surface shown in (A). The element is only approximately flat and approximately rectangular. Principal radii of curvature and force resultants are illustrated. (C) The normal and tangential surface tractions applied by the adjacent media to the differential surface element.

If we excise a small differential surface element (bounded by two meridional lines and two latitudinal circles), the force resultants and principal radii of curvature are as illustrated in Figure 3.5B. It is obvious that the element is only approximately flat and approximately rectangular. The small deviations from a flat plane and rectangular shape produce important contributions to the equations of equilibrium. The local spatial axes are chosen as x_1 tangent to a latitude circle (in a clockwise sense around the z

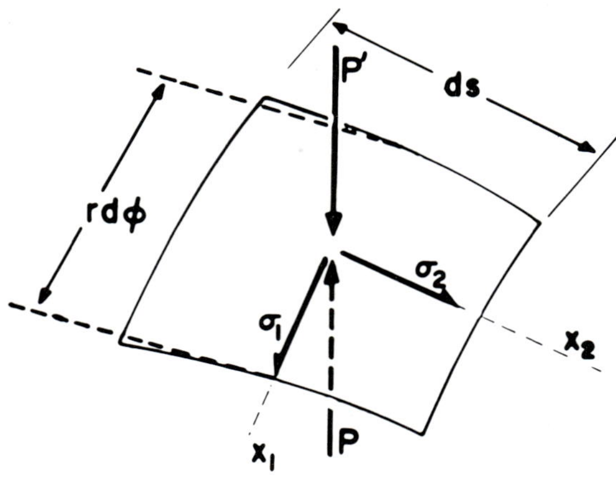

FIGURE 3.5 C

axis) and x_2 tangent to a meridian. Both axes are orthogonal to the surface normal. Consequently, the small surface element is described by differential coordinates (dx_1, dx_2), as projected onto the tangent plane, and by small deviations from the plane to the surface element along the normal to the surface. The projection of the differential surface element onto the tangent plane is approximately a trapezoid, illustrated in Figure 3.6A. The projected lengths of the latitude circles along the x_1 direction are different from one side of the element to the other as shown. The projected lengths of the bounding meridians are the same on both sides because of symmetry, but they deviate slightly from the x_2 direction. The differential coordinates (dx_1, dx_2) specify the mean of the base lengths and the altitude of the trapezoid, respectively. Therefore, the area of the element is $dx_1 \cdot dx_2$. The small angles at the base are given by half the rate of change of the differential length, dx_1, with respect to the x_2 direction, i.e., $\frac{1}{2}\,\partial(dx_1)/\partial x_2$ shown in Figure 3.6A. The deviations of the curved surface from the tangent plane are illustrated in Figure 3.6B. Since there are two principal radii of curvature that describe the surface (shown in Figure 3.5B), there will be two independent deviation angles, ($dx_1/2R_1$) and ($dx_2/2R_2$), both represented by ($dx/2R$) in Figure 3.6B.

Figure 3.5C shows the normal and tangential surface tractions (pressure and surface shear stress) applied by the adjacent environment. First, we will balance the forces in the x_1, x_2 directions, which define the plane tangent to the surface layer at the center of the chosen element. Then we will balance the forces normal to the tangential plane, along the vector **n**. Thus, Equations 3.4.1 will be satisfied for the surface element. In the x_1 direction, the difference in forces from one side of the element to the other includes contributions from the tension components, $d(T_{11}\,dx_2)$, the shear that acts on the edges formed by the latitude circles, $d(T_{12}\,dx_1)$, and the tangential surface traction, $\sigma_1\,dx_1\,dx_2$. In addition, there is a contribution of the shear on the edges formed by the meridians (ϕ = constant) because of the slightly trapezoidal shape of the element (schematically illustrated in Figure 3.6A. This is given by the sum,

$$\left[\left(T_{12} + \frac{\partial T_{12}}{\partial x_1}\,dx_1\right)dx_2 \cdot \tfrac{1}{2}\left(\frac{\partial(dx_1)}{\partial x_2}\right) + (T_{12})\,dx_2 \cdot \tfrac{1}{2}\left(\frac{\partial(dx_1)}{\partial x_2}\right)\right]$$

The trapezoidal shape is represented by the small angular deviations, $\frac{1}{2}[\partial(dx_1)/\partial x_2]$, from right angles at the bases formed by the latitude circles. There is no variation of

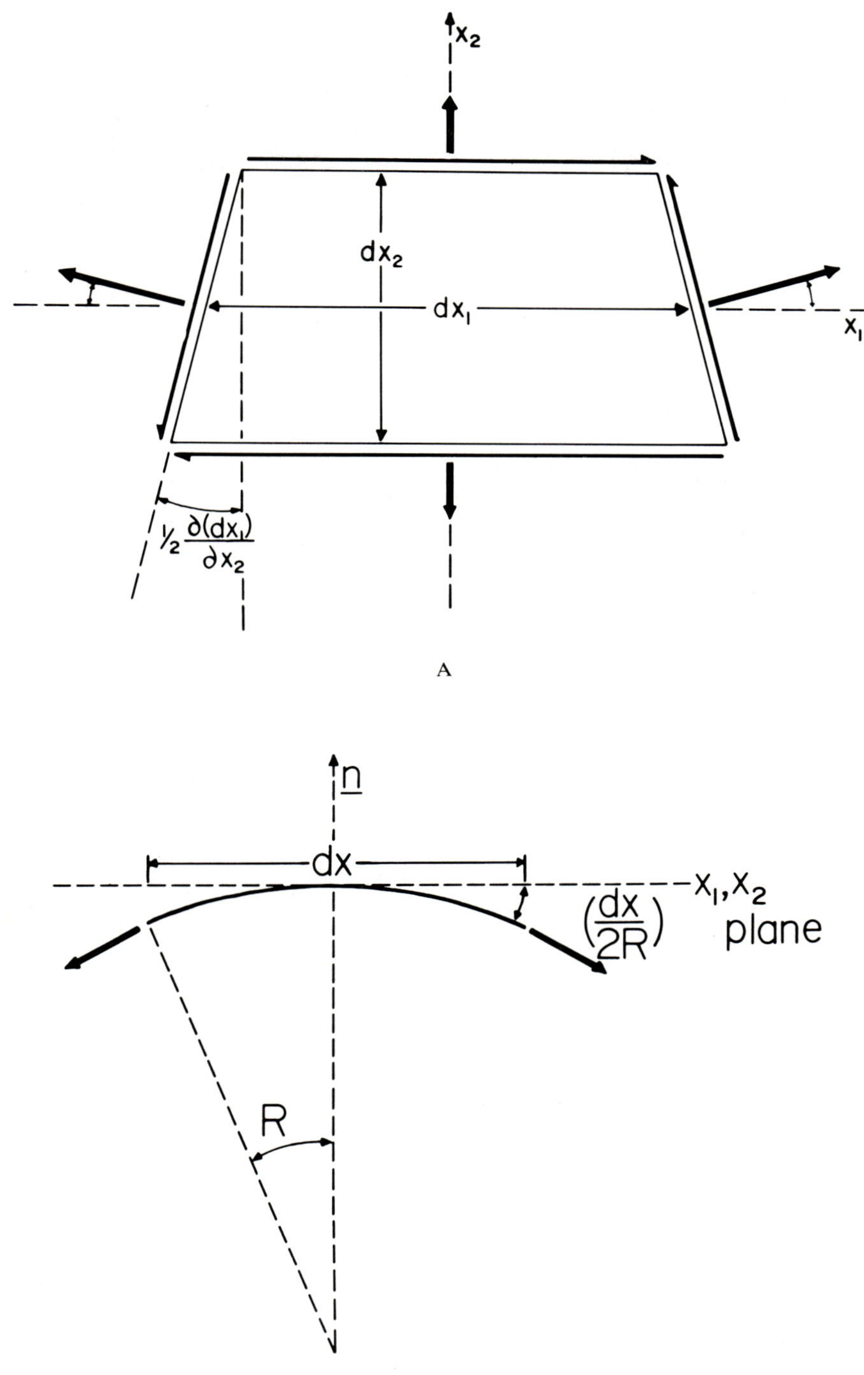

FIGURE 3.6.   (A) Trapezoidal projection of the curved surface element onto a plane tangent to the surface at its center. (B) Deviation of the curved surface from the tangent plane.

the edge length, dx_2, with respect to the x_1 direction because of symmetry. If we keep terms of first order, the force contributions are given by

$$d(T_{11}dx_2) = \frac{\partial(T_{11}dx_2)}{\partial x_1} dx_1 = \frac{\partial T_{11}}{\partial x_1} dx_2 dx_1$$

$$d(T_{12}dx_1) \doteq \frac{\partial(T_{12}dx_1)}{\partial x_2} dx_2$$

$$\left[\left(T_{12} + \frac{\partial T_{12}}{\partial x_1} dx_1\right) + T_{12}\right] \frac{dx_2}{2} \frac{\partial(dx_1)}{\partial x_2} \simeq T_{12} \frac{\partial(dx_1)}{\partial x_2} dx_2$$

(Note: As in the previous chapter, the differential operator, d( ), represents the local variation of a function over the scale of the differential coordinates, dx_1 and dx_2.)

Setting the sum of all forces equal to zero, we have

$$0 = \sigma_1 dx_1 dx_2 + \frac{\partial T_{11}}{\partial x_1} dx_2 dx_1 + \frac{\partial(T_{12}dx_1)}{\partial x_2} dx_2 + T_{12} \frac{\partial(dx_1)}{\partial x_2} dx_2 \qquad (3.5.1)$$

If we introduce curvilinear coordinates for the differential increments, dx_1 and dx_2,

$$dx_1 = r d\phi$$

$$dx_2 = ds$$

we can reduce Equation 3.5.1 to

$$0 = \sigma_\phi r + \frac{\partial T_{\phi\phi}}{\partial \phi} + \frac{\partial(T_{\phi m}r)}{\partial s} + T_{\phi m} \frac{\partial r}{\partial s} \qquad (3.5.2)$$

where we use the indices, m and ϕ, for components tangent and normal to the meridian, respectively. Equation 3.5.2 is the first equation of mechanical equilibrium.

For the x_2 direction, the force differences from one side of the element to the other include the tension component, $d(T_{22} dx_1)$, the shear on the edges formed by the meridian generators, $d(T_{12} dx_2)$, and the tangential surface traction, $\sigma_2 dx_1 dx_2$. Subtracted from this sum is a contribution from the other tension component that acts on the meridian edges because of the trapezoidal shape of the element (as shown in Figure 3.6A). The latter is

$$-\left[\left(T_{11} + \frac{\partial T_{11}}{\partial x_1} dx_1\right) dx_2 \cdot \tfrac{1}{2}\left(\frac{\partial(dx_1)}{\partial x_2}\right) + T_{11} dx_2 \cdot \tfrac{1}{2}\left(\frac{\partial(dx_1)}{\partial x_2}\right)\right]$$

The first order expressions for these contributions are

$$d(T_{22}dx_1) = \frac{\partial(T_{22}dx_1)}{\partial x_2} dx_2$$

$$d(T_{12}dx_2) = \frac{\partial(T_{12}dx_2)}{\partial x_1} dx_1 = \frac{\partial T_{12}}{\partial x_1} dx_2 dx_1$$

$$-\left[\left(T_{11} + \frac{\partial T_{11}}{\partial x_1} dx_1\right) + T_{11}\right] \frac{dx_2}{2} \frac{\partial(dx_1)}{\partial x_2} \simeq -T_{11} dx_2 \frac{\partial(dx_1)}{\partial x_2}$$

We set the sum of forces equal to zero,

$$0 = \sigma_2 \, dx_1 \, dx_2 + \frac{\partial(T_{22} \, dx_1)}{\partial x_2} \, dx_2 + \frac{\partial T_{12}}{\partial x_1} \, dx_2 \, dx_1 - T_{11} \, dx_2 \, \frac{\partial(dx_1)}{\partial x_2}$$

Again we use the curvilinear expressions for the differential increments, dx_1 and dx_2; the force balance reduces to

$$0 = \sigma_m r + \frac{\partial(T_{mm} r)}{\partial s} + \frac{\partial T_{\phi m}}{\partial \phi} - T_{\phi\phi} \, \frac{\partial r}{\partial s} \qquad (3.5.3)$$

which is the second equation of mechanical equilibrium with the subscript notation introduced previously for components normal and tangent to the meridian.

The last equilibrium equation is obtained from the balance of forces normal to the surface. The force differences are the net normal surface traction (pressure difference), $(P - P')dx_1 \, dx_2$, and the components of the tension resultants projected normal to the tangent plane (as illustrated in Figure 3.6B). These components are $-2(T_{11} \, dx_2) \, dx_1/2R_1$ and $-2(T_{22} \, dx_1) \, dx_2/2R_2$, where the small deviation angles between the force resultants and the tangent plane are given by the distance from the element center divided by the principal radius of curvature that describes the surface arc, e.g., $-dx_1/2R_1$ or $-dx_2/2R_2$. Consequently, curvature of the surface element creates force components to oppose the normal pressure difference,

$$0 = (P - P')dx_1 \, dx_2 - T_{11} \, dx_2 \left(\frac{dx_1}{R_1} \right) - T_{22} \, dx_1 \left(\frac{dx_2}{R_2} \right)$$

$$(3.5.4)$$

It is important to note that the radii of curvature are defined as positive for surface arcs which are concave toward the z axis.

Equations 3.5.2, 3.5.3, and 3.5.4 are the set of mechanical equilibrium equations for an axisymmetric membrane surface, without moments. These equations relate the intensive force resultants, T_{ij}, to the external forces on the membrane,

$$\frac{\partial T_{\phi\phi}}{\partial \phi} + \frac{\partial(T_{\phi m} r)}{\partial s} + T_{\phi m} \, \frac{\partial r}{\partial s} + \sigma_\phi r = 0 \qquad (3.5.5)$$

$$\frac{\partial(T_{mm} r)}{\partial s} + \frac{\partial T_{\phi m}}{\partial \phi} - T_{\phi\phi} \, \frac{\partial r}{\partial s} + \sigma_m r = 0 \qquad (3.5.6)$$

$$\frac{T_{mm}}{R_m} + \frac{T_{\phi\phi}}{R_\phi} = \Delta P \qquad (3.5.7)$$

where ΔP is the pressure difference across the membrane (inside minus outside pressure). The principal radii of curvature, R_m and R_ϕ, of the surface element are expressed in terms of the curvilinear coordinates by

$$R_m = - \frac{ds}{d\theta}$$

$$R_\theta = \frac{r}{\sin\theta} \qquad (3.5.8)$$

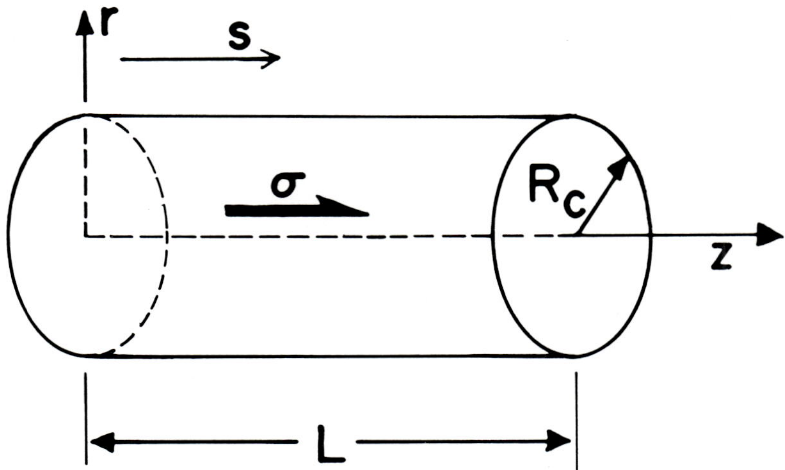

FIGURE 3.7.   Cylindrical surface fixed at one end and subject to surface traction, σ.

For axisymmetric loading of the surface, the shear stress, σ_ϕ, and the shear resultant, $T_{\phi m}$, are zero. Thus, the equilibrium equations are reduced to two equations involving the principal tensions, T_m and T_ϕ,

$$\frac{\partial(T_m r)}{\partial s} - T_\phi \frac{\partial r}{\partial s} + \sigma_m r = 0$$

$$\frac{T_m}{R_m} + \frac{T_\phi}{R_\phi} = \Delta P \tag{3.5.9}$$

(For a more extensive treatment of shells of revolution, see W. Flügge, 1973, *Stresses in Shells*.)[28]

Example 1

A sphere of radius, R_s, is subject to a uniform internal pressure, P_1, and an external pressure, P_2. The problem is to find the resultant tensions, T_m and T_ϕ.

Solving Equations 3.5.9 with $R_m = R_\phi = R_s$, we find that the tension is isotropic, $T_m = T_\phi$, and uniform, with $2T_m/R_s = \Delta P$. Hence $T_m = T_\phi = R_s \Delta P/2$.

Example 2

A cylinder of radius, R_c, and length, L, is capped at each end by a rigid plate. The internal pressure is P and there is a shear stress, σ, applied axially to the cylinder surface. The cylinder is held fixed at the end, $z = 0$. The tensions in the cylindrical surface are to be found. The coordinates are shown in Figure 3.7 with the fixed end at $z = 0$. At the free end, the pressure on the end plate must be balanced by the membrane resultant, T_m, times the circumference,

$$2\pi R_c T_m = \pi R_c^2 P$$

This gives a boundary condition on T_m at $s = L$,

$$T_m \Big|_{s=L} = \frac{R_c P}{2}$$

The equations of equilibrium for the cylindrical surface are,

$$\frac{T_\phi}{R_c} = P$$

$$R_c \frac{d}{ds}(T_m) + \sigma R_c = 0$$

These equations, together with the boundary conditions on T_m, give the distribution of principal tension along the cylinder,

$$T_m = \frac{R_c P}{2} + \sigma(L-s)$$

$$T_\phi = R_c P$$

Example 3

A membrane in the form of a truncated cone is capped at each end with a rigid plate. The conical membrane shell is subjected to an internal pressure, P. What are the tensions in the membrane? Let the radii of the ends be R_0 and R_L and the length of the cone be L as shown in Figure 3.8. The angle θ in this case is constant, $\theta = \pi/2 - \theta_c$, where θ_c is the half angle of the cone as shown. Equations 3.5.8 and 3.5.9 are given at any point by

$$\frac{T_\phi}{r} \sin\theta = P$$

$$\frac{d}{dr}(T_m r) - T_\phi = 0$$

The combination of these two relations yields an equation for the tension resultant along the meridian,

$$\frac{d}{dr}(T_m r) = \frac{Pr}{\sin\theta}$$

which can be integrated to give

$$T_m r = \frac{Pr^2}{2\sin\theta} + \text{constant}$$

At $r = R_L$, the tension is given by the axial force balance,

$$2\pi R_L \sin\theta\, T_m = \pi R_L{}^2 P$$

$$T_m = \frac{R_L P}{2\sin\theta}$$

Therefore, the tension along the meridian is given by

$$T_m = \frac{Pr}{2\sin\theta}$$

which is half of the circumferential tension, T_ϕ,

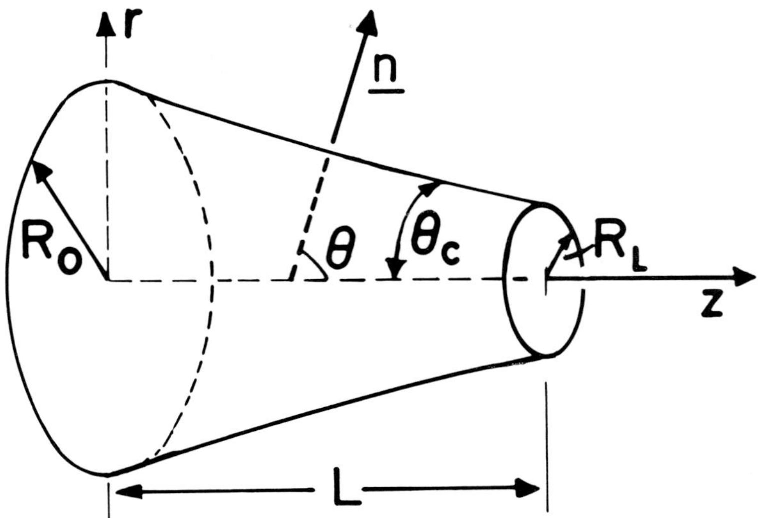

FIGURE 3.8.   Conical surface capped at both ends and subject to pressure difference, P.

$$T_\phi = \frac{Pr}{\sin\theta}$$

Example 4

Consider a flat annular disk with inner radius, r_1, and outer radius, r_2, that is rigidly supported at the outer radius. Assume that the inner circle, $r = r_1$, is subjected to radial tension, T_0. What is the distribution of principal tensions in the annular disk? The equilibrium equations (3.5.9) reduce to a single equation because both curvatures are zero,

$$\frac{d}{dr}(rT_m) - T_\phi = 0$$

Therefore, the situation is indeterminate unless another equation can be found to relate the principal tensions. In general, such additional equations are provided by the material response to the forces (i.e., constitutive relations). For instance, a liquid film has equal (isotropic) principal tensions, $\overline{T}$. In this case, the equation is

$$\frac{d\overline{T}}{dr} = 0$$

and $\overline{T}$ is a constant, equal to the radial tension at $r = r_1$. Another arbitrary example would be to set $T_\phi = 0$; the result with this prescription is

$$rT_m = \text{constant}$$

or

$$T_m = \frac{r_1 T_0}{r}$$

because $T_m = T_0$ at $r = r_1$.

From a general viewpoint, the principal tensions can be expressed in terms of isotropic and deviatoric (maximum shear) force resultants,

$$T_m \equiv \bar{T} + T_s$$

$$T_\phi \equiv \bar{T} - T_s$$

The single differential equation can be written in an alternative form,

$$\frac{dT_m}{dr} = \frac{2T_s}{r}$$

which can be integrated to give

$$T_m \Big|_{r_1}^{r_2} = 2 \int_{r_1}^{r_2} T_s \, d(\ln r)$$

This integral equation relates the boundary conditions on the meridional tension to the integral of the maximum shear force resultant in the surface. Consequently, the constitutive behavior for shear of the membrane determines the distribution of radial tension in a plane membrane surface.

Example 5

Consider an axially symmetric membrane envelope that supports a uniform axial force, F. This force must be balanced by the meridional tension at any transverse section of the surface according to

$$F = 2\pi r \cdot \sin \theta \, T_m$$

(The angle, θ, is defined in Figure 3.5A.) This relation can also be given in terms of the conical half-angle, θ_c, for a tangent to the meridian,

$$F = 2\pi r \cdot \cos \theta_c \cdot T_m$$

The equations of mechanical equilibrium were determined for a set of orthogonal axes (i.e., x_1, x_2, and the normal, n), defined by the surface location. These equations are

$$d(T_m r) - T_\phi \, dr = 0$$

$$\frac{T_m}{R_m} + \frac{T_\phi}{R_\phi} = \Delta P$$

In this case, $\Delta P = 0$, since there is no transmembrane pressure difference. At first glance it appears that the situation is overdetermined, i.e., three equations for the two unknowns, T_m and T_ϕ. However, we will demonstrate that if we choose either of the differential equations of equilibrium for the surface plus the axial force balance, we will identically satisfy the remaining equilibrium equation. For convenience, we define a function, $u \equiv \sin \theta$. The following identites can be expressed in terms of u, r, and θ (for the sense of the curvilinear coordinates shown in Figure 3.5A):

$$u \equiv \sin \theta$$

$$R_\phi = \frac{r}{u}$$

$$R_m = \frac{dr}{du}$$

$$\frac{dr}{ds} = -\sqrt{1 - u^2} \qquad\qquad (3.5.10)$$

Thus, the differential forms for equations of equilibrium are transformed to

$$d(T_m r) - T_\phi\, dr = 0$$

$$rT_m \cdot du + uT_\phi \cdot dr = r \cdot \Delta P\, dr = 0 \qquad\qquad (3.5.11)$$

(Again, note that $\Delta P = 0$ in this case.) The axial force balance becomes

$$F = 2\pi\, r\, u\, T_m \qquad\qquad (3.5.12)$$

First, we combine Equation 3.5.12 with the first of Equations 3.5.11 to give the result,

$$\frac{F}{2\pi} \cdot \frac{d}{dr}\left(\frac{1}{u}\right) = T_\phi$$

Substituting this relation in the second of Equations 3.5.11 with $\Delta P = 0$, we have

$$rT_m \cdot \frac{du}{dr} + \frac{F}{2\pi} \cdot u \cdot \frac{d}{dr}\left(\frac{1}{u}\right) = 0$$

This reduces to

$$\left(rT_m - \frac{F}{2\pi u}\right)\frac{du}{dr} = 0$$

which is consistent with Equation 3.5.12.

3.6 Mechanical Equilibrium of a Flat Membrane with Moment Resultants

As discussed in Section 3.1, moment and transverse shear resultants are induced by force resultant couples in membranes composed of associated molecular layers. Figure 3.9 illustrates the principal moment and transverse shear resultants experienced by a flat membrane element (the force resultants and surface tractions in Figure 3.4 also apply here, but are omitted for clarity). The sums of moments about the three orthogonal axes through the center of the membrane element must be zero, as are the sums of forces given in Equations 3.4.1,

$$\Sigma\,(M)_1 \equiv 0 \qquad \Sigma\,(M)_2 \equiv 0 \qquad \Sigma\,(M)_3 \equiv 0 \qquad\qquad (3.6.1)$$

Here, $\Sigma\,(M)_i$ represents the sum of moments about the ith axis through the centroid of the element. (Moments are defined as positive in the "right hand" sense of clockwise rotation about the axis, as viewed in the direction which the axis points.)

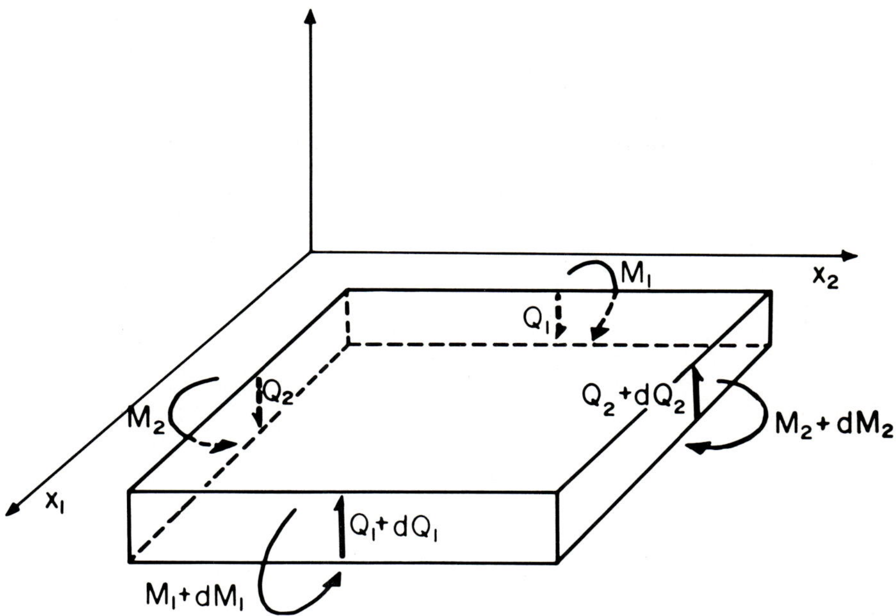

FIGURE 3.9.   The principal moment and transverse shear resultants supported by a flat membrane element. Note that only the force equilibrium equation in the direction normal to the surface will be affected by the transverse shear resultants.

From Figure 3.9, it is apparent that only the force equilibrium equation in the x_3 direction will be affected by the transverse shear resultants. The contribution comes from the force differences, $d(Q_1 dx_2)$ and $d(Q_2 dx_1)$, on the edges of the membrane element. With the first order expression for the transverse shear force differences, Equation 3.4.6 is modified by

$$0 = \Delta P \, dx_1 \, dx_2 \; + \; \frac{\partial Q_1}{\partial x_1} \; dx_2 \, dx_1 \; + \; \frac{\partial Q_2}{\partial x_2} \; dx_1 \, dx_2$$

without the body forces (as previously discussed). The third equation of force equilibrium, along with Equations 3.4.4 and 3.4.5, is

$$\Delta P \; + \; \frac{\partial Q_1}{\partial x_1} \; + \; \frac{\partial Q_2}{\partial x_2} = 0 \tag{3.6.2}$$

Therefore, gradients in transverse shear resultants can oppose a normal surface traction (pressure difference) in a flat region of the membrane. As we will show, the transverse shear resultant is related to gradients of the moment resultants. Consequently, the pressure difference in the equation of force equilibrium (3.6.2) will be related to the second derivatives of moment resultants.

Only two of the sums of the moments in Equations 3.6.1 need to be evaluated. The third, about the axis normal to the membrane element, was previously evaluated and resulted in the symmetry of the shear resultants, i.e., $T_{12} = T_{21}$. The sum of moments about the axis through the centroid, parallel to the x_1 material axis, is given to first order by the relation,

$$- d(M_2 \, dx_1) + 2(Q_2 \, dx_1) \; \frac{dx_2}{2} = 0$$

or,

$$-\frac{\partial M_2}{\partial x_2}\, dx_1\, dx_2 + Q_2\, dx_1\, dx_2 = 0$$

Therefore, the transverse shear is given by

$$Q_2 = \frac{\partial M_2}{\partial x_2} \tag{3.6.3}$$

Similarly, the sum of moments about the axis through the centroid, parellel to the x_2 material axis, is given to first order by the relation,

$$d(M_1\, dx_2) - 2(Q_1\, dx_2)\,\frac{dx_1}{2} = 0$$

or,

$$\frac{\partial M_1}{\partial x_1}\, dx_2\, dx_1 - Q_1\, dx_2\, dx_1 = 0$$

therefore,

$$\frac{\partial M_1}{\partial x_1} = Q_1 \tag{3.6.4}$$

The equation of force equilibrium in the normal direction, (3.6.2), may be written in terms of the moment resultants using Equations 3.6.3 and 3.6.4,

$$\Delta P + \frac{\partial^2 M_1}{\partial x_1{}^2} + \frac{\partial^2 M_2}{\partial x_2{}^2} = 0 \tag{3.6.5}$$

We have developed the sum of moment equations in a principal axes coordinate system for the moment resultants, M_1 and M_2. In general for nonprincipal axes systems, there will be two types of moment resultants: (1) moment resultants that act around the material edge (in the same sense as the principal moment resultants, M_1 and M_2), M_{11} and M_{22} and (2) moment resultant components which twist the element about the centroidal axes in the material plane (parallel to x_1 and x_2), M_{12} and M_{21}. The transverse shear resultants, Equations 3.6.3 and 3.6.4, must be modified to include the twist contributions (defined as positive clockwise around the centroidal axis parallel to x_2 and counterclockwise around the centroidal axis parallel to x_1),

$$\frac{\partial M_{11}}{\partial x_1} + \frac{\partial M_{12}}{\partial x_2} = Q_1 \tag{3.6.6}$$

$$\frac{\partial M_{21}}{\partial x_1} + \frac{\partial M_{22}}{\partial x_2} = Q_2 \tag{3.6.7}$$

These components can be written in matrix form with the summation rule,

$$\frac{\partial M_{ij}}{\partial x_j} = Q_i \tag{3.6.8}$$

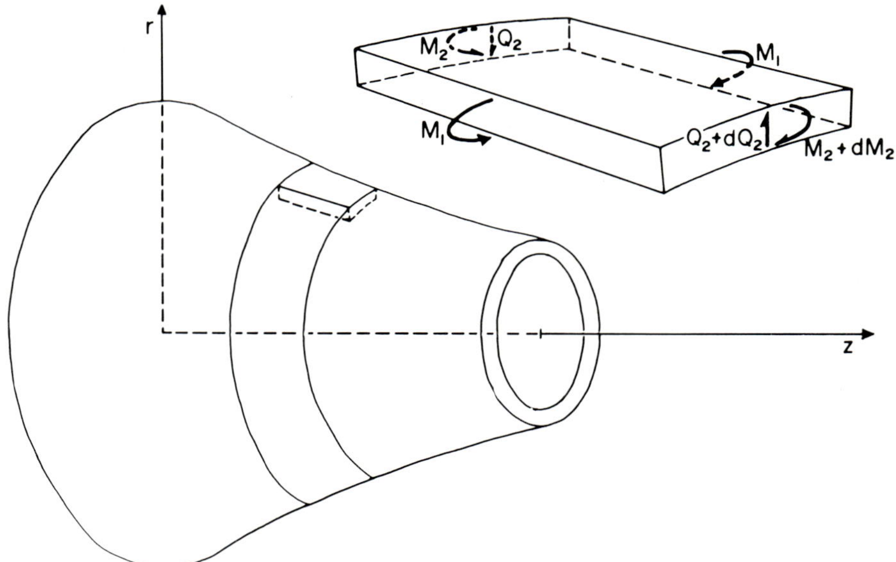

FIGURE 3.10.   A small element of an axisymmetric membrane with principal moment and transverse shear resultants displayed on exposed edges.

The direction of the induced transverse shear resultant is normal to the membrane plane. Equation 3.6.5 reflects this explicit feature and the general equation of force equilibrium in the normal direction for a plane membrane region is

$$\Delta P + \frac{\partial^2 M_{ij}}{\partial x_i \partial x_j} = 0 \tag{3.6.9}$$

3.7 Mechanical Equilibrium of an Axisymmetric Membrane with Moment Resultants

In the discussion of equilibrium for axisymmetric membranes that follows, axial symmetry will be assumed for the environmental tractions on the membrane surface. Thus, in curvilinear coordinates (s,θ,ϕ), only principal force resultants, tensions, and principal moment resultants will be involved in the equilibrium equations for the membrane. Normal and meridional surface tractions will act on the membrane element, i.e., ΔP and σ_m. The equations of equilibrium, without moment resultants, are Equations 3.5.9. Figure 3.10 illustrates the principal moment resultants and the induced transverse shear resultants. Again, we must consider the slightly trapezoidal shape and curvature of the element when we balance the forces and moments on the element.

Evaluating the sums of moments about axes through the centroid of the element and tangent to the material surface, we specify transverse shear resultants. Because of symmetry, the sum of moments about the centroidal axis, tangent to the meridian (x_2 coordinate), yields the trivial result that the transverse shear acting on the edges formed by the meridian segments is zero, i.e., $Q_1 = 0$, (since the moment resultant, M_1, does not vary across the element). The sum of moments about the centroidal axis, tangent to a latitude circle (x_1 coordinate), includes the induced transverse shear contributions, $2(Q_2\,dx_1)\,dx_2/2$, the difference in principal moments that act on the edges formed by latitude circles, $-\,d(M_2\,dx_1)$, plus the contributions of the moments about the edges formed by the meridian segments which are projected onto the centroidal axis. The latter are created by the slightly trapezoidal element shape: $2(M_1\,dx_2)\,\tfrac{1}{2}[\partial(dx_1)/\partial x_2]$. To first order, the sum of moments is given by

$$0 = Q_2 \, dx_1 \, dx_2 \; - \; \frac{\partial(M_2 \, dx_1)}{\partial x_2} \; dx_2 \; + \; M_1 \, dx_2 \; \frac{\partial(dx_1)}{\partial x_2}$$

With the curvilinear coordinate expressions for the differential increments, dx_1 and dx_2, the transverse shear resultant is obtained from the moment balance,

$$r Q_m \; = \; \frac{\partial(M_m r)}{\partial s} \; - \; M_\phi \frac{\partial r}{\partial s} \tag{3.7.1}$$

Now, we must modify the force balances in the directions normal to the surface and tangent to the surface meridian, Equations 3.5.4 and 3.5.3, respectively. We add the contributions from the transverse shear on the edges formed by the latitude circles. The contribution of the transverse shear to the force balance normal to the surface is

$$d(Q_2 \, dx_1) \; = \; \frac{\partial(Q_2 \, dx_1)}{\partial x_2} \; dx_2$$

thus, the sum of forces, Equation 3.5.4, becomes

$$0 = \Delta P dx_1 \, dx_2 \; - \; T_1 \, dx_2 \left(\frac{dx_1}{R_1} \right) \; - \; T_2 \, dx_1 \left(\frac{dx_2}{R_2} \right) \; + \; \frac{\partial(Q_2 \, dx_1)}{\partial x_2} \; dx_2$$

In curvilinear coordinates, the balance of normal forces yields

$$\frac{T_m}{R_m} + \frac{T_\phi}{R_\phi} \; - \; \frac{1}{r} \frac{\partial(Q_m r)}{\partial s} \; = \; \Delta P \tag{3.7.2}$$

The contribution to the force balance in the direction tangent to the meridian comes from the projection of the transverse shear forces onto the tangent plane. The projections are created by the curvature of the surface element with deviation angles, $- \, dx_2/2R_2$. The first order force addition is

$$2Q_2 \, dx_1 \left(\frac{dx_2}{2R_2} \right)$$

with the force balance, Equation 3.5.3, and curvilinear coordinates, the result is

$$0 = \sigma_m r \; + \; \frac{\partial(T_m r)}{\partial s} \; - \; T_\phi \frac{\partial r}{\partial s} \; + \; \frac{Q_m r}{R_m} \tag{3.7.3}$$

Equations 3.7.2 and 3.7.3 are the set of force equilibrium equations for the membrane. These equations are coupled to the moments by Equation 3.7.1 and can be expressed in terms of the principal moment resultants if we eliminate the transverse shear resultant.

$$0 = \sigma_m r \; + \; \frac{\partial(T_m r)}{\partial s} \; - \; T_\phi \frac{\partial r}{\partial s} \; + \; \frac{1}{R_m} \left[\frac{\partial(r M_m)}{\partial s} \; - \; M_\phi \frac{\partial r}{\partial s} \right]$$

$$\Delta P = \frac{T_m}{R_m} + \frac{T_\phi}{R_\phi} \; - \; \frac{1}{r} \frac{\partial^2(r M_m)}{\partial s^2} \; + \; \frac{1}{r} \frac{\partial}{\partial s} \left(M_\phi \frac{\partial r}{\partial s} \right)$$

$$\tag{3.7.4}$$

As previously discussed, the moment resultant contributions are usually small compared to the force resultants. However, Equations 3.7.4 demonstrate that there are situations where the small moment resultant contributions are required for membrane equilibrium (and stability). For instance, in a surface where the local curvatures are zero, the moment resultants are required to maintain equilibrium under local normal forces.

SECTION IV

4.1 Membrane Thermodynamics and Constitutive Relations

Scientific study of biological membrane structure and function requires the complementary use of biophysical and biochemical methodologies. The commonality between biophysical and biochemical approaches is provided by thermodynamics. Thermodynamics is composed of phenomenological laws that determine the behavior of a material in terms of a specific set of functions that characterize the state of the material. Thermodynamics is based on the hypothesis that the state functions are continuous from point to point in the material. In other words, the scale (in time and space) over which we measure state variables includes so many molecules that the eccentric behavior of individual molecules is averaged out. Thermodynamics, therefore, is made up of laws that represent the aggregate behavior of molecules. Departures from the expected group or ensemble behavior are exhibited as fluctuations.

The laws of thermodynamics provide empirical recipes for relating the mechanical work done on the material to the total energy and heat content in the material. The first law of thermodynamics is the postulate that energy is conserved, i.e., change in energy equals work plus heat addition. Work is defined quantitatively by the integration of the forces that are supported by the material times the displacements of the material. For a continuous material, forces and displacements are defined by macroscopic functions. Thus, along with work, they are predictable or deterministic variables. On the other hand, heat is a quantitative measure of the collective chaotic energy of the molecular constituents. Heat represents microscopic molecular processes that are random and incoherent. The diffuse nature of heat leads to a subjective relation called the second law, which describes the quality of the collective process that occurs during heat change, i.e., reversible or irreversible. For reversible processes, work may be attributed to conservative forces that are derivatives of potential functions. Heat is represented by a state function (entropy) and a proportionality factor (temperature). Irreversible processes, however, are less tangible. Some or all of the work can be due to nonconservative, or dissipative, forces. Heat exchange is not specified by changes in a state function. In contrast to the less satisfying second law of thermodynamics, the first law offers a secure refuge for the empirical scientist: conservation of energy. Heat and work are always accounted for in sum. Since we are neglecting inertia (kinetic energy) and the potential energy due to body forces, the total energy of the membrane is equal to the internal energy of the material. Internal energy is an abstract quantity which cannot be given an absolute value. Only changes in energy are observable effects, evidenced as heat exchange and work. Consequently, a reference state must always be assumed for thermodynamic state functions.

In biological membranes, we are faced with an anisotropic material which can be described as a continuum only in two dimensions (over large scale in the surface), but molecular discontinuity exists in the third dimension (thickness). We will conceptually treat the discontinuity in membrane composition from one side to the other by considering the membrane as a laminated structure made up of a strata of molecular layers.

The layers may be either coupled or uncoupled. This conceptual view does not restrict the application of the surface continuum approach only to membrane structures with distinguishable layers. Membranes with more homogeneous composition are represented equally well by the material properties that will be derived in this chapter. The thermodynamic variables for a membrane material are continuous functions of surface location, but are discrete functions of position within the thickness of the membrane. At a specific location on the membrane surface, the thermodynamic state represents a surface domain that contains a large number of molecules. Thus, we will have to respect the minimum scale on the surface above which this is a valid characterization (i.e., for which the fluctuations in mean thermodynamic functions are small).

We will be specifically interested in the thermodynamic changes that result from membrane deformation and rate of deformation as produced by external forces. Throughout the development, we will be considering the membrane material system as closed. In other words, there is no exchange of material or molecules between the membrane and its environment. This is true only for a limited period of time because the membrane may be remodeled by cellular processes or by exchange of lipids and proteins with the extracellular world. In general, these exchange processes take place over very long time periods compared with the mechanical experiments that are performed on the membrane. Therefore, our assumption is reasonable. We will develop membrane constitutive relations between the intensive distributions of membrane forces, deformation, and rate of deformation. These relations will be characterized by intrinsic material properties (e.g., coefficients for elasticity and viscosity) based on changes in thermodynamic state of the membrane.

The elastic constitutive relations that result from the development prescribe the mechanochemical equations of state for the membrane surface at constant temperature. These constitutive equations relate conservative membrane forces to the intensive deformation of the membrane. An example would be the change of surface pressure as a function of the change in the area in the membrane surface. Surface deformation can be decomposed into fractional changes in area of a surface element and extension of the element at constant area. Deformation of an individual molecular layer of the membrane depends only on these two parameters. Therefore, the elastic forces required to deform the molecular layer are functions of only these two parameters. (We assume that each molecular layer is isotropic in the plane of the layer.) Single molecular layers can only sustain forces that are locally tangent to the surface plane. The implication here is that single molecular layers exhibit a negligible energy change due to surface curvature change (bending resistance) unless the curvature change explicitly involves deformation of the surface element. The bending energy alone of single molecular layers is essentially negligible until the radii of curvature for the surface are on the order of molecular dimensions, at which point our continuum approach is no longer valid. We will concentrate first on the reversible thermodynamic behavior for each of the membrane molecular layers in response to surface deformation. Then, we will describe the composite behavior of the total membrane by summation over all layers.

Composite membrane structure introduces the possibility of coupled interactions between the layers. The surface deformation of the whole membrane will be experienced nearly the same by each layer if the surface radii of curvature are very much larger than the distance between the molecular layers of the composite. However, bending moments (couples) are produced by changes in the membrane curvature. These moments can be significant, especially if the mean surface deformation is small. The moments or force couples are due to small relative deformations between adjacent layers that create small energy variations (e.g., curvature can produce inner layer compression and outer layer expansion with no net change in the average membrane area). We will investigate the energy variations produced by changes in curvature of the membrane composite in order to determine bending resistance and moments. These curvature elastic effects will be treated subsequent to the analysis of membrane surface elasticity.

The elastic properties of a closed membrane system are associated with reversible thermodynamic changes in the membrane. These changes are produced by macroscopic deformation of the membrane structure. The elastic coefficients are derivatives of the free energy density (work per unit area of the membrane) taken with respect to intensive deformation at constant temperature. The free energy change is the sum of contributions from internal energy and heat content changes produced by deformation of the material. If the temperature of the material is changed, then internal energy and heat content also change. This process is accomplished by random exchanges of momentum between the material surface and the adjacent environment and by radiation absorption and molecular excitation. Temperature change produces small deformations in a free or unconstrained material. It can also produce forces in a material with constrained dimensions. For reversible processes, the thermoelastic alterations of material dimensions and forces combine with the elastic constitutive relations to give mechanochemical equations of state. These equations relate static material forces to deformation and temperature change. Furthermore, coefficients in the material equations of state specify the reversible changes of internal energy and heat content of the material that are produced by deformation at constant temperature. With this mechanical form of calorimetry, the material thermodynamic state can be investigated with temperature-dependent, mechanical experiments. Thus, thermoelasticity provides a direct thermodynamic probe of membrane structure *in situ,* which can be correlated with biochemical and ultrastructural studies. For example, it is possible to establish whether the configurational state of molecular complexes in the membrane is more or less ordered by a deformation and whether the energy of these complexes is appreciably changed by the deformation. For a biological membrane structure composed of amphiphilic components, thermoelasticity provides a direct way to assess the thermal repulsive forces in the membrane. In the natural state, without externally applied forces, thermal repulsive forces in the membrane are opposed by the hydrophobic interaction between the membrane surfaces and adjacent aqueous environment.

The elastic constitutive equations represent reversible thermodynamic changes produced by deformation. On the other hand, inelastic or nonconservative constitutive equations describe the dissipation of energy in the membrane produced by time rates of deformation. Nonconservative processes are thermodynamically irreversible and result in membrane forces that are proportional to rate of deformation, i.e., rate of area dilation and rate of extension at constant area of membrane material. In general, time-dependent response of a material to an applied force is complicated and can only be described by empirical equations. The relationship between the macroscopic behavior and the microscopic processes that occur at the molecular level is difficult to establish. However, it is possible to identify two characteristic types of irreversible phenomena that originate at the molecular level: (1) internal heat generation (molecular friction) superposed on the reversible thermodynamic behavior; (2) permanent deformation associated with structural changes produced by material forces. In the first case, time-dependent deformation creates forces in the material that exceed the elastic levels required to maintain static deformation. The additional force is required because mechanical work is dissipated by viscous friction produced by the nonzero rate of deformation. The dissipation occurs essentially in parallel with elastic energy changes. In the second case, permanent or plastic material deformation is proportional to the duration and strength of the forces supported by the material. Such a response can be considered to occur in series with the elastic mechanism. The two types of irreversible phenomena distinguish general regimes of material behavior: solid, semisolid, and liquid (plastic). Constitutive relations for viscoelastic deformation response and recovery represent the solid material regime. Relaxation and "creep" equations describe the semisolid transition from solid to plastic material regimes. The transition is especially significant because it represents molecular relaxation processes in structural elements of the membrane. Finally, viscoplasticity characterizes the liquid flow of membrane material. We will develop first order constitutive relations for each regime of material behavior. When appropriate, the equations will be given in both Lagrangian and Eulerian forms. In Section V, we will present applications of these relations to actual membrane mechanical experiments and will demonstrate the advantages peculiar to either the Lagrangian or Eulerian representation.

4.2 Thermodynamic Outline

Thermodynamics is based on two types of state variables: intensive and extensive. Extensive variables depend on the amount of material in the system (e.g., mass and energy); the intensive variables do not (e.g., pressure, density, temperature). Conservation principles start with extensive variables because they are concerned with the amount of something; intensive variables characterize the substance, not the amount. Such general attributes have been significant concerns in philosophy since early Greek considerations, almost three millenia in the past (Russell, 1972).[77] These philosophical interests have been coupled with Locke's concept of empirical truth to form the basis for natural science. Thermodynamics is a scientific formalism centered around these philosophical concepts. The two principal laws of thermodynamics deal with the conservation of energy (first law) and the character of the changes in the state of the system, i.e., reversible or irreversible (second law). (Katchalsky and Curran, *Nonequilibrium Thermodynamics in Biophysics*,[48] provides a good introduction; Morse's book, *Thermal Physics*,[62] contains historical notes and displays profound appreciation of the implications of developments in thermodynamics.)

The first law is a statement of conservation of energy: the sum of the heat exchange, ΔQ, from the exterior environment to the material (defined as positive into the material) and the work, ΔW, done by displacement of forces that act on the material must equal the change in internal energy of the material, ΔE,

$$\Delta E = \Delta Q + \Delta W \tag{4.2.1}$$

Here, the symbol, Δ, indicates a change or increment between beginning and end states. This may be written in integral form as

$$\int dE = \int (\delta Q + \delta W)$$

For infinitesimal increments, the law also applies,

$$dE = \delta Q + \delta W \tag{4.2.2}$$

Since energy is conserved by definition, the increment in energy is an exact differential, i.e., independent of the character of the process in going between the two states. However, the heat exchange and work increments may or may not be exact differentials. They depend on the character of the process which is used to go between the two states. "Process" is used in this context to represent the particular set of intermediate states between the initial and instantaneous state. For example, an isothermal reversible process is a particular procedure that produces conservative work on the material, independent of the path between states. Thus, it is only a function of the initial and final states. The incremental stages are given by an exact differential,

$$\delta W_T^{REV} = dW \tag{4.2.3}$$

where the subscript, T, implies constant temperature. From Equation 4.2.2, the incremental heat exchange in an isothermal, reversible process is also expressed as a total differential,

$$\delta Q_T^{REV} = dQ \tag{4.2.4}$$

Hence, by returning to the initial state through a reversible process, there will be no net heat exchange between the material and its environment. Each cyclic integral is zero,

$$\oint dE \equiv 0$$

$$\oint \delta W^{REV} \equiv 0$$

$$\oint \delta Q^{REV} \equiv 0 \tag{4.2.5}$$

(The second and third equations of (4.2.5) apply for a cyclic process that is only isothermal or adiabatic. They do not represent a cycle such as a Carnot cycle, which is a sequence of isothermal and adiabatic reversible processes between different temperatures.)

A perfectly elastic material obeys Equation 4.2.3 at constant temperature. External forces, which deform an elastic material, do work on the material that depends only on the initial and final state of deformation. The heat exchange produced during the deformation is completely recovered when the external forces are removed and the material returns to its initial state. However, deformation of any real material will take place over a finite period of time, and internal heat dissipation will occur due to the frictional interaction of molecules. In an actual process, irrecoverable heat exchange may result. Therefore, Equations 4.2.5 can only be approached in the limit, where dissipation is negligible. Another essential requirement for a reversible process is that there be no irreversible material alterations (e.g., molecular rearrangements). Permanent structural changes will occur in proportion to the magnitude and duration of the external forces. Consequently, the forces must be applied and removed such that the structure does not have time to relax to a new material configuration. It may be impossible for a given process to satisfy both requirements simultaneously, i.e., a slow enough deformation of the material to have negligible dissipation and fast enough to avoid structural change.

The second law of thermodynamics states that for all real processes the actual heat added, δQ, is less than the heat exchanged in a reversible process between the same states, i.e.,

$$\delta Q \leq \delta Q^{REV} \tag{4.2.6}$$

At constant temperature, Equation 4.2.6 states that there is thermal energy transfer to the environment in excess of the reversible amount. This is illustrated by the cyclic integral,

$$\oint \delta Q \leq 0$$

Empirically, it was recognized that an additional state variable is required to characterize the redistribution of energy which occurs within the material during the process. The total energy is a conserved quantity, but the partition of energy between the various possible molecular states is not specified. The ideal or reversible increments in the heat added to the system are determined by a second extensive state variable called entropy, S,

$$\delta Q^{REV} = TdS \tag{4.2.7}$$

where temperature is an integrating factor. The entropy increment, dS, is determined by the change in state of the system in any process. Equation 4.2.6 shows that the change in entropy of the system is greater than or equal to the heat added from external sources divided by the temperature,

$$dS \geq \frac{\delta Q}{T} \tag{4.2.8}$$

where the equality condition is an explicit statement of reversibility. In the 19th century, Boltzmann demonstrated that the entropy for a gas is determined by the distribution of molecules among possible states of the system. Subsequently, statistical mechanics has been built around this discrete concept as represented by

$$S = k \ln \Omega \tag{4.2.9}$$

where k is Boltzmann's constant, and Ω is the number of configurations in which the system can exist. Thus, entropy is a measure of the number of opportunities on a microscopic scale for distributing energy changes and work. It is an assessment of the molecular disorder or lack of coherence in the system. Equation 4.2.8 shows that dissipative processes tend to increase internal disorder. Deformation of a material may require work not only to change the internal energy of the material, but also to rearrange the internal structure (work in opposition to the natural, thermal randomization process).

4.3 Equilibrium Thermodynamics and Membrane Deformation

Equilibrium thermodynamics characterizes processes that take place in an essentially reversible manner. For deformations of a membrane surface that occur reversibly, we can express the total energy change of the membrane with the combined first and second laws of thermodynamics,

$$dE = TdS + dW$$

or

$$dW = dE - TdS \tag{4.3.1}$$

If we assume that the temperature is constant and that there is no irreversible alteration of the material and no viscous dissipation, then the work done on the membrane during deformation is reversible, i.e., dW in Equation 4.3.1 is an exact differential. Therefore, we will be able to identify an elastic, potential energy function which represents the work term.

As we said earlier, the first and second laws of thermodynamics involve state variables that are extensive properties of the system, i.e., they depend on the total amount of material in the system. We consider the membrane to be a continuous material with fixed mass. Thus, the value of an extensive variable for the whole system equals the sum of its incremental values for all small elements of the system. Each increment is the amount of the extensive variable contributed by the element, which does not depend on the overall size of the material. Consequently, it is possible to define a partial value of the extensive variable with respect to the size of a small element. The partial value becomes a density function for the extensive variable per unit area when the limit is taken where the element areas are arbitrarily small. For example, we can define area specific or density functions for internal energy and entropy per unit area of material surface,

$$\tilde{E} = \lim_{\Delta A_0 \to 0} \left(\frac{\Delta E}{\Delta A_0} \right)$$

$$\tilde{S} = \lim_{\Delta A_0 \to 0} \left(\frac{\Delta S}{\Delta A_0} \right) \tag{4.3.2}$$

The increments in the thermodynamic variables are represented by the amounts, $\Delta(\ )$, appropriate to the element area, ΔA_0. The thermodynamic distributions, $\tilde{E}$ and $\tilde{S}$, are defined relative to a fixed reference geometry, i.e., the initial or undeformed state that is represented by element area, ΔA_0. Consequently, these functions characterize the state of matter at point locations over the entire surface, regardless of the subsequent material deformation. The specific properties, $\tilde{E}$ and $\tilde{S}$, are in units of energy and energy per degree, respectively, both per unit initial area of membrane surface. (The definition per unit of initial area can also be interpreted as a per molecule basis if the membrane is a single component system.) Using the intensive representations for membrane forces and deformation that were developed in Sections II and III, we will be able to quantitatively describe the differential work done on a membrane element per unit area. The differential form for work in Equation 4.3.1 is related to the changes in energy and entropy density of the material.

$$d\tilde{W} = d\tilde{E} - Td\tilde{S} \tag{4.3.3}$$

A specific condition for which $d\tilde{W}$ in Equation 4.3.3 is a total differential is that the temperature of the material is held constant. This is the usual condition in a mechanical experiment. However, temperature change is required to determine how much of the isothermal work goes into energy change and how much into entropy change in the material. These thermoelastic effects will be considered in a later section. For constant temperature, the elastic potential energy function is the same as the Helmholtz free energy density of the material, $\tilde{F}$. The Helmholtz free energy is a thermodynamic potential defined by

$$\tilde{F} \equiv \tilde{E} - T\tilde{S}$$

and which is differentially expressed by

$$d\tilde{F} = d\tilde{E} - Td\tilde{S} - \tilde{S}\,dT$$

This potential function equals the conservative work on the system for the isothermal case,

$$(d\widetilde{F})_T = d\widetilde{E} - Td\widetilde{S} = d\widetilde{W}$$

[Note: Another condition under which $d\widetilde{W}$ in Equation 4.3.3 is a total differential is the adiabatic case (TdS ≡ 0), where

$$d\widetilde{W} = d\widetilde{E}$$

The internal energy itself is an elastic potential and obviously a conservative function.] The elastic potential, $\widetilde{F}$, at constant temperature is also referred to as a strain-energy function (see for example, Green and Adkins, *Large Elastic Deformations*).[35]

Since we treat the membrane as isotropic in the plane of the surface, the elastic potential energy density at constant temperature depends only on two independent variables which characterize the deformation. In Section 2.3 (see Figure 2.9) we demonstrated that these two variables are: (1) α, the fractional change in area of the surface element (isotropic dilation or condensation) and (2) β, the measure of extensional deformation of the surface element at constant surface area. (The parameter β represents the distortion of a circular region of the membrane surface to an ellipse.) Because the elastic potential, $\widetilde{F}$, at constant temperature is a function of only the linearly independent deformation variables, α and β, a differential form may be written in terms of contributions from each deformation parameter,

$$d\widetilde{F}_T = \left(\frac{\partial \widetilde{F}}{\partial \alpha}\right)_{\beta,T} d\alpha + \left(\frac{\partial \widetilde{F}}{\partial \beta}\right)_{\alpha,T} d\beta \tag{4.3.4}$$

(Subscripts indicate the independent variables that are held constant.) The assumption that the deformation is thermodynamically reversible implies that the elastic potential, $(\widetilde{F})_T$, is independent of the path taken between the initial and instantaneous states. Equation 4.3.4 is an explicit statement of the path independent deformation process with respect to α and β. The first term is the change in elastic potential energy density due to isotropic dilation or condensation of the surface keeping the shape of material elements unchanged. The second term is the change in elastic potential energy density contributed by the extensional deformation of the molecular structure at constant surface area. The deformation parameters must be linearly independent and rotationally invariant in order that Equation 4.3.4 be valid. Later examples will illustrate this principle, using two specific models for membrane materials.

4.4 Isothermal Constitutive Equations for Principal Axes

As previously discussed, mechanochemical equations of state at constant temperature describe the elastic constitutive behavior of a membrane. These equations are valid for reversible thermodynamic states where the viscous dissipation is negligible. Elastic constitutive equations are also called stress-strain relations in the literature of solid mechanics. For single molecular layers and thin membrane structures, the surface relations take the form of force resultants (e.g., tensions) as functions of strain or deformation. In order to establish the relationship of membrane force resultants to static membrane deformation (thermodynamic equilibrium), we equate the incremental change in Helmholtz free energy density (elastic potential) at constant temperature to the work done on an element of the material by the small displacements of forces supported by the material. We choose the convenient perspective provided by the prin-

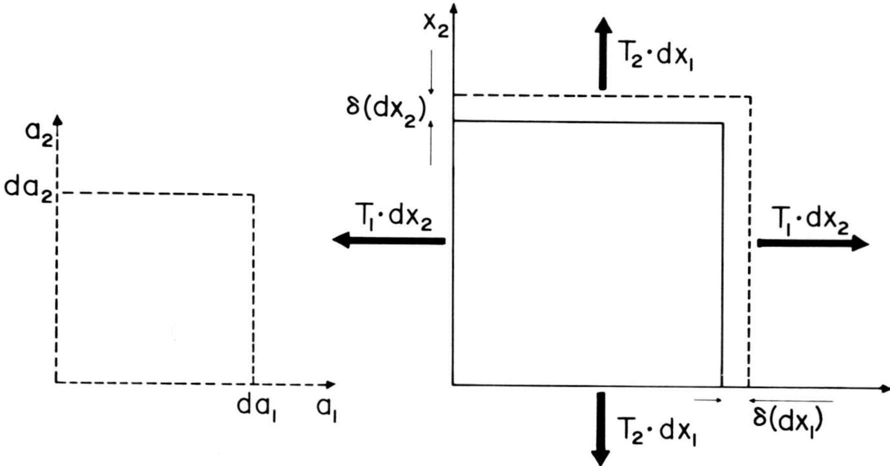

FIGURE 4.1. Incremental deformation of a differential element of membrane viewed in the principal axes system. The work done on the element is the product of the force times the displacement, summed for the x_1 and x_2 directions. T_1 and T_2 are the principal membrane tensions (intensive forces) described earlier. $T_1\lambda_2 da_2$ and $T_2\lambda_1 da_1$ are the forces acting along the element edge in the x_1 and x_2 directions, respectively, and $\delta(dx_1)$ and $\delta(dx_2)$ are the displacements which result from the deformation.

cipal axes system and then consider the principal axes deformation of a differential element of the membrane (Figure 4.1). The work done on the element is the product of the force times displacement, summed for the x_1 and x_2 directions. The principal forces on the element are approximately $T_1 \cdot dx_2$ and $T_2 \cdot dx_1$. These are displaced by small amounts, $\delta(dx_1)$ and $\delta(dx_2)$, respectively. The incremental work produced by these forces and small displacements is

$$\delta W = (T_1 \cdot dx_2)\, \delta(dx_1) + (T_2 \cdot dx_1)\, \delta(dx_2) \tag{4.4.1}$$

T_1 and T_2 are the principal membrane tensions (intensive forces) described in Section 3. The displacements, $\delta(dx_i)$, correspond to increments in the principal extension ratios,

$$\delta(dx_1) = da_1\, \delta\lambda_1$$

$$\delta(dx_2) = da_2\, \delta\lambda_2 \tag{4.4.2}$$

by noting that the initial element dimensions are constant. Equations 4.4.1 and 4.4.2 combine to give the incremental work per unit area of the undeformed membrane element ($dA_0 \equiv da_1 \cdot da_2$),

$$\delta W = \left[(T_1 \cdot \lambda_2)\delta\lambda_1 + (T_2 \cdot \lambda_1)\delta\lambda_2 \right] \cdot dA_0$$

The increments in extension ratios can be expressed as differential changes, which give a differential relation for the work per unit area. Using the definition for Helmholtz free energy density and the combined first and second laws of thermodynamics, Equation 4.3.3, we obtain the relation for change in elastic potential that is produced by intensive deformation,

$$(d\tilde{F})_T = (T_1 \lambda_2)\, d\lambda_1 + (T_2 \lambda_1)\, d\lambda_2 \tag{4.4.3}$$

The differential change in elastic potential, $(d\tilde{F})_T$, is only dependent on the differential changes in the independent deformation variables, α and β, i.e., it is equal to the sum of two contributions: one produced by fractional change in area and the other produced by extension at constant area. These geometric changes are explicitly given by the differential relations,

$$d\alpha = \left(\frac{\partial \alpha}{\partial \lambda_1}\right) d\lambda_1 + \left(\frac{\partial \alpha}{\partial \lambda_2}\right) d\lambda_2$$

and,

$$d\beta = \left(\frac{\partial \beta}{\partial \lambda_1}\right) d\lambda_1 + \left(\frac{\partial \beta}{\partial \lambda_2}\right) d\lambda_2$$

since they are functions of the principal extension ratios. We now take the chain rule expression for the elastic potential change, Equation 4.3.4, plus the differential relations above and equate the coefficients of $d\lambda_1$ and $d\lambda_2$ in the differential relations, Equation 4.4.3. This is valid because the extension ratios are linearly independent functions. The result is two independent equations,

$$\left[\left(\frac{\partial \tilde{F}}{\partial \alpha}\right)_{T,\beta} \frac{\partial \alpha}{\partial \lambda_1} + \left(\frac{\partial \tilde{F}}{\partial \beta}\right)_{T,\alpha} \frac{\partial \beta}{\partial \lambda_1}\right] = T_1 \lambda_2$$

$$\left[\left(\frac{\partial \tilde{F}}{\partial \alpha}\right)_{T,\beta} \frac{\partial \alpha}{\partial \lambda_2} + \left(\frac{\partial \tilde{F}}{\partial \beta}\right)_{T,\alpha} \frac{\partial \beta}{\partial \lambda_2}\right] = T_2 \lambda_1 \qquad (4.4.4)$$

We recall the definitions of α and β ($\alpha = \lambda_1\lambda_2 - 1$ and $\beta = (\lambda_1^2 + \lambda_2^2)/2\lambda_1\lambda_2 - 1$) to give

$$\frac{\partial \alpha}{\partial \lambda_1} = \lambda_2$$

$$\frac{\partial \alpha}{\partial \lambda_2} = \lambda_1$$

$$\frac{\partial \beta}{\partial \lambda_1} = \frac{1}{2\lambda_1^2\lambda_2} (\lambda_1^2 - \lambda_2^2)$$

$$\frac{\partial \beta}{\partial \lambda_2} = \frac{-1}{2\lambda_1\lambda_2^2} (\lambda_1^2 - \lambda_2^2)$$

We use these partial derivatives in Equations 4.4.4 to obtain the relationship between membrane tensions (intrinsic force resultants) and the changes in elastic potential energy density,

$$T_1 = \left[\left(\frac{\partial \tilde{F}}{\partial \alpha}\right)_{T,\beta} + \frac{1}{2\lambda_1^2\lambda_2^2} \left(\frac{\partial \tilde{F}}{\partial \beta}\right)_{T,\alpha} (\lambda_1^2 - \lambda_2^2)\right] \qquad (4.4.5)$$

$$T_2 = \left[\left(\frac{\partial \widetilde{F}}{\partial \alpha} \right)_{T,\beta} + \frac{1}{2\lambda_1^{\,2}\lambda_2^{\,2}} \left(\frac{\partial \widetilde{F}}{\partial \beta} \right)_{T,\alpha} (\lambda_2^{\,2} - \lambda_1^{\,2}) \right] \tag{4.4.5}$$

We see that both principal membrane tensions are associated with free energy changes which are produced by changes in surface area and surface extension at constant area.

In Section 3, we decomposed the principal tensions into an isotropic tension (that acts equally in all directions in the membrane surface) and a deviatoric resultant (which is the maximum shear force per unit length that acts in the membrane at ±45° to the directions of the principal axes). The isotropic tension is the mean of the principal components, and the maximum shear resultant is half the magnitude of the difference between principal tensions. Equations 4.4.5 give

$$\overline{T} = \tfrac{1}{2}(T_1 + T_2) = \left(\frac{\partial \widetilde{F}}{\partial \alpha} \right)_{T,\beta}$$

$$T_s = \tfrac{1}{2} \left| T_1 - T_2 \right| = \frac{1}{2\lambda_1^{\,2}\lambda_2^{\,2}} \left(\frac{\partial \widetilde{F}}{\partial \beta} \right)_{T,\alpha} \left| \lambda_1^{\,2} - \lambda_2^{\,2} \right| \tag{4.4.6}$$

These relations demonstrate that isotropic tension in the membrane is due to surface area change and that membrane shear is associated with surface extension at constant density. The membrane force resultants are proportional to the changes in free energy density that are produced by these geometric changes.

The maximum shear resultant can be written in the alternative form,

$$T_s = \frac{1}{2(1+\alpha)} \left(\frac{\partial \widetilde{F}}{\partial \beta} \right)_{T,\alpha} \left| \widetilde{\lambda}^{\,2} - \widetilde{\lambda}^{-2} \right| \tag{4.4.7}$$

where the scaled extension ratio, $\widetilde{\lambda}$ from Equation 2.4.7, for stretch at constant area is given by

$$\widetilde{\lambda} = \lambda_1 / \sqrt{\lambda_1 \lambda_2}$$

Therefore, the differential change in free energy density at constant temperature, Equation 4.4.3, can be expressed in terms of isotropic and deviatoric force resultants times fractional changes in surface area and surface extension,

$$(d\widetilde{F})_T = \overline{T}\, d\alpha + 2T_s (1 + \alpha)\, \frac{d\widetilde{\lambda}}{\widetilde{\lambda}} \tag{4.4.8}$$

This demonstrates the superposition of work produced by area change and work produced by extensional deformation at constant area.

The work done on the material is distributed per unit area of the surface in the initial, undeformed state. Thus, along with the elastic potential and the other thermodynamic state functions, it is a Lagrangian variable. The force resultants are determined in the instantaneous state, and yet Equations 4.4.6 relate the material forces to Lagrangian deformation variables. Subtle differences become apparent on closer inspection. For instance, the expression for the maximum Lagrangian shear strain is recalled as

$$\epsilon_s = \tfrac{1}{4}\ |\lambda_1{}^2 - \lambda_2{}^2|$$

The maximum shear resultant is given by

$$T_s = \frac{2}{\lambda_1{}^2 \lambda_2{}^2}\ \left(\frac{\partial \widetilde{F}}{\partial \beta}\right)_{T,\alpha} \epsilon_s$$

The shear resultant is not simply proportional to the Lagrangian shear strain (as in the classical theory of infinitesimal elasticity); it is inversely scaled by the area ratio (final to initial element area) squared. On the other hand, the maximum Eulerian shear strain is recalled as

$$e_s = \tfrac{1}{4}\ |\lambda_1{}^{-2} - \lambda_2{}^{-2}|$$

which gives a relation for the maximum shear resultant as a simple proportionality:

$$T_s = 2 \left(\frac{\partial \widetilde{F}}{\partial \beta}\right)_{T,\alpha} e_s$$

Even though the shear resultant is linear in the Eulerian shear strain, the finite deformation implicitly creates a nonlinear relationship to geometric extensions. The distinction between the magnitudes of the maximum Lagrangian and Eulerian shear strains disappears where the surface deformation occurs at constant area, i.e., $\lambda_1\lambda_2 \equiv 1$.

The partial derivative of the elastic energy with respect to the material stretch or element distortion at constant area completely specifies the shear rigidity of the membrane, μ,

$$\mu \equiv \left(\frac{\partial \widetilde{F}}{\partial \beta}\right)_{T,\alpha}$$

$$T_s = \frac{\mu}{2\lambda_1{}^2 \lambda_2{}^2}\ \left|\lambda_1{}^2 - \lambda_2{}^2\right| \tag{4.4.9}$$

The coefficient, μ, is a shear modulus, intrinsic to the membrane structure. The shear modulus represents the energy storage and static resistance to extensional deformations of the membrane surface. This is a property peculiar to solid materials; liquids have zero shear (elastic) moduli.

The Eulerian form for the shear resultant vs. shear strain is similar to the classical relation obtained from linear elasticity. However, it is important to recognize that Equation 4.4.9 is a nonlinear relation between extension and force resultant. In addition, an important difference can be recognized between the classical derivation of the shear modulus and the finite deformation approach that we have used. The shear modulus in the linear elastic theory for small deformations is obtained from the derivative of the elastic free energy density with respect to the square of the shear strain,

$$\mu \sim \left(\frac{\partial \widetilde{F}}{\partial e_s{}^2}\right)_T$$

(see Landau and Lifshitz, *Theory of Elasticity*).[51] For finite deformations, the square of the shear strain is on the order of the fourth power of the extension ratio,

$$(\epsilon_s^2) \sim \widetilde{\lambda}^4 + \widetilde{\lambda}^{-4}$$

as seen from Equation 2.4.1. This relation is essentially the square of the deformation variable, $(\beta + 1)^2$, i.e.,

$$(\beta + 1)^2 \sim \widetilde{\lambda}^4 + \widetilde{\lambda}^{-4}$$

which implies that the free energy density would depend on terms of order β^2 if the infinitesimal theory were to be used. This is not consistent with the results in Equation 4.4.7. Consequently, the infinitesimal theory cannot be simply extended to finite deformations with a free energy density that is proportional to the square of the shear strain. On the other hand, the nonlinear theory developed in this section can be shown to reduce to the linear elastic theory in the limit of small deformations. Note that for small deformations, the extension ratio, $\widetilde{\lambda}$, is given by

$$\widetilde{\lambda} = 1 + \delta\widetilde{\lambda}$$

where $\delta\widetilde{\lambda} \ll 1$. Therefore, the square of the shear strain is proportional to $(\delta\widetilde{\lambda})^2$, if we neglect terms of higher power,

$$\epsilon_s^2 \sim (\delta\widetilde{\lambda})^2$$

and, similarly, the deformation variable β is proportional to $(\delta\widetilde{\lambda})^2$. Therefore, for the case of small deformations, the shear modulus can be expressed by either

$$\mu \sim \left(\frac{\partial\widetilde{F}}{\partial\epsilon_s^2}\right)_T$$

or

$$\mu \sim \left(\frac{\partial\widetilde{F}}{\partial\beta}\right)_T$$

which is consistent with infinitesimal and finite deformation theories.

The functional dependence of the free energy density on the deformation parameter, β, can have physical significance. Elastic relations that are linear in β for large deformations are termed as "hyperelastic," a form of behavior exhibited by natural rubber in three dimensions. Later, we will illustrate the first order, hyperelastic behavior with a material model of a two-dimensional, loose network of long-chain flexible molecules. In this case, the Helmholtz free energy density is first order in the deformation parameter, β, and we can deduce a surface shear modulus for such "elastomeric" molecular layers in terms of parameters that describe the network configuration. In general, however, μ can be a function of the invariants, α, β, and temperature, T.

In contrast to membrane shear rigidity, the ability to support isotropic tension is characteristic of both solid and liquid membrane materials. Isotropic tension is produced by dilation or condensation of membrane surface area relative to the equilibrium state of the membrane. With a Taylor series expansion in the fractional change in area per molecule, α, relative to the equilibrium state, the elastic potential energy density is approximated by

$$(\widetilde{F} - \widetilde{F}_0)_T = \overline{T}_0\alpha + K\frac{\alpha^2}{2} + \ldots \tag{4.4.10}$$

where $(\tilde{F}_0)_T$ is the Helmholtz free energy density at the initial area. The coefficients, $\overline{T}_0$ and K, are defined by derivatives of the free energy density as

$$\overline{T}_0 \equiv \left(\frac{\partial \tilde{F}}{\partial \alpha} \right)_{T,\beta} \Bigg|_{\alpha=0}$$

$$K \equiv \left(\frac{\partial^2 \tilde{F}}{\partial \alpha^2} \right)_{T,\beta} \Bigg|_{\alpha=0}$$

The isotropic tension is obtained from Equation 4.4.6,

$$\overline{T} = \overline{T}_0 + K\alpha + \cdots$$

The coefficient, K, is the isothermal area compressibility modulus that relates changes in isotropic tension to small fractional changes in area relative to the initial state,

$$K \equiv \left(\frac{\partial \overline{T}}{\partial \alpha} \right)_T \Bigg|_{\alpha=0} \tag{4.4.11}$$

The constant, $\overline{T}_0$, is the initial isotropic tension in the membrane for the reference state. For instance, liquid films or interfaces (e.g., soap bubbles) maintain a constant tension, independent of the surface area. Commonly called "surface tension," this constant isotropic tension, $\overline{T}_0$, is characteristic of free interfaces where the surface is open to exchange of material with the environment or boundary regions of the film. However, closed interfaces or membranes (fixed mass) exhibit isotropic tension that depends on surface area changes as described by the elastic compressibility modulus, Equation 4.4.11. This behavior is called "Gibb's elasticity" for surfaces because of its original development by J. Willard Gibbs (1961)[33] in the nineteenth century.

For a closed membrane system such as a biological membrane, it is possible to define the initial, undeformed equilibrium state of the membrane as the state where the force resultants are zero. In addition, we will identify the natural state of the membrane as this force-free state. Therefore, relative to the natural state, the membrane isotropic tension and shear resultants are given by the following elastic constitutive equations,

$$\overline{T} = K\alpha + 0(\alpha^2) + \cdots$$

$$T_s = 2\mu e_s = \frac{2\mu e_s}{(1+\alpha)^2} = \frac{\mu}{2\lambda_1^{\,2}\lambda_2^{\,2}} \,|\,\lambda_1^{\,2} - \lambda_2^{\,2}\,| = \frac{\mu}{2(1+\alpha)} \,|\,\tilde{\lambda}^2 - \tilde{\lambda}^{-2}\,|$$

$$\tag{4.4.12}$$

where the fractional change in area, α, is taken to be small (remember that the shear resultant can be expressed in several alternative ways). The force resultants for any set of coordinate axes in the instantaneous state (x_i) can be obtained with the isotropic tension and maximum shear resultants and the rotation angle relative to the principal axes system as outlined in Sections 2 and 3. In the next section, the matrix form for the general equation which relates force resultants to elastic deformation will be derived for any instantaneous coordinate system. Figure 4.2 illustrates force resultants for the axes system in which maximum shear occurs. This is an axes system at $\pm45°$ with respect to the principal axes system. The elastic constitutive equations (4.4.12) are the mechanochemical equations of state for the membrane at constant temperature.

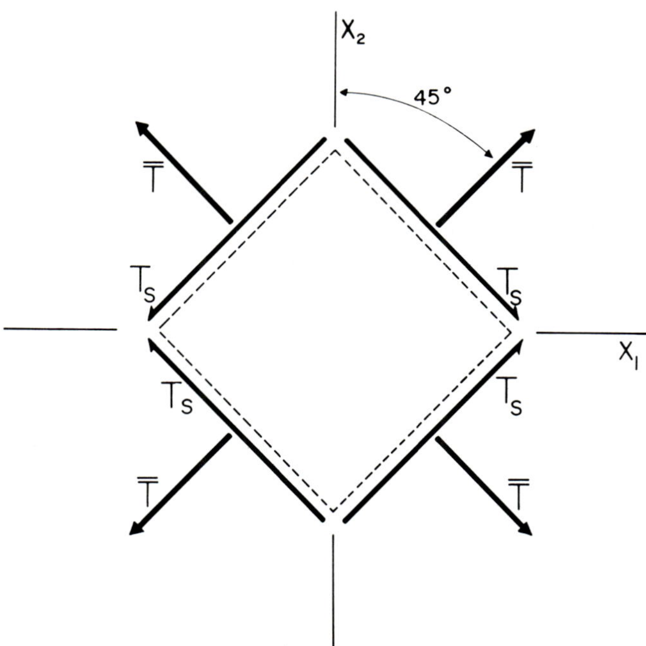

FIGURE 4.2. Force resultants for the axes system in which maximum shear occurs. This system is oriented at angles of ±45° with respect to the principal axes system.

There are two equations because the membrane may be a solid structure. We see that two intrinsic elastic moduli for thin membrane materials are required to relate the intensive force resultants in the material to the equilibrium surface deformation: (1) surface area compressibility modulus, K; (2) surface shear modulus, μ, for extension at constant area.

4.5 General Derivation of the Isothermal Constitutive Equations

In the previous section, we developed the isothermal mechanochemical equations of state from the work performed on a material element in the principal axes system. In this section, we derive these equations for arbitrary orientation of initial and instantaneous material coordinates. Both sets of coordinates are assumed to be Cartesian and referred to the same set of orthogonal axes.

At constant temperature, the rate of doing work on a material element per unit area is equal to the time rate of change of the elastic potential (Helmholtz free energy density),

$$\frac{\partial \widetilde{W}}{\partial t} = \left(\frac{\partial \widetilde{F}}{\partial t} \right)_T \tag{4.5.1}$$

Both variables in Equation 4.5.1 are Lagrangian in nature, since they are defined relative to a fixed reference state for the material. In the dynamic or instantaneous state, the rate of mechanical work is measured per unit area of the deformed material. This is called the instantaneous mechanical power density and is given by the scalar product of the Eulerian variables, rate of deformation times force resultant matrices, $T_{ij} \, V_{ij}$. The work per unit time on the material element is equal to the instantaneous power,

$$\left(\frac{\partial \widetilde{W}}{\partial t}\right) \cdot dA_0 = (T_{ij} V_{ij}) \cdot dA \tag{4.5.2}$$

where the initial and instantaneous element areas establish the extensive level of power produced by the element. The instantaneous mechanical power density can now be determined from the time rate of change of the Helmholtz free energy density,

$$T_{ij} V_{ij} = \frac{1}{\lambda_1 \lambda_2} \left(\frac{\partial \widetilde{F}}{\partial t}\right)_T \tag{4.5.3}$$

where the ratio of element areas is given by the product of principal extension ratios (rotationally invariant). The free energy density is a conservative function that depends only on the state of the deformation at any time. Therefore, we can write

$$\left(\frac{\partial \widetilde{F}}{\partial t}\right)_T = \left(\frac{\partial \widetilde{F}}{\partial \epsilon_{pq}}\right)_T \frac{\partial \epsilon_{pq}}{\partial t} \tag{4.5.4}$$

in terms of components of the Lagrangian strain matrix and their time derivatives. In Section II, we developed a relation between the rate of deformation matrix, V_{ij}, and the Lagrangian strain matrix, ϵ_{ij},

$$V_{ij} = \frac{\partial \epsilon_{pq}}{\partial t} \frac{\partial a_p}{\partial x_i} \frac{\partial a_q}{\partial x_j} \tag{4.5.5}$$

where

$$2\epsilon_{pq} = \left(\frac{\partial x_k}{\partial a_p} \frac{\partial x_k}{\partial a_q} - \delta_{pq}\right)$$

Therefore, Equations 4.5.3 through 4.5.5 combine to give the following coefficient for independent time derivatives of strain, $\partial \epsilon_{pq}/\partial t$,

$$\frac{1}{\lambda_1 \lambda_2} \left(\frac{\partial \widetilde{F}}{\partial \epsilon_{pq}}\right)_T = T_{ij} \left(\frac{\partial a_p}{\partial x_i} \frac{\partial a_q}{\partial x_j}\right)$$

or with the fact,

$$\frac{\partial a_p}{\partial x_i} \cdot \frac{\partial x_k}{\partial a_p} = \delta_{ki}$$

$$T_{ij} = \frac{1}{\lambda_1 \lambda_2} \left(\frac{\partial \widetilde{F}}{\partial \epsilon_{pq}}\right)_T \frac{\partial x_i}{\partial a_p} \frac{\partial x_j}{\partial a_q} \tag{4.5.6}$$

The differential form of Equation 4.3.4 is the expansion of the elastic potential in independent deformation parameters, α and β. Thus, the force resultant matrix or tensor is determined by

$$T_{ij} = \frac{1}{\lambda_1 \lambda_2} \frac{\partial x_i}{\partial a_p} \frac{\partial x_j}{\partial a_q} \left[\left(\frac{\partial \widetilde{F}}{\partial \alpha}\right)_{T,\beta} \frac{\partial \alpha}{\partial \epsilon_{pq}} + \left(\frac{\partial \widetilde{F}}{\partial \beta}\right)_{T,\alpha} \frac{\partial \beta}{\partial \epsilon_{pq}} \right]$$

$$(4.5.7)$$

The partial derivatives, $\partial \alpha / \partial \epsilon_{pq}$ and $\partial \beta / \partial \epsilon_{pq}$, are conveniently evaluated with a deformation matrix called Finger's strain tensor, B_{ij},

$$B_{ij} \equiv \frac{\partial x_i}{\partial a_\varrho} \frac{\partial x_j}{\partial a_\varrho}$$

This matrix may be regarded as a strain matrix oriented relative to the instantaneous coordinate system, although it is a Lagrangian type of variable because its magnitudes are measured relative to the initial state. It also is characterized by two invariants,

$$B_1 = B_{ii} = \lambda_1^2 + \lambda_2^2$$

$$B_2 = -\tfrac{1}{2}(B_{ij} B_{ji} - B_{ii} B_{jj}) = \lambda_1^2 \lambda_2^2$$

Hence, the required partial derivatives are given by

$$\frac{\partial \alpha}{\partial \epsilon_{pq}} = \frac{1}{2\lambda_1 \lambda_2} \frac{\partial B_2}{\partial \epsilon_{pq}}$$

$$\frac{\partial \alpha}{\partial \epsilon_{pq}} = \frac{1}{\lambda_1 \lambda_2} [\delta_{pq} - 2(\epsilon_{pq} - \delta_{pq} \epsilon_{kk})]$$

and

$$\frac{\partial \beta}{\partial \epsilon_{pq}} = \frac{1}{2\lambda_1 \lambda_2} \left[\frac{\partial B_1}{\partial \epsilon_{pq}} - \frac{B_1}{2\lambda_1^2 \lambda_2^2} \frac{\partial B_2}{\partial \epsilon_{pq}} \right]$$

$$\frac{\partial \beta}{\partial \epsilon_{pq}} = \frac{\delta_{pq}}{\lambda_1 \lambda_2} + \frac{B_1}{2\lambda_1^3 \lambda_2^3} \left[2(\epsilon_{pq} - \delta_{pq}\epsilon_{kk}) - \varepsilon_{pq} \right] = \frac{\delta_{pq}}{\lambda_1 \lambda_2} + \frac{B_1}{2\lambda_1^2 \lambda_2^2} \left(-\frac{\partial \alpha}{\partial \epsilon_{pq}} \right)$$

The final form of Equation 4.5.7 involves the contraction of matrix quantities over the indices, p and q. For this purpose, we develop the following identities,

$$\frac{\partial x_i}{\partial a_p} \frac{\partial x_j}{\partial a_q} \delta_{pq} \equiv B_{ij}$$

and

$$\frac{\partial x_i}{\partial a_p} \frac{\partial x_j}{\partial a_q} \left[2\epsilon_{pq} - (2\epsilon_{kk}+1) \delta_{pq} \right] = B_{ki} B_{kj} - B_1 B_{ij} = -\lambda_1^2 \lambda_2^2 \delta_{ij}$$

With these identities, Equation 4.5.7 is reduced to

$$T_{ij} = \left[\left(\frac{\partial \widetilde{F}}{\partial \alpha}\right)_{T,\beta} \delta_{ij} + \frac{1}{\lambda_1^2 \lambda_2^2} \left(\frac{\partial \widetilde{F}}{\partial \beta}\right)_{T,\alpha} \left(B_{ij} - \frac{B_1}{2} \delta_{ij} \right) \right]$$

$$(4.5.8)$$

The matrix, $B_{ij} - \frac{1}{2} B_1 \delta_{ij}$, is the deviator $\tilde{B}_{ij}$ of Finger's strain tensor. Thus, its trace is zero. We can regard $\tilde{B}_{ij}$ as a Lagrangian type of strain variable which is associated with changes in shape measured relative to the initial state. Here, it is convenient to introduce another strain variable $\hat{\varepsilon}_{ij} = \tilde{B}_{ij}/2$, i.e.,

$$2\hat{\epsilon}_{ij} \equiv \frac{\partial x_i}{\partial a_k} \frac{\partial x_j}{\partial a_k} - \frac{1}{2} \left(\frac{\partial x_\ell}{\partial a_k} \frac{\partial x_\ell}{\partial a_k} \right) \delta_{ij} \tag{4.5.9}$$

which has a zero first invariant, $\tilde{\varepsilon}_{ii} = 0$, and a second invariant which is the square of the maximum Lagrangian shear strain,

$$\epsilon_s^2 = -\hat{\epsilon}_{ij} \hat{\epsilon}_{ji} = -\hat{\epsilon}_1 \hat{\epsilon}_2$$

The principal values of $\hat{\varepsilon}_{ij}$ are

$$\hat{\epsilon}_1 = \left(\frac{\lambda_1^2 - \lambda_2^2}{4} \right)$$

$$\hat{\epsilon}_2 = \left(\frac{\lambda_2^2 - \lambda_1^2}{4} \right) \tag{4.5.10}$$

With $\hat{\varepsilon}_{ij}$ in Equation (4.5.8) a general constitutive relation for large elastic (reversible) deformations of the membrane surface is obtained,

$$T_{ij} = \left[\left(\frac{\partial \tilde{F}}{\partial \alpha} \right)_{T,\beta} \delta_{ij} + \frac{2}{\lambda_1^2 \lambda_2^2} \left(\frac{\partial \tilde{F}}{\partial \beta} \right)_{T,\alpha} \hat{\epsilon}_{ij} \right] \tag{4.5.11}$$

The principal tensions are exactly those derived in Section 4.4, Equation 4.4.5, based on displacements along the principal axes,

$$T_1 = \left[\left(\frac{\partial \tilde{F}}{\partial \alpha} \right)_{T,\beta} + \frac{1}{2\lambda_1^2 \lambda_2^2} \left(\frac{\partial \tilde{F}}{\partial \beta} \right)_{T,\alpha} (\lambda_1^2 - \lambda_2^2) \right]$$

$$T_2 = \left[\left(\frac{\partial \tilde{F}}{\partial \alpha} \right)_{T,\beta} + \frac{1}{2\lambda_1^2 \lambda_2^2} \left(\frac{\partial \tilde{F}}{\partial \beta} \right)_{T,\alpha} (\lambda_2^2 - \lambda_1^2) \right]$$

$$\tag{4.5.12}$$

The isotropic tension and maximum shear resultant are identical to the previous, principal axes development. Equations 4.5.11 make it possible, however, to determine immediately the force resultants for any choice of coordinate axes in the instantaneous system.

Because free energy is an extensive or additive property, we can cumulate the contributions from the several layers that make up a membrane. If these layers experience the same surface deformation, then the principal tensions (force resultants), elastic free energy density changes, and material properties are the sum of the individual layer properties since

$$\tilde{F} = \sum_{\ell=1}^{N} \tilde{F}_\ell$$

$$\overline{T} \equiv \left(\frac{\partial \widetilde{F}}{\partial \alpha}\right)_{T,\beta} = \sum_{\ell=1}^{N} \left(\frac{\partial \widetilde{F}_\ell}{\partial \alpha}\right)_{T,\beta}$$

$$T_s \equiv \frac{1}{2(1+\alpha)} \left(\frac{\partial \widetilde{F}}{\partial \beta}\right)_{T,\alpha} (\widetilde{\lambda}^2 - \widetilde{\lambda}^{-2}) = \frac{1}{2(1+\alpha)} \left[\sum_{\ell=1}^{N} \left(\frac{\partial \widetilde{F}_\ell}{\partial \beta}\right)_{T,\alpha}\right] (\widetilde{\lambda}^2 - \widetilde{\lambda}^{-2})$$

respectively, where

$$K \equiv \left(\frac{\partial^2 \widetilde{F}}{\partial \alpha^2}\right)_{T,\beta} = \sum_{\ell=1}^{N} \left(\frac{\partial^2 \widetilde{F}_\ell}{\partial \alpha^2}\right)_{T,\beta}$$

$$\mu \equiv \left(\frac{\partial \widetilde{F}}{\partial \beta}\right)_{T,\alpha} = \sum_{\ell=1}^{N} \left(\frac{\partial \widetilde{F}_\ell}{\partial \beta}\right)_{T,\alpha} \tag{4.5.13}$$

These material properties characterize the composite or total membrane structure and will be the measurable material constants in direct mechanical experiments.

4.6 Surface Pressure and the Tension-Free State

Biological cell membranes and artificial membrane vesicles (e.g., amphiphilic bilayer vesicles) behave as closed systems for time periods much longer than the time required for mechanical experiments. A closed membrane surface contains a fixed number of molecules, and as such, it can exist in a force-free equilibrium state (the force resultants or tensions are zero). In the force-free state, a membrane surface pressure that is produced by collisions and forces between molecules in the membrane surface can still exist. This surface pressure is the time average exchange of momentum in the plane of the membrane between molecular constituents of the membrane. In most solid and liquid materials, thermal repulsive forces between molecules are counteracted by cohesive or attractive forces. The level of stress, in the absence of external forces, is established only by the pressure of the environment. Membrane materials, on the other hand, strongly interact with the adjacent environment (usually a water solution containing ionic solutes and polar molecules). The interactions at the interfaces produce chemical forces that must be supported by the membrane. In the case of simple lipid bilayer systems, surface pressure can be opposed totally by cohesive effects due to the membrane interfaces with the aqueous media. Interfacial forces are related to the energy required to freely exchange water and solute molecules between the bulk aqueous phases and the interfaces. This exchange is an open process, in contrast to the closed system of membrane constituents. Expansion or condensation of the membrane will cause the interfaces to increase or decrease, which proportionately changes the interaction with water and solute material from the adjacent environment. Unlike the closed membrane system, an open or free membrane system (e.g., a soap film) can exchange membrane molecules with a bulk, fluid phase reservoir. As a result, it will reduce its exposed interfacial area by giving up material to the reservoir until restrained by external forces, such as the hydrostatic pressure difference across a soap bubble or oil drop interface. The reservoir may be provided by the system's ability to change thickness or by a build-up of material at a support boundary.

If we consider closed systems, the isotropic part of the membrane force resultants, $\overline{T}$, is the difference between the effective interfacial tension, γ, and the surface pressure, π,

$$\overline{T} = \gamma - \pi \tag{4.6.1}$$

The interfacial tension is the surface resultant of chemical forces created by the interactions at the membrane interfaces. Without membrane tension, $\overline{T} = 0$ and $\gamma = \pi_0$, where π_0 is the natural value of the surface pressure at force-free equilibrium. The interfacial tension, γ, is only a function of temperature for small changes in the area of the membrane. It follows that the isotropic tension is equal to the negative of the change in surface pressure,

$$\overline{T} = -\Delta\pi = \pi_0 - \pi \qquad (4.6.2)$$

Therefore, the work done by the isotropic membrane force resultant, $\overline{T}$, during an increase in area, dA, is

$$dW = \overline{T} \cdot dA = (\pi_0 - \pi)\, dA$$

Hence, per unit initial element area, the free energy density change is

$$(d\widetilde{F})_T = (\pi_0 - \pi)\, d\alpha$$

The relationship between surface pressure, membrane area, and temperature may be termed an internal equation of state. A conceptual isotherm of an internal equation of state is illustrated in Figure 4.3 (See Davies and Rideal, *Interfacial Phenomena*,[12] for an extensive treatment of equations of state for monomolecular layers.) In the unstressed or force-free state, $\overline{T} = 0$, the surface pressure, π_0, is balanced by the interfacial chemical force resultant or tension, γ.

At an interface between two immiscible fluids, the surface free energy of an insoluble monolayer is defined as the surface free energy in the presence of the monolayer minus the surface free energy of the liquid-liquid substrate as the monolayer density approaches zero (see Defay and Prigogine, *Surface Tension and Adsorption*, 1966.)[13] Surface free energy density for other closed membranes that separate aqueous phases can be defined in a manner similar to the case of an insoluble monolayer. The Helmholtz free energy density is partitioned into two contributions: (1) $\widetilde{F}_p$, for the interactions between molecules of the membrane which are assumed to be excluded from exchange with the aqueous media and (2) $\widetilde{F}_w$, for the interactions between the aqueous media and the interfacial groups of the membrane. With this definition, the total Helmholtz free energy density of the membrane may be written as

$$\widetilde{F} \equiv \widetilde{F}_p + \widetilde{F}_w \qquad (4.6.3)$$

By definition, then, the interaction between membrane molecules specifies surface pressure according to the partial derivative,

$$-\pi \equiv \left(\frac{\partial \widetilde{F}_p}{\partial \alpha}\right)_T \qquad (4.6.4)$$

Therefore, surface pressure explicitly represents the time average momentum exchange in the plane of the membrane between the membrane molecules. The partial derivative of the total Helmholtz free energy density includes the contribution, γ, for the interaction of the aqueous media at the membrane interfaces.

$$\left(\frac{\partial \widetilde{F}}{\partial \alpha}\right)_T = \gamma - \pi \qquad (4.6.5)$$

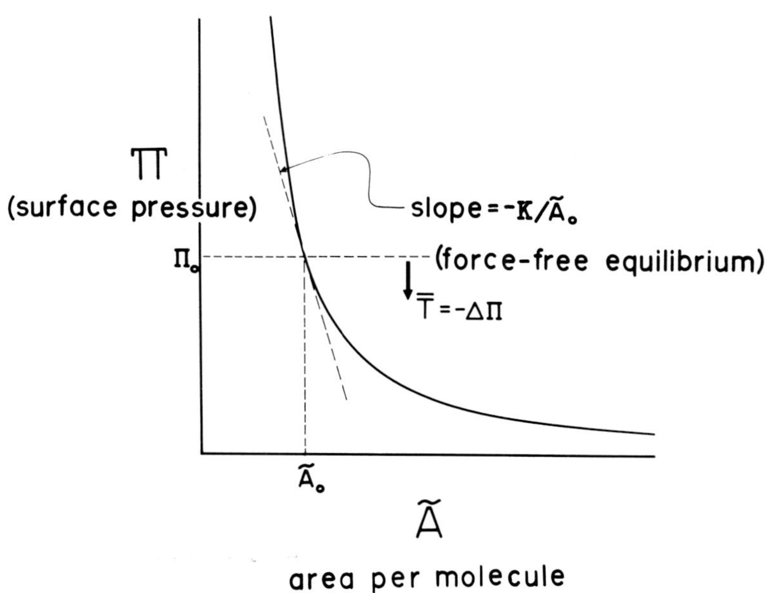

FIGURE 4.3. A conceptual isotherm of surface pressure vs. membrane area equation of state. In the force-free natural state, the membrane isotropic tension is zero, and the surface pressure, π_0, exactly opposes the interfacial chemical force resultant, γ. For any departure from the force-free state, $\overline{T} = -\Delta\pi$.

where

$$\gamma \equiv \left(\frac{\partial \tilde{F}_w}{\partial \alpha}\right)_T \qquad (4.6.6)$$

Therefore, we choose to call γ an "interfacial" free energy density or "tension" for interaction between the aqueous environment and membrane molecules.

An equivalent approach is to consider the chemical equilibrium between water at the membrane interface and the bulk aqueous phase with which it is free to exchange. This approach leads to the concept of surface osmotic stress, as introduced by Defay and Prigogine (1966).[13] Since the polar regions of membrane molecules occupy space in the aqueous interfaces, they exclude water from the interfaces. Thus, the activity of water in the interfaces is lower than in the bulk phase, which creates a surface osmotic stress. There is another metaphor that is equivalent to the concept of surface osmotic stress, which is: the interface is a surface solution of water and polar regions of membrane molecules. Consequently, the partial pressure of water is reduced in the interfacial phases and is balanced by the surface pressure of the membrane molecules. If we consider a short-time mechanical experiment, the membrane system is unable to exchange molecules with the aqueous environment. However, water is free to exchange between the bulk aqueous phases and the membrane interfaces. If we consider this exchange to be an equilibrium process, then the chemical potential for water in the interfacial phases, ϕ_w^s, is equal to the chemical potential in the bulk phase, ϕ_w^0,

$$\phi_w^s = \phi_w^0$$

The interfacial phases contribute to the total membrane tension; the component from each interface is $\overline{T}^i$, where $i = 1, 2$. These components are determined by chemical equilibrium. For a multilayer membrane, the total membrane equilibrium will implicitly represent the interfacial contact between layers and the distribution of tension between the various layers. As we will show in Section 4.10, small changes in the chemical equilibrium of membrane layers can induce large scale curvature alterations of the surface envelope. For equilibrium of the interfacial phases, the chemical potential for water in the interface is determined by the isotropic tension and the activity of water in the interfacial phase, $a_w{}^s$ (see Defay and Prigogine, 1966),[13]

$$\phi_w^{s_i} = \phi_w^{0,s_i} + \widetilde{R}T \ln a_w^{s_i} - \overline{T}^i / \widetilde{n}_w^{s_i} \tag{4.6.7}$$

where $\phi_w^{0,s}$ is the standard chemical potential (reference state) for water at the interface, $\widetilde{R}$ is the gas constant, and $\widetilde{n}_w^s$ is the number of moles of water per square centimeter at the interface. The terms in Equation 4.6.7 are thermodynamic definitions: (1) the standard chemical state for water in the interfacial phase, $\phi_w^{0,s}$, is defined by the conditions where the tension contributed by the interface is zero and the activity of water is unity; (2) relative to the reference state just chosen, the activity of water in the interfacial phase, a_w^s, is reduced in proportion to the tension. Thus, at constant temperature, the tension is directly related to the activity of water at the interface. The partial molar density of water in the interfacial phase, $\widetilde{n}_w^s$, is assumed to be constant, i.e., the area occupied by water molecules in the interfacial phase is unchanged by addition or removal of water. The chemical equilibrium expressions combine to give the relation between the tension contributed to the membrane by the interfaces, the difference in standard chemical potentials for water in bulk and interfacial phases, and the activity of water in the interfacial phase,

$$\overline{T} = \sum_{i=1}^{2} \left[\widetilde{n}_w^{s_i} (\phi_w^{0,s_i} - \phi_w^0) + \widetilde{n}_w^{s_i} \widetilde{R}T \ln a_w^{s_i} \right]$$

or

$$\overline{T} = \gamma - \pi$$

From this relation, the surface pressure in the membrane, π, and the interfacial free energy density, γ, are defined by the sum over both interfaces,

$$\pi \equiv \sum_{i=1}^{2} -\widetilde{n}_w^{s_i} \widetilde{R}T \ln a_w^{s_i}$$

$$\gamma \equiv \sum_{i=1}^{2} \widetilde{n}_w^{s_i} (\phi_w^{0,s_i} - \phi_w^0) \tag{4.6.8}$$

With this approach, we determine that the interfacial free energy density is the "unitary" free energy of transfer of water to the interface. This free energy will largely be hydrophobic in character if the hydration of the polar head groups is constant. The surface pressure is related to the activity of water at the interface in a manner analogous to osmotic pressure of water in a bulk solution. The surface activity represents the work that is required to move membrane molecules closer together or farther apart as the water interface decreases or increases in area. If the simplifying assumptions of Defay and Prigogine are used, the activity of water may be related to the mole fraction of membrane polar head groups. From this approach, a surface pressure vs. area equation of state may be derived.

The interaction between membrane and the aqueous interfaces consists of polar head group interaction plus the hydrophobic interaction with the aqueous media. If we assume that the hydration of the polar head groups changes negligibly for small variations in surface area per molecule, the change in interfacial interaction will be due to changes in the interfacial contact between the hydrophobic groups and water. According to Tanford (1973, 1974)[88,89] and Reynolds et al. (1974),[75] the interaction between hydrocarbon and water, i.e., the hydrophobic effect, is proportional to the area of contact. For small changes in area per molecule in the surface, it is reasonable to assume that the contact area between hydrophobic groups and the aqueous phase increases in proportion to the change in area per molecule. Therefore,

$$\tilde{F}_w \cong \tilde{F}_w^0 + \left(\frac{\partial \tilde{F}_H}{\partial \alpha}\right)_T \alpha \tag{4.6.9}$$

where $\tilde{F}_w^0$ is the free energy reference of the aqueous interface for the force-free, natural state; $(\partial \tilde{F}_H / \partial \alpha)_T$ is the interfacial free energy density of hydrophobic interaction. From Equation 4.6.6, we see that γ is the interfacial free energy density of the hydrophobic interaction,

$$\gamma = \left(\frac{\partial \tilde{F}_H}{\partial \alpha}\right)_T \tag{4.6.10}$$

γ depends only on temperature and other properties of the aqueous environment. The hydrophobic interaction is determined by the geometry of the cavities between molecules. These cavities form the interstitial area of contact between aqueous phase and the hydrophobic region of the molecule.* The definition of natural state prescribes a surface pressure, π_0, where $\overline{T} = 0$. This is obtained from Equation 4.6.1,

$$\overline{T} = \gamma - \pi$$

when the tension is zero,

$$\pi_0 = \gamma \tag{4.6.11}$$

In this state, the potential loss of repulsive energy from expansion of the membrane is opposed by the potential increase in energy required to enlarge the interfacial contact between hydrophobic groups and the aqueous environment. Unlike the free surface of a soap bubble, the vesicle or cell membrane surface can exist at zero tension if the hydrostatic pressure inside the vesicle or cell is equal to the outside pressure. In other types of amphiphilic bilayer membranes, such as planar bilayer films which contain hydrocarbon solvent, the exchange of material with the boundary "torus" creates a tension at the boundary. The tension is uniform throughout the surface film and is called the surface tension, γ_σ, of the planar bilayer,

$$\overline{T} = \gamma_\sigma$$

Hence, a planar bilayer membrane would have a surface pressure equal to

---

* Any changes in water interaction with the polar head groups which are produced by small changes in area per molecule would be included to first order in the linear expansion, Equation 4.5.9.

$$\pi = \gamma - \gamma_\sigma$$

The previous discussion and development of surface pressure and interfacial free energy density can be represented by an empirical hypothesis for general membrane materials. The hypothesis is simply that the isotropic tension in the membrane is related to two variables: one is a function of both temperature and surface density, labeled $\pi(\alpha, T)$, and the other is only a function of temperature, labeled by $\gamma(T)$. The function, $\pi(\alpha, T)$, is what we will call surface pressure and is an equation of state internal to the membrane. Even though the surface pressure cannot be directly determined in a closed membrane system, we can relate the internal equation of state, $\pi = f(\alpha, T)$, to the mechanochemical equation of state, $\overline{T} = f(\alpha, T)$, through investigation of effects of changes in area and temperature. We will develop these thermoelastic relations in Section 4.11. In the following examples, we will illustrate the nature of the interfacial free energy density, γ, and a surface equation of state for a phospholipid bilayer membrane.

Example 1

Phospholipid molecules are amphiphilic species. They are characterized by a polar head group which has a high affinity for aqueous solutions and two acyl chains, attached to the glycerol "backbone" of the polar end, which are essentially insoluble in aqueous media. Consequently, these molecules can form a stable bilayer configuration with the hydrocarbon polymer chains (acyl chains) in the middle and the polar ends fixed at the aqueous interfaces of the membrane (Tanford, 1973).[88] For a closed bilayer, we cannot experimentally determine the interfacial free energy density of hydrophobic interaction, γ, and the associated surface pressure, π_0. However, we can consider a simple model for the interstitial hydrocarbon regions between polar head groups. This model consists of surface regions of CH_2 groups in contact with water. Reynolds et al. (1974)[75] have shown that the "unitary" free energy density of transfer of hydrocarbon to an aqueous solution is about 25 cal/mol/$\mathring{A}^2$ at 25°C. In other words, each $\mathring{A}^2$ of hydrocarbon-water interface for a single molecule requires 25 cal for 1 mol of hydrocarbon molecules. To calculate the free energy density in erg/cm², we must determine the number of moles of CH_2 groups per cm² of surface plus the interfacial contact area of a CH_2 group in $\mathring{A}^2$. A CH_2 group is approximately 20 $\mathring{A}^2$ in cross section, giving about 10^{-9} mol/cm². If the interfacial contact surface of a single CH_2 group with water is approximated by a hemisphere, the contact area per group would be twice its cross sectional area, i.e., 40 $\mathring{A}^2$. This gives a free energy difference on a per mole basis of 1000 cal/mol. Per square centimeter of contact area, the interfacial free energy density, γ, would be

$$\gamma \sim 0.84 \times 10^{-6} \text{ cal/cm}^2$$

or

$$\gamma \sim 35 \text{ dyn/cm}$$

As is expected, this value is the approximate surface free energy density for a paraffin oil-water interface. Consequently, the energy expense of changing the exposed hydrocarbon regions in a bilayer would be twice this value or 70 dyn/cm, and the surface pressure, π_0, would be 70 dyn/cm in the unstressed bilayer membrane.

Example 2

To illustrate the dependence of surface pressure on area or surface density in a phospholipid bilayer membrane, we use the two-dimensional Van der Waal's model of a condensed gas without attractive forces, i.e., a collection of hard cylinders in random thermal motion. This is an approximate empirical model for a limited range of surface pressure vs. area per molecule behavior of lipid monolayers in the fluid state (above the order-disorder transition temperature for the hydrocarbon chains (Davies and Rideal, 1961).[12] Such an equation of state for a phospholipid bilayer would be represented by

$$\pi(\tilde{A} - \tilde{A}_e) = c \cdot T \tag{4.6.12}$$

where $\tilde{A}$ is the area per molecule (in each monolayer); $\tilde{A}_e$ is the excluded area per molecule. The constant, c, is theoretically equal to $4 \cdot k$, where k is Boltzmann's constant (the factor of "4" is for a phospholipid bilayer with four acyl chains per area, $\tilde{A}$). The area compressibility modulus is obtained using Equations 4.4.11 and 4.6.2,

$$K = \left(\frac{\partial \bar{T}}{\partial \alpha}\right)_T \Bigg|_{\alpha \to 0} = -\left(\frac{\partial \pi}{\partial \alpha}\right)_T \Bigg|_{\alpha \to 0}$$

We define the fractional change in area as

$$\alpha = \frac{\tilde{A} - \tilde{A}_0}{\tilde{A}_0}$$

Therefore, the area compressibility modulus is related to the surface pressure for the natural state,

$$K = \pi_0 \left(\frac{\tilde{A}_0}{\tilde{A}_0 - \tilde{A}_e}\right)$$

where $\tilde{A}_0$ is the area per molecule in the initial, undeformed state.

Data for lecithin monolayers spread at oil-water interfaces (Yue et al., 1975)[98] can be approximated by values of 38 Å^2 per molecule for the excluded area and 0.7×10^{-15} erg/K for the constant, c. At the surface pressure, π_0, equal to 70 dyn/cm which we estimated from the previous example, the equilibrium area per molecule, $\tilde{A}_0$, would be 68 Å^2 in the natural state.* Therefore, the elastic area compressibility modulus for the lecithin bilayer is expected to be of the order,

$$K = 70 \cdot \frac{68}{30} \simeq 160 \text{ dyn/cm}$$

4.7 Shear Hyperelasticity: Example of a 2-D Elastomer

In this section, we will present a thermodynamic model for an "elastomeric" surface. A material of this type is hyperelastic, i.e., capable of finite elastic deformations. In addition, the material model exhibits first order dependence on the geometric deformation variable, β. The example will illustrate the statistical nature of entropy and how the surface elastic shear modulus, a property of the continuum, is related to the microstructure of this particular material. In the example, we will consider the work

---

* It is interesting to note that the predicted value of 68 Å^2 per molecule for lecithin molecules in an unstressed bilayer is equal to the area per molecule determined from X-ray diffraction measurements of multilamellar phases of egg yolk lecithin and water for the fully hydrated state (Reiss-Husson, 1967; Luzzati, 1968).[56,72]

of extension of the surface material at constant surface density. From this, then, we can determine a surface elastic shear modulus according to Equation 4.4.9.

The thermodynamic basis for elastic behavior is determined by measuring the reversible heat exchange produced by the deformation, either calorimetrically or from thermoelastic properties. Entropic elastomers are not the only forms of hyperelastic materials, although they are commonly identified as such materials. We will show that the hyperelastic behavior of red cell membrane is not derived from negative entropy changes that are produced by extension (Section V). Unlike entropic elastomers, extension of the red cell membrane appears to increase the entropy of the material, i.e., the reversible heat of extension is positive. This means that the material takes up heat from the environment as it is extended, whereas an entropic elastomer gives up heat to the environment when extended.

At constant temperature, the elastic potential energy density is equal to the Helmholtz free energy density and is composed of two contributions: internal energy changes and changes in the distribution of energy between random molecular states (i.e., temperature times change in entropy),

$$(d\widetilde{F})_T = d\widetilde{E} - T\,d\widetilde{S}$$

An elastomer is made up of a loose network of randomly arranged, flexible molecules. The major contribution to changes in free energy density come from the work required to order the network molecular chains (if the material density is not required to change). (See Meares, *Polymers: Structure and Bulk Properties*, 1965,[59] for a discussion of this kind of elastomeric material.) For such an elastomer, little change in internal energy is produced by extensional deformation. Therefore, we can associate the change in Helmholtz free energy density created by deformation with the entropy contribution,

$$(d\widetilde{F})_T \simeq -T\,d\widetilde{S} \qquad\qquad (4.7.1)$$

With Boltzmann's equation (4.2.9), the entropy is given by the distribution of molecules among possible network configurations,

$$dA_0 \cdot \widetilde{S} = k\,\ell n\,\Omega \qquad\qquad (4.7.2)$$

In general, it is difficult to specify the possible molecular arrangements in an exact, quantitative manner. Therefore, we characterize the structure by statistical parameters that are derived from the nature of network molecules. Each network molecule is assumed to be highly flexible. Thus, the small segments of the molecule will be randomly arranged and oriented as illustrated in Figure 4.4. The two statistically significant parameters will be the end-to-end distance of the whole molecule and the variance of this distance (i.e., the mean square distance). For an individual molecular chain which contains a large number, n, of small segments of equal length, ℓ_0, we can specify the probability that one end of the molecule will lie in the range of (x_1, x_2) to $(x_1 + dx_1, x_2 + dx_2)$ relative to the other end. It is given by the probability density function $P(x_1, x_2)$ times the differential area, $dx_1 \cdot dx_2$. The probability density function is defined by the density of possible configurations local to the point (x_1, x_2) such that the integral of the probability density over the surface is unity,

$$\iint P(x_1, x_2)\,dx_1\,dx_2 \equiv 1$$

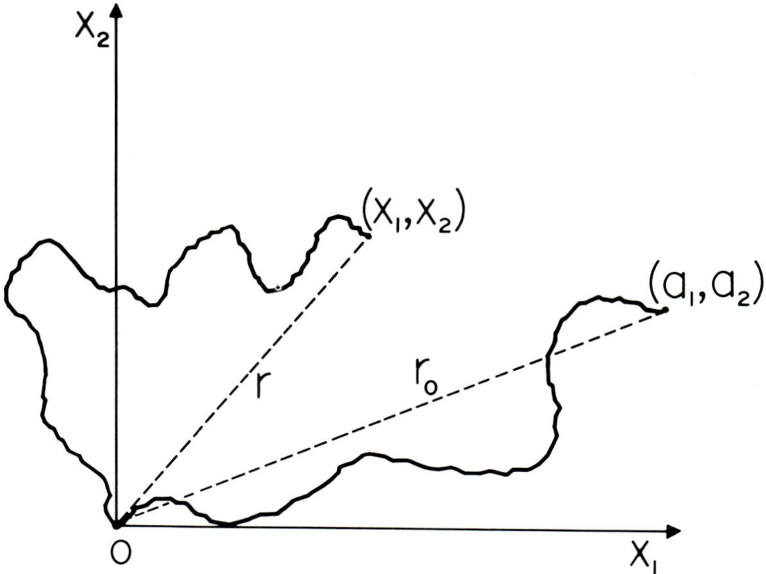

FIGURE 4.4.   The initial and deformed configurations of a randomly arranged, flexible molecule. The end-to-end distance of the whole molecule and the variance of this distance are the statistically significant parameters. End locations of molecules are specified by (a_1,a_2) for the initial molecular configuration and by (x_1,x_2) for the deformed state.

We assume that end locations of molecules are randomly distributed about an expected position given by coordinates $\bar{x}_1$ and $\bar{x}_2$ relative to the origin, as illustrated in Figure 4.4. The probability distribution function is chosen as a Gaussian (normal) distribution with a variance, σ^2.

$$P(x_1,x_2) = \left(\frac{1}{2\pi\sigma_1\sigma_2} \right) \exp\left[-\frac{(x_1-\bar{x}_1)^2}{2\sigma_1^2} \right]$$

$$\exp\left[-\frac{(x_2-\bar{x}_2)^2}{2\sigma_2^2} \right]$$

The density of possible configurations, $P(x_1, x_2)$, of an individual molecule specifies the distribution function for an individual molecule. The network distribution function for the differential element involves $\bar{N}$ molecular chains each of which has an individual distribution function that may have different average end locations $(\bar{x}_1, \bar{x}_2)$ and variances, σ_1^2, σ_2^2, depending on the network cross-linking and molecular composition.

Because of the assumption of isotropy and the assumption that each molecule contains a large number of small segments, the average end-to-end positions are given by the sum of the projections of each segment onto the x_1 and x_2 axes. These projections are oriented at angles, ϕ, that are uniformly distributed (equally probable) about the origin. Hence, the mean location of one end relative to the other is given by

$$\bar{x}_1 = \frac{n\ell_0}{2\pi} \int_0^{2\pi} \cos\phi \, d\phi = 0$$

$$\overline{x}_2 = \frac{n\ell_0}{2\pi} \int_0^{2\pi} \sin\phi\, d\phi = 0$$

Therefore, the average distance from end to end of the molecular chain is negligible, but the variance is not. The variance is the mean square deviation from the expected location,

$$\sigma_1^2 = \overline{(x_1 - \overline{x}_1)^2} = \overline{(x_1^2)} = \frac{n\ell_0^2}{2\pi} \int_0^{2\pi} \cos^2\phi\, d\phi = n\ell_0^2/2$$

$$\sigma_2^2 = \overline{(x_2 - \overline{x}_2)^2} = \overline{(x_2^2)} = \frac{n\ell_0^2}{2\pi} \int_0^{2\pi} \sin^2\phi\, d\phi = n\ell_0^2/2$$

As we see, this is consistent with isotropy, i.e., $\sigma_1^2 = \sigma_2^2 \equiv \sigma^2$. The isotropic variance, σ^2, is a measure of the area occupied by the molecule.

For a particular chain in the network, say the Nth molecule, the distribution of the metric for distances between network points (end to end) is given by the Gaussian distribution,

$$\Omega_N = \left(\frac{1}{2\pi\sigma_N^2}\right) e^{-\frac{r^2}{2\sigma_N^2}} \tag{4.7.3}$$

where $r^2 = x_1^2 + x_2^2$ is the metric or measure of end-to-end distance, and Ω_N is the density of molecular configurations for the Nth molecule with end-to-end distances in the range $(r, r + dr)$.

For a loose network, the total distribution function, Ω, is the product of the individual functions, Ω_N, since each molecule is assumed to be free to arrange itself independent of the others. Thus,

$$\Omega = \Omega_1 \cdot \Omega_2 \cdots = \prod_{N=1}^{\overline{N}} \Omega_N \tag{4.7.4}$$

where the symbol $\prod_{N=1}^{\overline{N}}$ represents the product of $\overline{N}$ terms. The configurational entropy is obtained with Equations 4.7.2, and 4.7.4,

$$dA \cdot \widetilde{S}_0 = k \sum_{N=1}^{\overline{N}} \ell n\, \Omega_N$$

The logarithm converts the product to a sum. Therefore, with Equation 4.7.3, the configurational entropy is given by

$$dA \cdot \widetilde{S}_0 = k \left[\text{constant} - \sum_{N=1}^{\overline{N}} \frac{r^2(N)}{2\sigma_N^2} \right] \tag{4.7.5}$$

Equation 4.7.5 is independent of the choice of axes for coordinates local to a molecule. The molecules are randomly oriented and so are the vectors that define the end-to-end positions. Consequently, only a radial distribution can be specified, which represents the mean square distance from end to end. The statistics "smear out" angular dependence because of the large number of subunits and flexibility in the molecular constituents. Consequently, deformation produces a change in entropy of the material element that is given by the initial mean square distance, r_0^2, and the instantaneous mean square distance, r^2, which describe the molecular distributions. With Equation 4.7.5, we define the entropy change by

$$d\,A_0\,(\widetilde{S} - \widetilde{S}_0) = -\frac{k}{2} \sum_{N=1}^{\overline{N}} \left[\frac{r^2(N)}{\sigma_N^2} - \frac{r_0^2(N)}{\sigma_{N_0}^2} \right] \tag{4.7.6}$$

We have allowed the statistical variances, σ_N^2 and $\sigma_{N_0}^2$, in the distributions to be different for the instantaneous and initial configurations.

We now make a crucial assumption: the deformation is experienced in a geometrically similar manner at the molecular level as on the continuum scale. In other words, molecular coordinates are scaled in the same proportions as those of the macroscopic element. For time scales over which the material cannot thermally relax, this assumption is reasonable. Therefore, if the metrics, r^2 and r_0^2, are the instantaneous and initial radial distances of one molecular end from the other, they are given by the ratio of sum of squares of characteristic lengths for each state (taken as unity for the initial state),

$$\frac{r^2}{r_0^2} = \frac{\lambda_1^2 + \lambda_2^2}{2} \tag{4.7.7}$$

The variances are isotropic properties that are scaled by the ratio of instantaneous to initial area,

$$\frac{\sigma_N^2}{\sigma_{N_0}^2} = \frac{dA}{dA_0} = \lambda_1 \lambda_2 \tag{4.7.8}$$

Now, the change in configurational entropy is obtained in terms of the element deformation extension ratios and network molecular properties,

$$d\,A_0\,(\widetilde{S} - \widetilde{S}_0) = -k \sum_{N=1}^{\overline{N}} \frac{r_0^2(N)}{\sigma_{N_0}^2} \left[\frac{1}{\lambda_1 \lambda_2} \left(\frac{\lambda_1^2 + \lambda_2^2}{2} \right) - 1 \right]$$

We see that the entropy change is simply proportional to the extensional deformation variable, β,

$$d\,A_0\,(\widetilde{S} - \widetilde{S}_0) = -k\,\beta\,\overline{N} \left\langle \frac{r_0^2}{\sigma_0^2} \right\rangle$$

where

$$\beta = \tfrac{1}{2} \left(\frac{\lambda_1}{\lambda_2} + \frac{\lambda_2}{\lambda_1} \right) - 1$$

as it was defined in Section II for stretch of the material at constant area. The network properties have been represented by

$$\bar{N} \left\langle \frac{r_0^2}{\sigma_0^2} \right\rangle_{T,A_0} \equiv \sum_{N=1}^{\bar{N}} \frac{r_0^2(N)}{\sigma_{N_0}^2}$$

where the ensemble average is taken at constant temperature and initial area. The ensemble average $\langle r_0^2/\sigma_0^2 \rangle$ gives the expected value of the quadratic variation in molecular chain length for the whole network normalized by the "free" variation of an individual chain. This average represents the percentage of active chains out of the total number. If all were equally active and identical in composition, the factor would be unity. We can also introduce the surface density, ϱ_0; the total number of molecules times the average molecular weight, $\langle M_w \rangle$, equals the total mass of the element,

$$\rho_0 \cdot dA_0 = \frac{\bar{N}}{N_A} \langle M_W \rangle$$

(N_A is Avogadro's number.) Therefore, the entropy change can be reduced to a change in entropy density,

$$(\tilde{S} - \tilde{S}_0) = - \frac{\rho_0 \tilde{R}\beta}{\langle M_W \rangle} \left\langle \frac{r_0^2}{\sigma_0^2} \right\rangle \tag{4.7.9}$$

where $\tilde{R} = kN_A$ is the gas constant. Since the free energy density depends linearly on β, as seen from Equations 4.7.1 and 4.7.9, the surface elastic shear modulus is determined by the configuration parameters of the molecular network and the absolute temperature. First, we recall the relation for the surface shear modulus,

$$\mu = \left(\frac{\partial \tilde{F}}{\partial \beta} \right)_{T,\alpha} = -T \left(\frac{\partial \tilde{S}}{\partial \beta} \right)_{T,\alpha}$$

then from Equation 4.7.9, we obtain the result,

$$\mu = \frac{\rho_0 \tilde{R} T}{\langle M_W \rangle} \left\langle \frac{r_0^2}{\sigma_0^2} \right\rangle \tag{4.7.10}$$

Now, we can see an interesting property of entropic elastomers: the elastic shear modulus increases with temperature. The work of extension for an elastomeric surface is opposed by the natural, thermal randomization processes, which occur at the molecular level. This is an expression of the second law of thermodynamics: work is required to increase order in the molecular world.

From Equation 4.7.10, we also see that the reversible heat exchange for extension of the surface is given by the surface shear modulus, which has units of energy per unit area (erg/cm²). We can represent the reversible heat exchange per mole of molecular chains in the surface network,

$$\frac{\langle M_W \rangle}{\rho_0} \cdot \mu = \tilde{R} T \left\langle \frac{r_0^2}{\sigma_0^2} \right\rangle$$

which has units of energy per mole (erg/mol). The heat content or thermal energy for a degree of molecular freedom is on the order of the gas constant times temperature, $\tilde{R}T$, for a mole of the substance. Therefore, we see directly that the reversible heat of extension of an entropic elastomer represents the reduction in average degrees of freedom of the molecular chains.

Example 1

As an illustration, what magnitude of elastic shear modulus would we expect for a membrane that is supported by a filamentous network of flexible molecules with the size and numbers per unit area of the red cell membrane protein, spectrin? The idealized view for the red cell membrane is shown Figure 1.1 with spectrin underneath a lipid and protein mixture (bilayer). Based on work of biochemists, beginning with Marchesi et al. (1969, 1970),[57,58] the molecular weight of spectrin units is on the order of 2×10^5 daltons and the surface density is about 10^{-7} g/cm^2, which implies a molar density of about 5×10^{-13} mol/cm^2. At a temperature of 300 K and with the gas constant equal to 8.3×10^7 erg/K/mol, the shear modulus of a surface elastomer of molecules with the above weight and density properties would be on the order of

$$\mu \sim 10^{-2} \quad \text{dyn/cm (erg/cm}^2\text{)}$$

The measured value for the red cell membrane shear modulus is also on the order of 10^{-2} dyn/cm (see Evans and Hochmuth, 1978, for discussion).[22] However, recent measurements (Waugh, 1977; Waugh and Evans, 1978)[91,93] have shown that red cell membrane shear modulus decreases as temperature increases. Consequently, the entropy change from membrane extension is opposite to that expected for a simple elastomer (Evans and Waugh, 1978).[26] It is significant, though, that the magnitude of the red cell membrane shear modulus is on the order calculated from entropy considerations for spectrin. If we assume that spectrin is the supporting structural matrix (as all evidence to date seems to imply), subtle changes in molecular arrangement and configuration of the spectrin matrix are important and must be studied further.

A fundamental aspect of material behavior for cell membranes is apparent when we note that the values for membrane shear moduli can be so small. These values are about four orders of magnitude lower than the elastic area compressibility modulus ($\sim 10^2$ dyn/cm) estimated for a phospholipid bilayer and actually measured for a red cell membrane (Evans and Waugh, 1977).[25] This implies that membrane composites greatly resist intensive changes in surface density or area, but can easily deform by surface extension at constant area. Thus, in most deformations, area changes are very small and often negligible, i.e., the surface can be considered as two-dimensionally incompressible. In contrast to this behavior, we will give an example in Section V for the membrane-cortex of the sea urchin egg where the elastic area compressibility modulus is much smaller than for an amphiphilic bilayer or red cell membrane. Here, surface area changes of 20 to 30% occur easily. However, this membrane is actually a ruffled structure at the submicroscopic level with considerable material associated with its subsurface structure (it is on the order of 1 to 3 μm thick, 100 times larger than most molecules).

4.8 Hyperelastic Membrane with Small Area Compressibility

The examples given in Sections 4.6 and 4.7 illustrate the main features of the observed material behavior of many cell membranes, i.e., large resistance to area dilation and low resistance to extension or shear at constant area (Skalak, 1973; Evans, 1973).[16,85] This behavior may be described to first approximation by the free energy density, $\tilde{F}$, relative to the tension free state, $\tilde{F}_0$,

$$(\widetilde{F} - \widetilde{F}_0)_T \cong K \frac{\alpha^2}{2} + \mu\beta \tag{4.8.1}$$

where the deformation variables, α and β, quantify the intensive area change and surface extension. From this expression for free energy density, the force resultants are derived with Equation 4.5.11,

$$T_{ij} = K\alpha\,\delta_{ij} + \frac{2}{\lambda_1^{\,2}\lambda_2^{\,2}}\,\mu\,\hat{\epsilon}_{ij} \tag{4.8.2}$$

where the isotropic part of T_{ij} is $K \cdot \alpha$. Equation 4.8.2 applies for any choice of local coordinates and degree of extension.

In the case of small area compressibility, it is implied that K is large relative to μ and that α will be small. The product, $K \cdot \alpha$, may be comparable in magnitude to the second term in Equation 4.8.2, or even dominate it. Since α is small, the approximation of constant area may be used, i.e.,

$$\lambda_1\lambda_2 \simeq 1 \tag{4.8.3}$$

Therefore Equation 4.8.2 is reduced to

$$T_{ij} = K\alpha\,\delta_{ij} + 2\mu\,\hat{\epsilon}_{ij} \tag{4.8.4}$$

In the principal axes system, the principal tensions are essentially determined by the small fractional change in area, α, and a single extension ratio, e.g., λ_1,

$$T_1 = K\alpha + \mu(\lambda_1^2 - \lambda_1^{-2})/2$$

$$T_2 = K\alpha - \mu(\lambda_1^2 - \lambda_1^{-2})/2 \tag{4.8.5}$$

It is understood from Equation 4.8.3 that $\lambda_2 \cong \lambda_1^{-1}$.

When the approximation (4.8.3) is adopted, α can no longer be computed from the extension ratios, but must be regarded as an independent variable. The product, $K \cdot \alpha$, can be represented by a single variable, analogous to the hydrostatic pressure in the theory of incompressible fluids. We replace $K \cdot \alpha$ by an isotropic tension, $\overline{T}$, which may be a function of time and position. The isotropic tension is determined from the equations of motion and the boundary conditions. Thus Equations 4.8.4 and 4.8.5 become

$$T_{ij} = \overline{T}\,\delta_{ij} + 2\mu\,\hat{\epsilon}_{ij} \tag{4.8.6}$$

and in the principal axes systems,

$$T_1 = \overline{T} + \mu(\lambda_1^2 - \lambda_1^{-2})/2$$

$$T_2 = \overline{T} - \mu(\lambda_1^2 - \lambda_1^{-2})/2 \tag{4.8.7}$$

From a physical viewpoint, the isotropic part of the force resultants may be attributed to a change in surface pressure, π, which is approximated by

$$\overline{T} = -\Delta\pi \cong K \cdot \alpha \tag{4.8.8}$$

With the constant area approximation, Equations 4.8.3, 4.8.6, or 4.8.7 are a convenient and sufficiently accurate basis for the solution of most problems that deal with

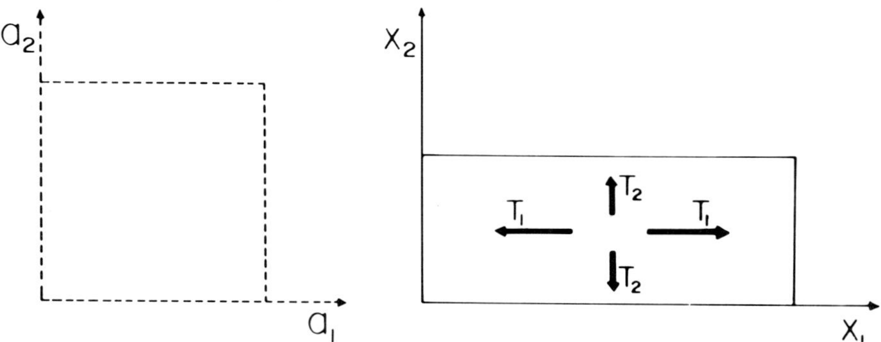

FIGURE 4.5. Elastic extension of a square element into a rectangle at nearly constant area. The principal tensions, T_1 and T_2, are given by the constitutive relations, Equations 4.8.7.

the deformation of membranes which encapsulate cells. The preponderance of experimental work has been done on red blood cell and lipid bilayer membranes. However, our development of hyperelastic constitutive relations for membranes is directly applicable to more complicated membrane structures. In Section V, we will apply the elastic constitutive relations in examples of membrane mechanical experiments. Here, we consider some elementary aspects of the elastic constitutive relation which are appropriate to extension and simple shear deformation of material elements.

Example 1

Consider the elongation of a square element into a rectangle at nearly constant area. The principal tensions are given by Equations 4.8.7 and are illustrated in Figure 4.5,

$$T_1 = \bar{T} + \frac{\mu}{2} (\lambda_x{}^2 - \lambda_x{}^{-2})$$

$$T_2 = \bar{T} - \frac{\mu}{2} (\lambda_x{}^2 - \lambda_x{}^{-2})$$

The shear resultants, T_{12} and T_{21}, are zero; the isotropic tension replaces the surface pressure change, $-K \cdot \alpha$. The isotropic tension must be determined by a separate statement, e.g., by the specific value of the small fractional area change or by the specification of the value of one of the tensions through a boundary condition. For instance, if the element is deformed by simple, uniaxial tension (i.e., $T_2 \equiv 0$), then the force resultants (principal tensions) are given by

$$T_2 = 0$$

and

$$T_1 = \mu(\lambda_x^2 - \lambda_x{}^{-2})$$

The isotropic tension and change in surface pressure are related to the uniaxial tension by

$$\bar{T} = -\Delta\pi = \frac{T_1}{2}$$

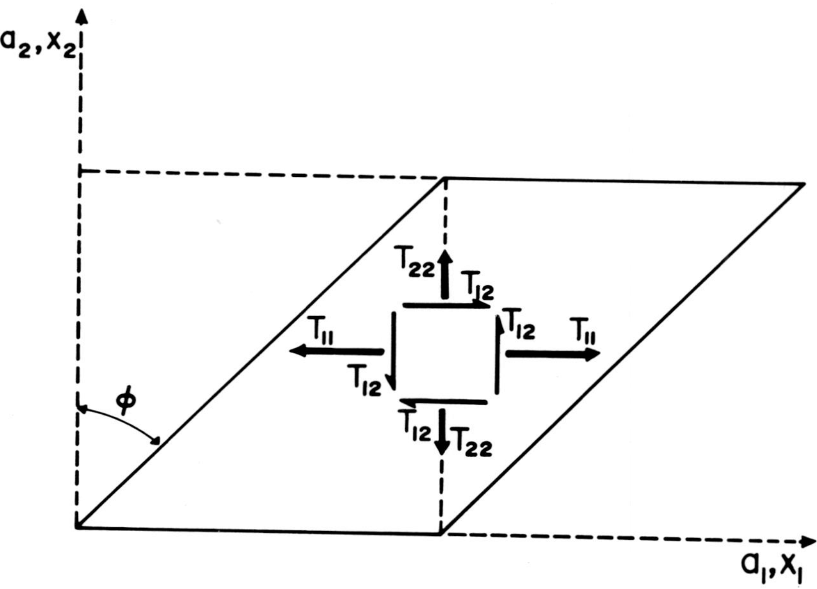

FIGURE 4.6.   Simple shear deformation of a square element into a parallelogram at constant area. ϕ is the deformation angle and is used to determine the components of the force resultant matrix. For small deformation, $\phi \ll 1$, the principal axes are simply oriented at 45° to the instantaneous coordinate axes.

Example 2

Consider the simple shear deformation of a square element into a parallelogram at constant area. The force resultants are illustrated in Figure 4.6. We want to find the values for the force resultants in terms of the shear deformation angle, ϕ, and to find the principal axes system for the force resultants. We use Equation 4.8.6,

$$T_{ij} = \overline{T}\,\delta_{ij} + 2\mu\hat{\epsilon}_{ij}$$

First, we write out the components of the strain representation, $\hat{\epsilon}_{ij}$, from Equation 4.5.9. The principal components are equal in magnitude, opposite in sign,

$$\hat{\epsilon}_{11} = \tfrac{1}{4}\left(\frac{\partial x_1}{\partial a_1}\frac{\partial x_1}{\partial a_1} + \frac{\partial x_1}{\partial a_2}\frac{\partial x_1}{\partial a_2} - \frac{\partial x_2}{\partial a_1}\frac{\partial x_2}{\partial a_1} - \frac{\partial x_2}{\partial a_2}\frac{\partial x_2}{\partial a_2}\right)$$

$$\hat{\epsilon}_{22} = \tfrac{1}{4}\left(\frac{\partial x_2}{\partial a_1}\frac{\partial x_2}{\partial a_1} + \frac{\partial x_2}{\partial a_2}\frac{\partial x_2}{\partial a_2} - \frac{\partial x_1}{\partial a_2}\frac{\partial x_1}{\partial a_2} - \frac{\partial x_1}{\partial a_1}\frac{\partial x_1}{\partial a_1}\right)$$

i.e.,

$$\hat{\epsilon}_{22} \equiv -\hat{\epsilon}_{11}$$

The shear components are

$$\hat{\epsilon}_{12} = \hat{\epsilon}_{21} = \tfrac{1}{2}\left(\frac{\partial x_1}{\partial a_1}\frac{\partial x_2}{\partial a_1} + \frac{\partial x_1}{\partial a_2}\frac{\partial x_2}{\partial a_2}\right)$$

For this particular problem, the components are calculated with the coordinate transformations given in Example 2 of Section 2.2,

$$\hat{\epsilon}_{11} = -\hat{\epsilon}_{22} = \frac{\tan^2 \phi}{4}$$

$$\hat{\epsilon}_{12} = \frac{\tan \phi}{2}$$

Thus, the force resultant components are determined to be

$$T_{11} = \overline{T} + \frac{\mu}{2} \tan^2 \phi$$

$$T_{22} = \overline{T} - \frac{\mu}{2} \tan^2 \phi$$

$$T_{12} = \mu \tan \phi$$

Additional information on the boundary conditions must be provided in order to specify the isotropic tension.

The orientation of principal axes for the force resultant matrix is given by counterclockwise rotation of ϕ_R radians in the instantaneous coordinate system. The rotation angle is determined from the relation,

$$\tan 2\phi_R = \frac{2\hat{\epsilon}_{12}}{\hat{\epsilon}_{11} - \hat{\epsilon}_{22}} = 2(\tan \phi)^{-1}$$

For small deformation, $\phi \ll 1$, the principal axes are simply oriented at 45° to the instantaneous coordinate axes, and the first order values for the force resultants indicate the simple shear process,

$$T_{11} \simeq \overline{T}$$

$$T_{22} \simeq \overline{T}$$

and,

$$T_{12} \simeq \mu\phi$$

4.9 Membrane Bending Moments for Coupled Molecular Layers

In the introduction, we mentioned that curvature elasticity or bending resistance is essentially negligible in a single molecular layer until the radii of curvature for the layer are on the order of molecular dimensions; then the continuum hypothesis is no longer valid. Membranes made up of two or more molecular layers, which are free to slide relative to one another (uncoupled) without restrained edges, also would not develop moments due to changes of curvature. Thus, until now, we have only needed to consider membrane deformations that do not involve radii of curvature, i.e., plane deformations of differential surface elements. Now, we will consider composite membrane structures and introduce the possibility of coupled interactions between the layers of the composite. These interactions are usually second order compared with the energy changes produced by the force resultants and displacements already discussed. (Fung (1966)[29] pointed out that bending resistance is most often negligible compared with membrane force resultants (tension). However, Fung also noted that many situations arise where force resultants are small or do not contribute to the local membrane stability. (Here, small bending moments are significant.) We will be primarily concerned in this section with coupled strata and, in the next section, with uncoupled layers. We assume that coupled layers are rigidly connected so that no relative sliding of adjacent layers can occur. If the surface radii of curvature are much larger than the distance between the molecular layers of the composite, the surface deformation of the whole membrane will be experienced nearly the same by each layer. However, from layer to layer, there can be small relative deformations that can produce bending moments. The bending moments result from the free energy variations caused by the small relative deformations.

Curvature changes produce small relative deformation of one layer compared to another; Figure 4.7 illustrates the effect for a double layer. If the two elements of surface were originally the same size when the double layer was flat, then the change in curvature has left the outer layer slightly expanded relative to the inner layer for the convex shape shown in Figure 4.7. We assume that the distance between layers is fixed and that the distance is much smaller than the principal radii of curvature, R_1 and R_2, illustrated in Figure 4.7. Thus, deformation of the total membrane structure is described by the mean surface deformation, i.e., fractional change in area and extension at constant area, with the relative deformation of individual layers described by small variations from the mean. The mean deformation characterizes the shape change of an element of surface which is defined by an abstract "neutral" layer or "neutral" surface. This surface is the plane of action for the total membrane force resultants. For the thin membrane material, the radii of curvature for the neutral surface are good approximations for any location within the membrane. Therefore, we simply identify the change in curvature of the membrane and the component layer as

$$C_1 \equiv \Delta \left(\frac{1}{R_1} \right)$$

$$C_2 \equiv \Delta \left(\frac{1}{R_2} \right)$$

These curvature changes are defined as positive for convex changes in the sense of the outward surface normal (e.g., a positive curvature change is shown in Figure 4.7). We assume that the principal axes for deformation of the total membrane structure coincide with the orthogonal planes that contain the surface arcs generated by the principal curvatures, as in axisymmetric problems. For rigidly coupled layers, the relative deformation of an individual layer compared to the deformation of the neutral surface is proportional to the layer's distance, h_ℓ, from the neutral surface times the change in curvature that has occurred in conjunction with the membrane deformation. The first order approximations for the variation in extension ratios define the deformation relative to the neutral surface,

$$\delta \lambda_1^\ell \simeq \lambda_1 \, h_{\ell_1} \, C_1$$

$$\delta \lambda_2^\ell \simeq \lambda_2 \, h_{\ell_2} \, C_2 \qquad\qquad (4.9.1)$$

where λ_1 and λ_2 are the principal extension ratios that represent the mean deformation of the membrane (i.e., neutral surface). The ℓth layer is located at a distance, h_ℓ, which is considered positive in the direction outward from the center of curvature.

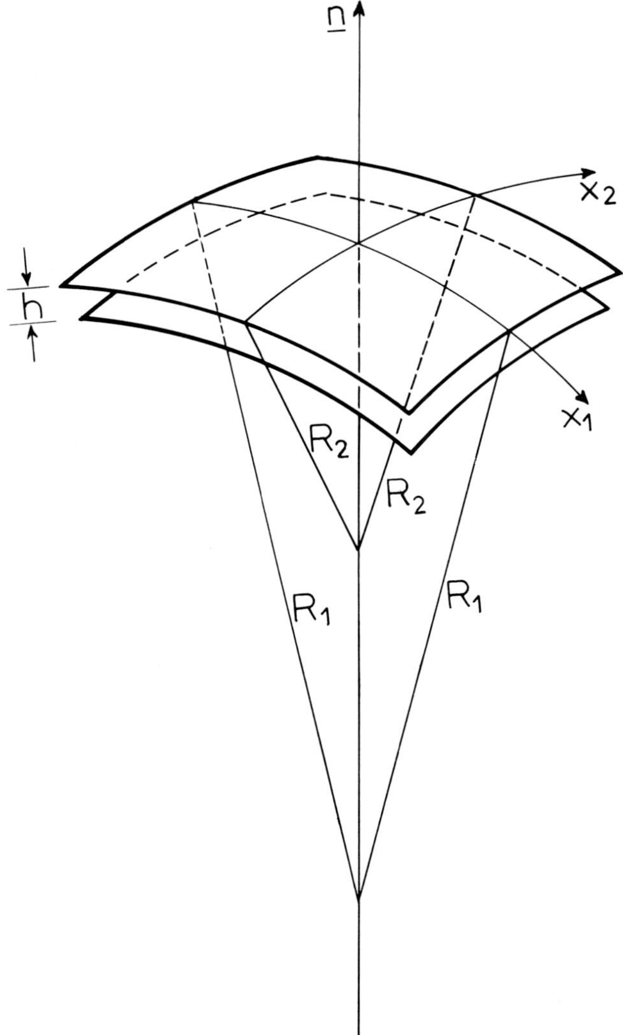

FIGURE 4.7. Effects of curvature changes for a double layer structure. Note that the outer layer is slightly expanded relative to the inner layer for this positive curvature change. The principal axes for deformation of the composite coincide with the orthogonal planes that contain the surface arcs generated by the principal curvatures. R_1 and R_2 are the principal radii of curvature.

The deformation of each constituent layer of the membrane is given, therefore, by the approximation,

$$\lambda_1^\ell \simeq \lambda_1 + \delta \lambda_1^\ell$$

$$\lambda_2^\ell \simeq \lambda_2 + \delta \lambda_1^\ell$$

The subscripts, 1 and 2, for the distances from the neutral surface represent a subtlety that is generally not significant for thin membrane materials: the lines of action for principal forces resultants may not be coplanar. In other words, the center of force within the membrane may be slightly different for the tensions of layers that act in one principal direction than for those that act in the other principal direction. We will eventually eliminate this distinction and will continue to refer to the neutral surface as if lines of action of membrane force resultants are coplanar.

The change in free energy of the membrane, which is produced by deformation, is the sum of free energy changes for the component layers. We assume that single layers support force resultants in the tangent plane, but no bending moments. The differential work per unit area was given by Equation 4.4.3 for an incremental deformation of a single layer. For a membrane composed of N coupled layers, the differential change in elastic free energy is the sum of the differential work produced in each layer by the deformation,

$$\sum_{\ell=1}^{N} (d\widetilde{F}_{\ell})_T = \sum_{\ell=1}^{N} [\lambda_2 (T_1^{\ell} d\lambda_1^{\ell}) + \lambda_1 (T_2^{\ell} d\lambda_2^{\ell})] \tag{4.9.2}$$

all per unit initial area of the membrane element. Since we assume that the layers experience nearly the same deformation with only a small variation between the layers produced by curvature, we express Equation 4.9.2 as contributions associated with the deformation of the neutral surface (the surface which contains the lines of action of the total force resultants) and the variations relative to this surface.

$$\sum_{\ell=1}^{N} (d\widetilde{F}_{\ell})_T = \sum_{\ell=1}^{N} \left[\lambda_2 \left(T_1^{\ell}(\alpha,\beta) + \delta T_1^{\ell} \right) \left(d\lambda_1 + d(\delta\lambda_1^{\ell}) \right) + \right.$$
$$\left. \lambda_1 \left(T_2^{\ell}(\alpha,\beta) + \delta T_2^{\ell} \right) \left(d\lambda_2 + d(\delta\lambda_2^{\ell}) \right) \right]$$

where the deformation of the neutral surface is defined by the mean fractional area change and extension at constant area, α and β, with principal extension ratios, λ_1 and λ_2. The variations, $\delta T_1^{\ell}, \delta\lambda_1^{\ell}$, etc., represent the differences between values in the ℓth layer and the neutral surface values. Similarly, the differential of the free energy density can be expressed as the contribution of the deformation of the neutral surface and the free energy variations relative to the neutral surface,

$$\sum_{\ell=1}^{N} (d\widetilde{F}_{\ell})_T = \left(d\widetilde{F}(\alpha,\beta) \right)_T + \sum_{\ell=1}^{N} \left[\left(\frac{\partial\widetilde{F}_{\ell}}{\partial\alpha} \right)_{T,\beta} d(\delta\alpha_{\ell}) \right.$$
$$\left. + \left(\frac{\partial\widetilde{F}_{\ell}}{\partial\beta} \right)_{T,\alpha} d(\delta\beta_{\ell}) \right] \tag{4.9.3}$$

The deformation of the neutral surface produces a differential change in free energy density given by

$$\left(d\widetilde{F}(\alpha,\beta) \right)_T = \left[\sum_{\ell=1}^{N} \left(\frac{\partial\widetilde{F}_{\ell}}{\partial\alpha} \right)_{T,\beta} \right] d\alpha + \left[\sum_{\ell=1}^{N} \left(\frac{\partial\widetilde{F}_{\ell}}{\partial\beta} \right)_{T,\alpha} \right] d\beta$$

We now equate Equations 4.9.2 and 4.9.3, and identify the part of the free energy change due to the total membrane force resultants, i.e., resultants that are associated with the neutral surface deformation. The terms associated with the neutral surface deformation are

$$\left(d\widetilde{F}(\alpha,\beta) \right)_T = \lambda_2 \left(\sum_{\ell=1}^{N} T_1^{\ell}(\alpha,\beta) \right) d\lambda_1 + \lambda_1 \left(\sum_{\ell=1}^{N} T_2^{\ell}(\alpha,\beta) \right) d\lambda_2$$

(4.9.4)

or

$$\left(d\widetilde{F}(\alpha,\beta) \right)_T = \lambda_2 T_1 d\lambda_1 + \lambda_1 T_2 d\lambda_2$$

T_1 and T_2 are the total resultants for the entire membrane which act in the plane of the neutral surface. In the absence of bending effects, the free energy density change is characterized by the cumulated material properties given in Equation 4.5.13. When bending is present, the variation of the free energy density associated with curvature is given by the difference between Equations 4.9.3 and 4.9.4,

$$\sum_{\ell=1}^{N} d(\delta\widetilde{F}_{\ell})_T \equiv \sum_{\ell=1}^{N} (d\widetilde{F}_{\ell})_T - \left(d\widetilde{F}(\alpha,\beta) \right)_T$$

Thus, the variation in work per unit area due to curvature change is equal to the product of variations in tension times variation in extension ratio,

$$\sum_{\ell=1}^{N} d(\delta\widetilde{F}_{\ell})_T = \sum_{\ell=1}^{N} \left[\lambda_2 \, \delta T_1^{\ell} d(\delta\lambda_1^{\ell}) + \lambda_1 \, \delta T_2^{\ell} d(\delta\lambda_2^{\ell}) \right]$$

(4.9.5)

where the following identities have been employed,

$$\sum_{\ell=1}^{N} \delta T_1^{\ell} = 0, \quad \sum_{\ell=1}^{N} \delta T_2^{\ell} = 0$$

and

$$\sum_{\ell=1}^{N} T_1^{\ell}(\alpha,\beta) \, d(\delta\lambda_1^{\ell}) \equiv 0$$

$$\sum_{\ell=1}^{N} T_2^{\ell}(\alpha,\beta) \, d(\delta\lambda_2^{\ell}) \equiv 0$$

(4.9.6)

The first pair of the identities (4.9.6) are equivalent to the definition of δT_i^{ℓ} as variations from the mean force resultants. The second pair of identities are statements that define the location of the neutral surface, i.e., the center of force within the membrane. This is demonstrated by the fact that the variations in extension ratios relative to the neutral surface are proportional to the distance from the neutral surface, h_ϱ, Equation 4.9.1. Therefore, the relations,

$$\sum_{\varrho=1}^{N} T_1^\varrho (\alpha,\beta) h_{\varrho_1} \equiv 0$$

$$\sum_{\varrho=1}^{N} T_2^\varrho (\alpha,\beta) h_{\varrho_2} \equiv 0 \tag{4.9.7}$$

state that the sum of moments of the force resultants relative to the neutral surface are zero when each layer experiences the same deformation (α,β). In other words, uniform deformation of the layered composite does not produce any net moment resultants in the membrane. Equations 4.9.7 provide the definition of the neutral surface. [Note: We have assumed that the principal radii of curvature describe arcs that coincide with the principal axes of the deformation, Figure 4.7. This will be the case for axisymmetric deformations. For more general expressions, see Fung and Sechler (1960) and Naghdi (1972).][30,63]

We now introduce the first order approximation for the variations in extension ratios, Equation 4.9.1, into Equation 4.9.5. This gives the variation in work per unit area as a function of differential changes in curvature,

$$\sum_{\varrho=1}^{N} d(\delta\widetilde{F}_\varrho)_T = \lambda_1\lambda_2 \left[\left(\sum_{\varrho=1}^{N} \delta T_1^\varrho \cdot h_{\varrho_1} \right) \cdot dC_1 + \left(\sum_{\varrho=1}^{N} \delta T_2^\varrho \cdot h_{\varrho_2} \right) \cdot dC_2 \right] \tag{4.9.8}$$

Equation 4.9.8 is the differential change of curvature or bending elastic energy density in the membrane. The principal moment resultants in the membrane are defined by the sums of torques about the center of force (neutral surface), which are produced by the force resultant variations,

$$M_1 \equiv \sum_{\varrho=1}^{N} \delta T_1^\varrho h_{\varrho_1}$$

$$M_2 \equiv \sum_{\varrho=1}^{N} \delta T_2^\varrho h_{\varrho_2} \tag{4.9.9}$$

Therefore, the curvature elastic energy density is associated with the work of membrane moment resultants produced by membrane bending,

$$\sum_{\varrho=1}^{N} d(\delta\widetilde{F}_\varrho)_T = \lambda_1\lambda_2 (M_1 dC_1 + M_2 dC_2) \tag{4.9.10}$$

In order to determine the free energy variation, we will use the first order elastic constitutive behavior, Equation 4.8.1, to represent the free energy change of each layer. Thus, we can expand the total membrane free energy density, Equation 4.9.3, in terms of the mean deformation variables, α and β, and the variations, $\delta\alpha_\varrho$ and $\delta\beta_\varrho$, from layer to layer. The mean deformation produces the free energy density change of the neutral surface given by

$$\sum_{\varrho=1}^{N} \left(d\widetilde{F}_\varrho(\alpha,\beta)\right)_T = \sum_{\varrho=1}^{N} (K_\varrho \, \alpha \, d\alpha + \mu_\varrho \, d\beta)$$

and the free energy density variation is expressed in terms of the variations, $\delta\alpha_\varrho$ and $\delta\beta_\varrho$

$$\sum_{\varrho=1}^{N} d(\delta\widetilde{F}_\varrho)_T = \sum_{\varrho=1}^{N} [K_\varrho(\alpha + \delta\alpha_\varrho)d(\delta\alpha_\varrho) + \mu_\varrho d(\delta\beta_\varrho)]$$

$$(4.9.11)$$

The variations in deformation parameters are derived from the variations in extension ratios as

$$\delta\alpha_\varrho = \frac{\partial\alpha}{\partial\lambda_1}\,\delta\lambda_1^\varrho + \frac{\partial\alpha}{\partial\lambda_2}\,\delta\lambda_2^\varrho$$

$$\delta\beta_\varrho = \frac{\partial\beta}{\partial\lambda_1}\,\delta\lambda_1^\varrho + \frac{\partial\beta}{\partial\lambda_2}\,\delta\lambda_2^\varrho$$

which can be expressed in terms of curvature changes by

$$\delta\alpha_\varrho \simeq \lambda_1\lambda_2\,(h_{\varrho_1} C_1 + h_{\varrho_2} C_2)$$

$$\delta\beta_\varrho \simeq \frac{(\lambda_1^2 - \lambda_2^2)}{2\lambda_1\lambda_2}\,(h_{\varrho_1} C_1 - h_{\varrho_2} C_2)$$

With these variations, the free energy variation is given by four sums,

$$\sum_{\varrho=1}^{N} d(\delta\widetilde{F}_\varrho)_T = \lambda_1^2\lambda_2^2 \left[dC_1 \sum_{\varrho=1}^{N} h_{\varrho_1} K_\varrho(h_{\varrho_1} C_1 + h_{\varrho_2} C_2) \right.$$

$$\left. + \, dC_2 \sum_{\varrho=1}^{N} h_{\varrho_2} K_\varrho(h_{\varrho_1} C_1 + h_{\varrho_2} C_2) \right]$$

$$+ \, \lambda_1\lambda_2 \left[dC_1 \sum_{\varrho=1}^{N} T_1^\varrho(\alpha,\beta)h_{\varrho_1} + dC_2 \sum_{\varrho=1}^{N} T_2^\varrho(\alpha,\beta)h_{\varrho_2} \right]$$

The last two sums are zero because of the neutral surface definition, Equations 4.9.7. Therefore, the free energy density variation associated with bending or curvature change is given by

$$\sum_{\ell=1}^{N} d(\delta \widetilde{F}_\ell)_T = \lambda_1{}^2 \lambda_2{}^2 \left[dC_1 \sum_{\ell=1}^{N} h_{\ell_1} K_\ell (h_{\ell_1} C_1 + h_{\ell_2} C_2) + dC_2 \sum_{\ell=1}^{N} h_{\ell_2} K_\ell (h_{\ell_1} C_1 + h_{\ell_2} C_2) \right]$$

$$(4.9.12)$$

The integrated form of Equation 4.9.12 is the bending or curvature elastic energy density,

$$\sum_{\ell=1}^{N} (\delta \widetilde{F}_\ell)_T = \frac{\lambda_1{}^2 \lambda_2{}^2}{2} \left[\sum_{\ell=1}^{N} K_\ell (h_{\ell_1} C_1 + h_{\ell_2} C_2)^2 \right]$$

$$(4.9.13)$$

In general, bending free energy variations are contributed by second or higher derivatives of the free energy density of a single layer with respect to α and β,

$$\sum_{\ell=1}^{N} (\delta \widetilde{F}_\ell)_T = \sum_{\ell=1}^{N} \left[\left(\frac{\partial^2 \widetilde{F}_\ell}{\partial \alpha^2} \right)_{T,\beta} \frac{\delta \alpha_\ell{}^2}{2} + \cdots + \left(\frac{\partial^2 \widetilde{F}_\ell}{\partial \beta^2} \right)_{T,\alpha} \frac{\delta \beta_\ell{}^2}{2} + \cdots \right]$$

where the partial derivatives are evaluated at the neutral surface. The first order contribution is zero due to the definition of the location of the force resultant centroid, Equations 4.9.7. Thus the shear modulus for the hyperelastic layers does not appear explicitly in the thin membrane representation, Equation 4.9.13. We assume that the free energy density is separable into functions of α and β. Otherwise, it would be necessary to include cross derivatives. Moment resultants are determined by the partial derivatives of the free energy density with respect to curvature changes as is observed in Equation 4.9.10,

$$M_1 = \frac{1}{\lambda_1 \lambda_2} \frac{\partial}{\partial C_1} \left[\sum_{\ell=1}^{N} (\delta \widetilde{F}_\ell)_T \right]$$

$$M_2 = \frac{1}{\lambda_1 \lambda_2} \frac{\partial}{\partial C_2} \left[\sum_{\ell=1}^{N} (\delta \widetilde{F}_\ell)_T \right]$$

$$(4.9.14)$$

This yields the curvature elastic constitutive relations,

$$M_1 = \lambda_1 \lambda_2 \left[\left(\sum_{\ell=1}^{N} K_\ell \cdot h_{\ell_1}{}^2 \right) C_1 + \left(\sum_{\ell=1}^{N} K_\ell \cdot h_{\ell_1} \cdot h_{\ell_2} \right) C_2 \right]$$

$$M_2 = \lambda_1 \lambda_2 \left[\left(\sum_{\ell=1}^{N} K_\ell \cdot h_{\ell_1} \cdot h_{\ell_2} \right) C_1 + \left(\sum_{\ell=1}^{N} K_\ell \cdot h_{\ell_2}{}^2 \right) C_2 \right]$$

$$(4.9.15)$$

Therefore, the bending moments are given by material properties times the changes in principal curvatures,

$$M_1 = \lambda_1 \lambda_2 (B_{11} \cdot C_1 + B_{12} \cdot C_2)$$

$$M_2 = \lambda_1 \lambda_2 (B_{21} \cdot C_1 + B_{22} \cdot C_2)$$

where the matrix properties are curvature elastic moduli or bending moduli for the membrane.

In order to evaluate these properties, it is necessary to establish the location of each layer relative to the neutral surface, as defined by Equations 4.9.7. As discussed, the neutral surface may not be a single surface but, instead, two separate lines of action (one normal to each principal axis direction). In other words, the total force resultants may not lie in the same plane. This difference will be negligible for certain kinds of membrane strata, e.g., for membranes with very small area compressibility so that shear terms can be neglected in Equations 4.9.7. In the case of small area compressibility, the bending moments are equal, i.e., isotropic,

$$M_1 = M_2 = \lambda_1 \lambda_2 (C_1 + C_2) B$$

and the curvature elastic, or bending, modulus is simply,

$$B \equiv \sum_{\ell=1}^{N} h_\ell^2 K_\ell$$

(B is also referred to as the coefficient of bending resistance or rigidity.) The curvature elastic energy density (bending energy) is given by

$$\sum_{\ell=1}^{N} (\delta \widetilde{F}_\ell)_T = \frac{\lambda_1^2 \lambda_2^2}{2} (C_1 + C_2)^2 \, B \tag{4.9.16}$$

In the case of isotropic bending moments, the location of the neutral surface (center of force) within the membrane is determined from Equations 4.9.7 with isotropic force resultants for the layers,

$$0 = \sum_{\ell=1}^{N} \overline{T}^\ell \cdot h_\ell = \sum_{\ell=1}^{N} (K_\ell \cdot h_\ell)\alpha$$

where the mean deformation has been factored out of the equation. The curvature elastic modulus (bending modulus) can now be expressed in terms of the fixed distances between membrane layers. We define

$$h_{m\ell} = h_m - h_\ell$$

the curvature elastic modulus, B, is given by the progressive sum,

$$B = \sum_{\ell=1}^{N} h_\ell^2 K_\ell = \frac{1}{K} \sum_{\ell > m = 1}^{N} (h_{m\ell})^2 K_m K_\ell \tag{4.9.17}$$

For two molecular layers separated by a distance, h, the coefficient of bending resistance is

$$B = h^2 \left(\frac{K_1 K_2}{K_1 + K_2} \right)$$

and for three layers separated by h_1 and h_2 distances, respectively,

$$B = \frac{(h_1^2 K_1 K_2 + h_2^2 K_2 K_3 + (h_1 + h_2)^2 K_1 K_3)}{(K_1 + K_2 + K_3)}$$

and so on . . .

Isotropic bending is also the result obtained in the classical problem of bending of a thin shell (Landau and Lifshitz, *Theory of Elasticity,* 1970)[51] and is often referred to as cylindrical bending. The free energy variation produced by bending moments, Equation 4.9.16, is similar to the empirical curvature elasticity developed by Helfrich (1973).[37] Zarda (1974)[99] used a macroscopic material model where the bending moduli are given by, $B_{22} = B_{11}$ and $B_{12} = v \cdot B_{11}$, i.e., two independent curvature elastic constants. In 1970, Canham[5] minimized the surface integral of the square of the curvature as a model for the natural, biconcave shape of the red blood cell. Canham described his model as a minimization of the bending energy of the membrane. Subsequently, Helfrich and Dueling (1975)[39] have based numerical solutions on minimization of their curvature elasticity model in similar studies. In this section, we have established a thermodynamic basis for the coefficient of the bending energy, which is related to membrane structure (Evans, 1974).[17]

In the following examples, we will assess the contribution of curvature free energy change, Equation 4.9.16, in comparison to the free energy change that results from shear deformation of the neutral surface, Equation 4.6.4. The examples will also provide estimates of the magnitude of bending coefficients for a lipid bilayer and for a trilamellar structure which idealizes a red cell membrane.

Example 1

First, we will consider a lipid bilayer that is about 30×10^{-8} cm thick. X-ray diffraction data on lipid bilayers (Luzzati, 1968)[56] definitely show a region of low electron density midway between the lipid layers. Thus, these results indicate that the layers are uncoupled and free to slip relative to one another. Consequently, the lipid bilayer itself does not present any appreciable resistance to curvature change. However, if the boundary of the lipid bilayer is constrained so that the layers cannot slip, then we may consider the bilayer as coupled. Such is the case for long lipid bilayer cylinders of fixed length. The bending of these cylinders produces relative changes in area per molecule of lipids in one layer compared to the other. Servuss et al. (1976)[80] studied the bending fluctuations of lecithin bilayer cylinders and deduced a value for the curvature elastic modulus of 2.3×10^{-12} erg. We will demonstrate that this value is consistent with the value predicted by Equation 4.9.17 with the elastic area compressibility of a lecithin monolayer used for each layer of the bilayer. The coefficient of bending resistance is given by

$$B = h^2 \frac{K_1 K_2}{K_1 + K_2} = h^2 \frac{K_1}{2}$$

for the bilayer where the layers have the same area compressibility modulus. We will use the lecithin monolayer value of $K_1 \simeq 80$ dyn/cm and $h = 3 \times 10^{-7}$ cm. The result is

$$B = 3.6 \times 10^{-12} \quad \text{(dyn-cm or ergs)}$$

which is consistent with the value obtained by Servuss et al.(1976).[80] For uncoupled bilayers, the bending or curvature elastic modulus would be negligible by comparison.

Example 2

Consider the trilamellar membrane illustrated by Figure 1.1 in Section I. Assume that the lipid and protein bilayer is uncoupled and that the network underlayer is rigidly coupled to the lower lipid layer. Also, assume that the distance between the polar head groups of the lower lipid layer and the network is about 10×10^{-8} cm. If we arbitrarily assume that the total membrane area compressibility modulus can be partitioned equally between layers, $K_1 = K_2 = K_3$, then $K_1 \simeq 150$ dyn/cm based on the area compressibility modulus of the red cell membrane (Evans and Waugh, 1977).[25] The coefficient of bending resistance is

$$B = h^2 \left(\frac{K_1 K_2}{K_1 + K_2} \right) \simeq 7 \times 10^{-13} \ (\text{ergs})$$

For the red blood cell membrane, there may be some coupling of the two lipid layers by the large globular proteins which penetrate them, but the extent of such coupling is difficult to evaluate at this time. Measurement of red cell membrane "flicker" has indicated a bending modulus on the order of 1 to 3×10^{-13} erg (Servuss et al. 1976).[80]

Example 3

Consider the trilamellar composite in the previous example. What is the relative magnitude of the curvature elastic energy compared to the elastic free energy change produced by the shear deformation of a flat disk to a hemisphere? We assume that the membrane deforms at constant area. Therefore, the area of the disk is equal to the area of the hemisphere,

$$\pi r_0^2 = 2\pi R_s^2$$

where r_0 is the radius of the disk, R_s is the radius of the hemisphere. If λ is the principal extension ratio along the meridian of the hemisphere, the extension ratio is identically one at the pole of the hemisphere, and at the base of the hemisphere is given by

$$\lambda^2 = \left(\frac{r_0}{R_s} \right)^2 = 2$$

The average deformation parameter, which characterizes shear, $\langle \beta \rangle$, is approximately two thirds of the value at the equator,

$$\langle \beta \rangle \simeq \frac{2}{3} \left(\frac{\lambda^2 + \lambda^{-2} - 2}{2} \right) = \frac{1}{6}$$

Hence, the free energy density change due to shear deformation at constant area is approximately

$$(\Delta \widetilde{F}_s)_T \simeq \mu \langle \beta \rangle = \frac{\mu}{6}$$

The bending or curvature free energy density is given by

$$(\Delta \tilde{F}_B)_T \approx \frac{B}{2}\left(\frac{2}{R_s}\right)^2 = \frac{2B}{R_s^2}$$

The ratio of the two free energy density contributions is on the order of

$$\frac{(\Delta \tilde{F}_B)_T}{(\Delta \tilde{F}_s)_T} \sim \frac{10B}{\mu R_s^2}$$

With conservative estimates for red cell membrane properties, e.g., the value for B from Example 2 and $\mu \sim 10^{-2}$ dyn/cm, this ratio depends approximately on the radius of curvature of the hemisphere as

$$\frac{(\Delta \tilde{F}_B)_T}{(\Delta \tilde{F}_s)_T} \sim \frac{7 \times 10^{-10}}{R_s^2}$$

Thus, in order to neglect the curvature free energy contribution (bending resistance), the radius of the hemisphere would have to satisfy the condition that

$$R_s > 2.5 \times 10^{-5} \quad \text{(cm)}$$

The curvature elastic energy contribution would be about 25% for a radius of curvature of 0.5×10^{-4} cm (0.5 μm).

For red cell membrane changes in curvature on the order of 10^4 cm^{-1}, the material shear elastic energy is dominant if the surface extension ratios are 1.1 or greater. Such is the case in micropipet aspiration experiments on flaccid red cell disks where the extension ratio ranges from 1.5 to 3. On the other hand, for red cell deformations like osmotic sphering where the material stretch is negligible over large regions of the cell membrane, the curvature or bending energy will be the dominant rigidity mechanism (Zarda, 1974; Zarda et al., 1977).[99,100] See Section V for details.

4.10 Chemically Induced Curvature and Moments

In the last section, we developed the elastic free energy variation that is produced by bending of coupled or constrained membrane strata. This variation in membrane free energy is based on fixed membrane properties. Thus, the chemical state is assumed to remain unchanged. Normally, the chemical equilibrium of the membrane does not change during the course of mechanical experiments, but, often, important biological phenomena arise when the external environment or chemical equilibrium of the membrane is altered. As outlined in Section 4.6, the equilibrium surface pressure in the membrane is determined by the interfacial free energy density, γ. If the chemical state of a membrane interface is changed, then γ is changed, which disturbs the natural, force-free state of the membrane. Such chemical alterations may induce moments or produce curvature of the membrane as the constituent layers of the composite attempt to condense or expand relative to the original equilibrium state. Thus, the elastic free energy differential, Equation 4.9.3, must include an additional term to account for the change in chemical equilibrium as well as deformation. This term is given by

$$\sum_{\ell=1}^{N} \delta\gamma_\ell \, d(\alpha + \delta\alpha_\ell)$$

(4.10.1)

Here, we use $\delta\gamma_\ell$ to represent any variation in free energy density of the chemical reference state that is not attributable to deformation of the ℓth layer (as a closed system). The isotropic terms in Equation 4.9.3 can be interpreted as the results of surface pressure changes in each layer relative to the original equilibrium or initial state. Equation 4.10.1 consists of two contributions: a change in the free energy density due to deformation of the neutral surface given by

$$\sum_{\ell=1}^{N} \delta\gamma_\ell \, d\alpha \tag{4.10.2}$$

plus the sum of free energy density variations that are produced by relative deformations between layers, similar to Equation 4.9.11,

$$\sum_{\ell=1}^{N} \delta\gamma_\ell \, d(\delta\alpha_\ell) \tag{4.10.3}$$

For coupled layers, the free energy variation, (Equation 4.10.3), gives a result like that of Equation 4.9.10,

$$\lambda_1\lambda_2 \sum_{\ell=1}^{N} \delta\gamma_\ell (h_{\ell_1} \, dC_1 + h_{\ell_2} \, dC_2) \tag{4.10.4}$$

where the expression for area variations in terms of curvature increments has been used. Equation 4.10.4 implies that the membrane can approach a new equilibrium state by curvature change. The total variation of the elastic free energy that is associated with curvature changes is given by the sum of bending terms, Equation 4.9.12, and the chemically induced variation, Equation 4.10.4,

$$
\sum_{\ell=1}^{N} d(\delta\widetilde{F}_\ell)_T = \lambda_1{}^2\lambda_2{}^2 \left[dC_1 \sum_{\ell=1}^{N} h_{\ell_1} K_\ell (h_{\ell_1} C_1 + h_{\ell_2} C_2) \right.
$$

$$
\left. + \, dC_2 \sum_{\ell=1}^{N} h_{\ell_2} K_\ell (h_{\ell_1} C_1 + h_{\ell_2} C_2) \right]
$$

$$
+ \lambda_1\lambda_2 \left[dC_1 \sum_{\ell=1}^{N} \delta\gamma_\ell h_{\ell_1} + dC_2 \sum_{\ell=1}^{N} \delta\gamma_\ell h_{\ell_2} \right] \tag{4.10.5}
$$

This equation can be written in integrated form as

$$
\sum_{\ell=1}^{N} (\delta\widetilde{F}_\ell)_T = \frac{\lambda_1{}^2\lambda_2{}^2}{2} \left[\sum_{\ell=1}^{N} K_\ell (h_{\ell_1} C_1 + h_{\ell_2} C_2)^2 \right]
$$

$$
+ \lambda_1\lambda_2 \left[\sum_{\ell=1}^{N} \delta\gamma_\ell (h_{\ell_1} C_1 + h_{\ell_2} C_2) \right] \tag{4.10.6}
$$

Therefore, we can identify moment resultants which are induced by the change in chemical state of the membrane,

$$M_1^c \equiv \sum_{\ell=1}^{N} \delta\gamma_\ell h_{\ell_1}$$

$$M_2^c \equiv \sum_{\ell=1}^{N} \delta\gamma_\ell h_{\ell_2} \tag{4.10.7}$$

These moment contributions do not depend on membrane curvature changes. The total moment resultant is the sum of the bending moment, Equation 4.9.15, plus the chemically induced moment,

$$M_1 = M_1^B + M_1^c$$

$$M_2 = M_2^B + M_2^c$$

We again assume that we can neglect shear elasticity in determining the location of the neutral surface, i.e., we can use the location established by the isotropic force resultants,

$$h_{\ell_1} = h_{\ell_2} \quad \text{and} \quad \sum_{\ell=1}^{N} h_\ell K_\ell = 0$$

As a result, the induced moment contributions, Equations 4.10.7, are isotropic,

$$M_1^c = M_2^c = \Gamma$$

and we can define the induced moment coefficient, Γ, by the progressive sum,

$$\Gamma = \frac{1}{K} \sum_{\ell > m = 1}^{N} \left(\frac{\delta\gamma_\ell}{K_\ell} - \frac{\delta\gamma_m}{K_m} \right) K_\ell K_m h_{\ell m} \tag{4.10.8}$$

(It is important that the order of the layers be specified from number one on the inside to N on the outside in the sense of positive curvature, i.e., convex surfaces.) For two molecular layers separated by a distance, h, the induced moment coefficient is

$$\Gamma = \left(\frac{\delta\gamma_2}{K_2} - \frac{\delta\gamma_1}{K_1} \right) \frac{K_1 K_2 h}{(K_1 + K_2)}$$

For three layers separated by distances, h_1 and h_2, respectively, the coefficient is

$$\Gamma = \frac{1}{K} \left[\left(\frac{\delta\gamma_3}{K_3} - \frac{\delta\gamma_2}{K_2} \right) K_3 K_2 h_2 + \left(\frac{\delta\gamma_3}{K_3} - \frac{\delta\gamma_1}{K_1} \right) K_3 K_1 (h_1 + h_2) \right.$$

$$\left. + \left(\frac{\delta\gamma_2}{K_2} - \frac{\delta\gamma_1}{K_1} \right) K_1 K_2 h_1 \right]$$

where,

$$K = K_1 + K_2 + K_3$$

In terms of the curvature elastic modulus and the induced moment coefficient, the moment resultants are given by

$$M_1 = M_2 = \lambda_1 \lambda_2 (C_1 + C_2) B + \Gamma \qquad (4.10.9)$$

and the curvature elastic free energy density (bending energy plus chemical induction) is

$$\sum_{\ell=1}^{N} (\delta \tilde{F}_\ell)_T = \frac{\lambda_1^2 \lambda_2^2}{2} (C_1 + C_2)^2 B + \lambda_1 \lambda_2 (C_1 + C_2) \Gamma \qquad (4.10.10)$$

If membrane force resultants are negligible, then a new equilibrium state will be achieved by alteration of the curvature in the membrane. The new state is determined by the minimization of the free energy density with respect to the mean curvature, $C_1 + C_2$,

$$\frac{\partial}{\partial(C_1 + C_2)} \left[\sum_{\ell=1}^{N} (\delta \tilde{F}_\ell)_T \right] = 0$$

The new equilibrium curvature is obtained using Equation 4.10.10,

$$C_0 \equiv (C_1 + C_2)_0 = \frac{-\Gamma}{\lambda_1 \lambda_2 B}$$

The bending energy density can now be given relative to the new equilibrium configuration,

$$\sum_{\ell=1}^{N} (\delta \tilde{F}_\ell)_T = \frac{\lambda_1^2 \lambda_2^2}{2} (C_1 + C_2 - C_0)^2 B \qquad (4.10.11)$$

where C_0 is the chemically induced curvature (Evans, 1974).[17] This is the counterpart of the "spontaneous" curvature described by Helfrich (1974).[38]

As we discussed previously, changes of curvature of uncoupled molecular layers produce negligible moment resultants because the layers are free to shear or slip relative to one another, although repulsive forces maintain fixed distances between the layers. If the edges of the membrane strata are constrained or if they are closed as in a cellular envelope, there may be a variation in free energy due to membrane curvature changes. The free energy change is the effect of variations in free energy of the component layer integrated over the entire surface and is given by

$$\sum_{\ell=1}^{N} \iint (\delta \tilde{F}_\ell)_T da_1\, da_2 = \sum_{\ell=1}^{N} \left[K_\ell \iint \frac{(\delta \alpha_\ell)^2}{2} da_1\, da_2 + \delta \gamma_\ell \iint \delta \alpha_\ell\, da_1\, da_2 \right]$$

We can normalize this expression by the total surface area. We obtain the average over the surface (indicated by $< >$),

$$\sum_{\ell=1}^{N} <(\delta \tilde{F}_\ell)_T> = \sum_{\ell=1}^{N} \left[K_\ell \frac{\left< \delta \alpha_\ell^2 \right>}{2} + \delta \gamma_\ell <\delta \alpha_\ell> \right]$$

(We assume that membrane properties are uniform over the whole surface.) Since the layers are uncoupled, the variation in area between the layers and the neutral surface will be uniform over the surface. Therefore,

$$<\delta \alpha_\ell^2> = <\delta \alpha_\ell>^2$$

Consequently, the free energy variation is related to the average membrane curvature change, i.e., a nonlocal effect,

$$\sum_{\ell=1}^{N} <(\delta \tilde{F}_\ell)_T> = \frac{B}{2} <\lambda_1 \lambda_2 (C_1 + C_2)>^2 + \Gamma <\lambda_1 \lambda_2 (C_1 + C_2)>$$

This is the surface average free energy variation due to membrane curvature for uncoupled layers. The sum of this expression plus the free energy due to the in-plane deformations of the membrane must be minimized for the entire membrane surface to obtain the equilibrium geometry. Thus, the minimization is a global procedure.

Example 1

Consider a flat trilamellar composite that we used in Example 2 of Section 4.9. What is the chemical free energy change required to produce a hemisphere with the radius of 1×10^{-4} cm, first for a chemical change in the inner network layer, second in the outer lipid layer? We have chosen a radius of curvature such that the bending resistance can be neglected. Therefore, we must balance the shear elastic effects against the chemical changes, i.e.,

$$(\Delta \tilde{F}_s)_T + (\Delta \tilde{F}_c)_T \sim 0$$

We use the result from the example in Section 4.9 to give

$$(\Delta \tilde{F}_s)_T \sim 2 \times 10^{-3} \ \text{erg/cm}^2$$

The induced free energy density is represented by Equation 4.10.10,

$$(\Delta \tilde{F}_c)_T \sim \frac{2\Gamma}{R_s} = 2 \times 10^4 \cdot \Gamma \ \text{erg/cm}^2$$

which is combined with the shear elastic energy to provide an estimate of the induced moment coefficient,

$$\Gamma \sim -\ 10^{-7} \ \text{erg/cm}$$

The induced moment coefficient, Γ, is given for the two situations by Equation 4.10.8. For chemical change in the inner network, we have

$$(1) \ \Gamma = -\frac{\delta \gamma_1}{K} \left(K_3 (h_1 + h_2) + K_2 h_1 \right)$$

and for the outer lipid layer, we have

$$(2)\quad \Gamma = \frac{\delta\gamma_3}{K}\left(K_2 h_2 + K_1(h_1 + h_2)\right)$$

If we make the same assumptions as in the first two examples of Section 4.9, we can use $K_1 = K_2 = K_3 \simeq 150$ dyn/cm and $K \simeq 450$ dyn/cm with interlayer distances $h_1 = 10^{-7}$ cm and $h_2 = 3 \times 10^{-7}$ cm.

For a change in the inner network layer, case (1), the induced moment and change in free energy density would approximately be given by

$$\Gamma \sim -\delta\gamma_1 \cdot (10^{-7})$$

and

$$\delta\gamma_1 \sim 1 \text{ erg/cm}^2 \text{ (dyn/cm)}$$

For the red cell membrane, the number of mol/cm^2 of network material (spectrin) is about 0.5×10^{-12} mol/cm^2. Therefore, this change would be equivalent to 2×10^{12} erg/mol or 48 kcal/mol of spectrin molecules.

For a change in outer lipid layer, case (2), the induced moment and change in free energy density would be given by

$$\Gamma \sim \delta\gamma_3 \cdot (10^{-7})$$

and

$$-\delta\gamma_3 \sim 10^{-1} \text{ (erg/cm}^2 \text{ or dyn/cm)}$$

From Example 2 of Section 4.6, the chemical change, $\delta\gamma_3$, is less than 1% of the interfacial free energy density of the water-hydrocarbon interaction (natural state surface pressure) of the outer layer (40 dyn/cm). In calories per mole of phospholipid, the change is only about 10 cal/mol out of 2 kcal/mol total.

Note that reducing the free energy density of the outer layer can produce the same curvature as increasing the free energy density of the inner layer. Obviously, small changes in chemical equilibrium can produce major membrane surface deformations and curvature changes. This is relevant to the shape transformations induced in red cell membranes by chemical agents, described by Bessis (1973)[3] and Ponder (1971)[68] as the discocyte-echinocyte-stomatocyte shape change. Sheetz and Singer (1975)[81] postulated a membrane "couple" to account for red cell shape change, but provided no detailed thermodynamic basis for the hypothesis. A variety of shapes have been computed by Dueling and Helfrich on the basis of spontaneous curvature changes, again without detailed molecular basis. They also neglect the elastic energy due to deformation in the plane of the membrane. In view of the estimates above, neglecting the shear rigidity of the membrane is questionable.

4.11 Thermoelasticity

As we discussed in the introduction to this chapter, conservation of energy is an unchallenged principle. However, energy can be changed by heat exchange and mechanical work. The thermodynamic state of the substance may be characterized by two extensive state functions: internal energy and entropy. The mechanical work done on a material by displacement of forces can alter the energy and/or heat content of the material. For reversible processes at constant temperature, the work done on a material element is represented by an elastic, conservative potential function. From this function, we have derived elastic constitutive relations. We recognize that these changes in potential embody both changes in energy of the material and reversible changes in heat content. The latter are reversible heats of deformation. The heats of deformation are given by the isothermal changes in entropy of the material that are produced by the deformation and scaled by temperature. Entropy changes represent alterations in configuration of the material microstructure, which are reversible in this case. The conversion of work to heat has a counterpart: the conversion of heat to work, which is evidenced as thermoelastic effects. This is the thermoelastic behavior of the mechanochemical equations of state; it complements the isothermal, elastic constitutive behavior. Coupled with the first law of thermodynamics, thermoelasticity will be shown to represent the reversible heat exchange that occurs in an isothermal deformation process. With thermoelasticity, we will be able to decompose the elastic free energy change that is produced by deformation into contributions of internal energy and entropy of the material (Evans and Waugh, 1977).[24]

We have developed isothermal elastic constitutive relations in terms of the partial derivatives of the Helmholtz free energy density at constant temperature. These equations relate isotropic tension and the maximum shear resultant to the deformation of the membrane and as such are isotherms of mechanochemical equations of state for a membrane. In general, equations of state depend on temperature as well as on the deformation parameters. Now, we examine the temperature dependence of the membrane force resultants. The first mechanochemical state equation is for the isotropic tension,

$$\bar{T} = f_\beta(\alpha, T) \tag{4.11.1}$$

This can be written in differential form as

$$d\bar{T} = \left(\frac{\partial \bar{T}}{\partial \alpha}\right)_{T,\beta} d\alpha + \left(\frac{\partial \bar{T}}{\partial T}\right)_{\alpha,\beta} dT \tag{4.11.2}$$

where the shape of the material element is held constant, i.e., β = constant; only uniform area expansion or condensation is performed. Next, if we hold the area of a material element constant and express the temperature dependence of the shear resultant, we have the second mechanochemical equation of state,

$$T_s = f_\alpha(\beta, T) \tag{4.11.3}$$

Since β is a function only of the scaled extension ratio, Equation 2.3.21, this can be expressed equivalently as

$$T_s = f_\alpha(\tilde{\lambda}, T)$$

The differential form is given by

$$dT_s = \left(\frac{\partial T_s}{\partial \tilde{\lambda}}\right)_{T,\alpha} d\tilde{\lambda} + \left(\frac{\partial T_s}{\partial T}\right)_{\beta,\alpha} dT \tag{4.11.4}$$

where the area of the material element is held constant, i.e., α = constant; only deformation at constant area is performed. Consequently, we see that a solid material must be characterized by at least two equations of state, Equations 4.11.2 and 4.11.4. The elastic constitutive behavior is identified in Equations 4.11.2 and 4.11.4 as the isothermal response. The thermoelastic behavior is given by the changes in material force resultants that are produced by temperature change, with the dimensions of the material element fixed.

As discussed in Section 4.6, there is also an isotropic equation of state characterizing the surface pressure internal to the material,

$$\pi = f(\alpha, T) \tag{4.11.5}$$

This internal equation of state is closely related to the mechanochemical equation for isotropic tension. Equations 4.11.1 and 4.11.5 provide an important link between the thermodynamic effect of the interfacial interactions (e.g., hydrophobicity, surface charge, adsorbed components, etc.) and the membrane thermoelasticity. The differential expressions for both isotropic equations of state, (4.11.1) and (4.11.5), are given by

$$d\bar{T} = \left(\frac{\partial \bar{T}}{\partial \alpha}\right)_T d\alpha + \left(\frac{\partial \bar{T}}{\partial T}\right)_\alpha dT$$

$$d\pi = \left(\frac{\partial \pi}{\partial \alpha}\right)_T d\alpha + \left(\frac{\partial \pi}{\partial T}\right)_\alpha dT \tag{4.11.6}$$

in terms of the intensive variables (α, T). These two equations are related by

$$\bar{T} = \gamma - \pi \tag{4.11.7}$$

At constant temperature, the change in isotropic tension is equal and opposite in sign to the change in surface pressure since by definition γ depends only on temperature and other properties of the aqueous environment. Therefore, the isothermal modulus of compressibility is identical for both, Equations 4.11.1 and 4.11.5, i.e., the slope of the surface pressure isotherm gives the isothermal modulus of compressibility,

$$K = \left(\frac{\partial \bar{T}}{\partial \alpha}\right)_T \Bigg|_{\alpha \to 0} = -A\left(\frac{\partial \pi}{\partial A}\right)_T = -\left(\frac{\partial \pi}{\partial \alpha}\right)_T \Bigg|_{\alpha \to 0} \tag{4.11.8}$$

This isothermal modulus of membrane area compressibility can be obtained directly by measuring the change in membrane isotropic tension vs. fractional area change. For instance, the area compressibility modulus for the red blood cell membrane has been measured to be about 450 dyn/cm at 25°C (Evans and Waugh, 1977).[25] The compressibility modulus for a phospholipid bilayer is as yet unmeasured. We can estimate the area elastic modulus by idealizing the bilayer as two monolayers. Then, for example, the value for a lecithin bilayer would be on the order of 160 dyn/cm as we discussed in Example 2 of Section 4.6.

The coefficients of the temperature differentials in Equations 4.11.6 are the isotropic, thermoelastic coefficients. We can use a mathematical corollary to express these coefficients in terms of changes in area and temperature. The general rule which will be employed is that the partial derivatives of any analytic function, $f(X,Y,Z)$, of three variables are related by

$$\left(\frac{\partial X}{\partial Z}\right)_Y = - \left(\frac{\partial X}{\partial Y}\right)_Z \cdot \left(\frac{\partial Y}{\partial Z}\right)_X$$

Therefore, the isotropic thermoelastic coefficients are given by the area compressibility modulus times the fractional change in area with temperature at constant surface pressure or isotropic tension,

$$\left(\frac{\partial \pi}{\partial T}\right)_\alpha = - \left(\frac{\partial \pi}{\partial \alpha}\right)_T \left(\frac{\partial \alpha}{\partial T}\right)_\pi = K \left(\frac{\partial \alpha}{\partial T}\right)_\pi \qquad (4.11.9)$$

or

$$\left(\frac{\partial \bar{T}}{\partial T}\right)_\alpha = - \left(\frac{\partial \bar{T}}{\partial \alpha}\right)_T \left(\frac{\partial \alpha}{\partial T}\right)_{\bar{T}} = -K \left(\frac{\partial \alpha}{\partial T}\right)_{\bar{T}} \qquad (4.11.10)$$

The fractional change in area with temperature at constant isotropic tension can be measured by observing the change in membrane surface area vs. the change in temperature. This is called the thermal area expansivity for the membrane. The thermoelastic change in membrane tension at constant area is the product of the thermal area expansivity and the elastic area compressibility modulus. It is apparent from Equations 4.11.9 and 4.11.10 that there are two different measures of thermal area expansivity: one at constant surface pressure,

$$\left(\frac{\partial \alpha}{\partial T}\right)_\pi = \frac{1}{A} \left(\frac{\partial A}{\partial T}\right)_\pi \qquad (4.11.11)$$

and the other at constant tension,

$$\left(\frac{\partial \alpha}{\partial T}\right)_{\bar{T}} = \frac{1}{A} \left(\frac{\partial A}{\partial T}\right)_{\bar{T}} \qquad (4.11.12)$$

If we take the partial derivative of the isotropic tension, Equation 4.11.7, with respect to temperature at constant area, we see that the interfacial interactions contribute to the thermoelastic change,

$$\left(\frac{\partial \bar{T}}{\partial T}\right)_\alpha = \frac{d\gamma}{dT} - \left(\frac{\partial \pi}{\partial T}\right)_\alpha = \frac{d\gamma}{dT} - K \left(\frac{\partial \alpha}{\partial T}\right)_\pi \qquad (4.11.13)$$

where the interfacial free energy density, γ (or natural state surface pressure), is only a function of temperature. We will show later on that the difference between Equations 4.11.9 and 4.11.10 can be attributed to the heat of expansion of the interfaces. The relationship between the measured thermal area expansivity (i.e., at constant tension) and the thermal area expansivity at constant surface pressure is established by Equation 4.11.13,

$$\left(\frac{\partial \alpha}{\partial T}\right)_{\bar{T}} = \left(\frac{\partial \alpha}{\partial T}\right)_\pi - \frac{1}{K} \frac{d\gamma}{dT} \qquad (4.11.14)$$

Therefore, the differential equation of state (based on surface pressure) is given by

$$d\pi = -K \, d\alpha + \left[K \left(\frac{\partial \alpha}{\partial T} \right)_{\overline{T}} + \frac{d\gamma}{dT} \right] dT$$

$$(4.11.15)$$

which can be called an internal equation of state. This equation of state is indeterminate, since the change in interfacial interaction with temperature cannot be measured directly.

On the other hand, the mechanochemical equation of state,

$$d\overline{T} = K \, d\alpha - K \left(\frac{\partial \alpha}{\partial T} \right)_{\overline{T}} dT$$

$$(4.11.16)$$

is the form which can be directly determined by experiments. Equations 4.11.13 and 4.11.14 demonstrate that the observed temperature effects in membranes involve inseparable changes in surface pressure and interfacial interaction.

The second mechanochemical state equation, (4.11.3), describes behavior that is specific to solid membrane materials, whereas the isotropic behavior is appropriate for both liquid and solid surfaces. We can use the mathematical corollary to obtain

$$\left(\frac{\partial T_s}{\partial T} \right)_{\beta,\alpha} = - \left(\frac{\partial T_s}{\partial \tilde{\lambda}} \right)_{T,\alpha} \left(\frac{\partial \tilde{\lambda}}{\partial T} \right)_{T_s,\alpha}$$

$$(4.11.17)$$

which gives the thermoelastic coefficient of the shear resultant in terms of elastic properties and thermal extensibility (i.e., fractional change in extension with temperature). The shear resultant is given by the elastic constitutive relation,

$$T_s = \frac{\mu}{2(1+\alpha)} (\tilde{\lambda}^2 - \tilde{\lambda}^{-2})$$

where $\tilde{\lambda}$ is chosen greater than one. Therefore, the coefficients in the differential form of the mechanochemical state equation are obtained by taking the following partial derivatives:

$$\left(\frac{\partial T_s}{\partial \tilde{\lambda}} \right)_{T,\alpha} = \left(\frac{\partial \mu}{\partial \tilde{\lambda}} \right)_{T,\alpha} \frac{(\tilde{\lambda}^2 - \tilde{\lambda}^{-2})}{2(1+\alpha)} + \frac{\mu}{(1+\alpha)} (\tilde{\lambda} + \tilde{\lambda}^{-3})$$

$$\left(\frac{\partial T_s}{\partial T} \right)_{\beta,\alpha} = \left(\frac{\partial \mu}{\partial T} \right)_{\beta,\alpha} \frac{(\tilde{\lambda}^2 - \tilde{\lambda}^{-2})}{2(1+\alpha)}$$

$$(4.11.18)$$

Now, Equation 4.11.17 may be rearranged with the coefficients, Equations 4.11.18, to give the thermal extensibility as a function of the temperature gradient of the elastic shear modulus,

$$\left(\frac{\partial \tilde{\lambda}}{\partial T} \right)_{T_s,\alpha} = \frac{- \left(\frac{\partial \ln \mu}{\partial T} \right)_{\beta,\alpha} \cdot (\tilde{\lambda}^2 - \tilde{\lambda}^{-2})}{\left(\frac{\partial \ln \mu}{\partial \tilde{\lambda}} \right)_{T,\alpha} \cdot (\tilde{\lambda}^2 - \tilde{\lambda}^{-2}) + 2(\tilde{\lambda} + \tilde{\lambda}^{-3})}$$

$$(4.11.19)$$

If we assume that the membrane is a first order, hyperelastic surface, i.e., $(\partial \mu / \partial \tilde{\lambda})_T \equiv 0$, then we obtain,

$$\left(\frac{\partial \ell n\, \tilde{\lambda}}{\partial T} \right)_{T_s, a} = - \frac{(\tilde{\lambda}^2 - \tilde{\lambda}^{-2})}{2(\tilde{\lambda}^2 + \tilde{\lambda}^{-2})} \cdot \left(\frac{\partial\, \ell n\, \mu}{\partial T} \right)_{\beta, \alpha} \qquad (4.11.20)$$

The differential form for the second equation of state becomes

$$dT_s = \frac{\mu(\tilde{\lambda}^2 + \tilde{\lambda}^{-2})}{(1 + \alpha)} \left[d\,(\ell n\, \tilde{\lambda}) - \left(\frac{\partial\, \ell n\, \tilde{\lambda}}{\partial T} \right)_{T_s, a} dT \right]$$

$$dT_s = \frac{2\mu(\beta + 1)}{(1 + \alpha)} \left[d(\ell n\, \tilde{\lambda}) - \left(\frac{\partial\, \ell n\, \tilde{\lambda}}{\partial T} \right)_{T_s, \alpha} dT \right] \qquad (4.11.21)$$

The temperature dependence of mechanochemical equations of state provides not only the thermoelastic description of the membrane but also the dependence of the thermodynamic state variables, E and S, on membrane deformation. Since internal energy and entropy are state functions, differential changes in their values are the superposition of increments that are taken along independent pathways, i.e., state functions are exact differentials that can be expressed by the chain rule of differentiation,

$$d\tilde{E} = \left(\frac{\partial \tilde{E}}{\partial \alpha} \right)_{T, \beta} \cdot d\alpha + \left(\frac{\partial \tilde{E}}{\partial \ell n\, \tilde{\lambda}} \right)_{T, \alpha} \cdot d(\ell n\, \tilde{\lambda}) + \left(\frac{\partial \tilde{E}}{\partial T} \right)_{\alpha, \beta} dT$$

$$d\tilde{S} = \left(\frac{\partial \tilde{S}}{\partial \alpha} \right)_{T, \beta} \cdot d\alpha + \left(\frac{\partial \tilde{S}}{\partial \ell n\, \tilde{\lambda}} \right)_{T, \alpha} \cdot d(\ell n\, \tilde{\lambda}) + \left(\frac{\partial \tilde{S}}{\partial T} \right)_{\alpha, \beta} dT$$

$$(4.11.22)$$

An exact differential is a characteristic feature of an analytic function, say f(x,y). This means that cross derivatives with respect to the independent variables are equal. In other words, the expansion of an exact differential,

$$df = \frac{\partial f}{\partial x} \cdot dx + \frac{\partial f}{\partial y} \cdot dy$$

implies that

$$\frac{\partial^2 f}{\partial x \partial y} \equiv \frac{\partial^2 f}{\partial y \partial x}$$

Thus, the cross derivatives of internal energy and entropy with respect to temperature and deformation variables must be equal. For instance,

$$\left(\frac{\partial}{\partial \alpha} \right)_{T, \beta} \cdot \left(\frac{\partial \tilde{E}}{\partial T} \right)_{\alpha, \beta} = \left(\frac{\partial}{\partial T} \right)_{\alpha, \beta} \left(\frac{\partial \tilde{E}}{\partial \alpha} \right)_{T, \beta} \qquad (4.11.23)$$

or

$$\left(\frac{\partial}{\partial \ell n\, \tilde{\lambda}}\right)_{T,\alpha} \cdot \left(\frac{\partial \tilde{E}}{\partial T}\right)_{\alpha,\beta} = \left(\frac{\partial}{\partial T}\right)_{\alpha,\beta} \left(\frac{\partial \tilde{E}}{\partial \ell n\, \tilde{\lambda}}\right)_{T,\alpha}$$

In order to take advantage of these analytic properties of the state functions, we use the combined first and second laws of thermodynamics for reversible processes,

$$d\tilde{W} = d\tilde{E} - T\, d\tilde{S}$$

and the expression for the mechanical work that is produced by area dilation and by extension of the material element,

$$d\tilde{W} = \bar{T} \cdot d\alpha + 2T_s \cdot (1 + \alpha) \cdot d(\ell n\, \tilde{\lambda}) \tag{4.11.25}$$

We see that Equations 4.11.22, 4.11.24, and 4.11.25 allow us to relate the partial derivatives of internal energy and entropy for the differential changes in intensive variables: fractional change in area, $d\alpha$, fractional extension at constant area, $d(\ell n\, \tilde{\lambda})$, and temperature change, dT. For instance, the partial derivatives of the entropy density are given by

$$\left(\frac{\partial \tilde{S}}{\partial \alpha}\right)_{T,\beta} = \frac{1}{T}\left[\left(\frac{\partial \tilde{E}}{\partial \alpha}\right)_{T,\beta} - \bar{T}\right]$$

$$\left(\frac{\partial \tilde{S}}{\partial \ell n\, \tilde{\lambda}}\right)_{T,\alpha} = \frac{1}{T}\left[\left(\frac{\partial \tilde{E}}{\partial \ell n\, \tilde{\lambda}}\right)_{T,\alpha} - 2(1 + \alpha)\cdot T_s\right]$$

$$\left(\frac{\partial \tilde{S}}{\partial T}\right)_{\alpha,\beta} = \frac{1}{T}\left(\frac{\partial \tilde{E}}{\partial T}\right)_{\alpha,\beta} \tag{4.11.26}$$

If we take cross derivatives with respect to temperature and deformation variables, we may use Equations 4.11.23 to obtain the internal energy density change that is produced by isothermal deformation,

$$\left(\frac{\partial \tilde{E}}{\partial \alpha}\right)_{T,\beta} = \bar{T} - T \cdot \left(\frac{\partial \bar{T}}{\partial T}\right)_{\alpha,\beta}$$

$$\left(\frac{\partial \tilde{E}}{\partial \ell n\, \tilde{\lambda}}\right)_{T,\alpha} = 2(1 + \alpha) \cdot \left[T_s - T \cdot \left(\frac{\partial T_s}{\partial T}\right)_{\alpha,\beta}\right] \tag{4.11.27}$$

We also obtain the heats of deformation with Equations 4.11.27,

$$T \cdot \left(\frac{\partial \tilde{S}}{\partial \alpha}\right)_{T,\beta} = -T \cdot \left(\frac{\partial \bar{T}}{\partial T}\right)_{\alpha,\beta}$$

$$T \cdot \left(\frac{\partial \tilde{S}}{\partial \ell n\, \tilde{\lambda}}\right)_{T,\alpha} = -2(1 + \alpha) \cdot T \cdot \left(\frac{\partial T_s}{\partial T}\right)_{\alpha,\beta} \tag{4.11.28}$$

Thus, the reversible heats of deformation are exhibited as thermoelastic effects. If we first consider the isotropic equation of state, we may express the internal energy and heat of expansion in terms of the first order material properties,

$$\left(\frac{\partial \widetilde{E}}{\partial \alpha}\right)_{T,\beta} = K \cdot \alpha + T \cdot K \cdot \left(\frac{\partial \alpha}{\partial T}\right)_{\overline{T}}$$

$$T \cdot \left(\frac{\partial \widetilde{S}}{\partial \alpha}\right)_{T,\beta} = T \cdot K \cdot \left(\frac{\partial \alpha}{\partial T}\right)_{\overline{T}} \qquad (4.11.29)$$

where we have assumed that these changes are measured relative to the tension-free or natural state. In the natural state, we know that the free energy must be minimum,

$$\left. (\delta \widetilde{F})_T \right|_{\alpha \to 0} = 0$$

which implies that

$$\left. \left(\frac{\partial \widetilde{F}}{\partial \alpha}\right)_T \right|_{\alpha \to 0} = 0$$

We see that this condition is satisfied by Equations 4.11.29. Furthermore, we note that there exists a natural heat of expansion in the tension-free condition,

$$T \cdot \left(\frac{\partial \widetilde{S}}{\partial \alpha}\right)_{T,\beta} = T \cdot K \cdot \left(\frac{\partial \alpha}{\partial T}\right)_{\overline{T}=0} \qquad (4.11.30)$$

which represents the thermal repulsive forces internal to the material. The natural heat of expansion is opposed by a cohesive energy density in the material of equal magnitude,

$$\left(\frac{\partial \widetilde{E}}{\partial \alpha}\right)_{T,\beta} = T \cdot K \cdot \left(\frac{\partial \alpha}{\partial T}\right)_{\overline{T}=0}$$

The thermal area expansivity at zero tension is related to the thermal area expansivity at nonzero tension by

$$\left(\frac{\partial \alpha}{\partial T}\right)_{\overline{T}} = \left(\frac{\partial \alpha}{\partial T}\right)_{\overline{T}=0} - \frac{\overline{T}}{K} \cdot \left(\frac{\partial \ell n\, K}{\partial T}\right)$$

Thus, the heat of expansion is equal to

$$T \left(\frac{\partial \widetilde{S}}{\partial \alpha}\right)_{T,\beta} = T \cdot K \cdot \left(\frac{\partial \alpha}{\partial T}\right)_{\overline{T}=0} - T \cdot \left(\frac{\partial \ell n\, K}{\partial T}\right) \cdot \overline{T}$$

$$(4.11.31)$$

where the second term is often negligible compared to the natural heat of expansion.

The free energy decomposition into internal energy and heat of expansion is expressed in terms of measurable variables: elastic compressibility modulus and thermal area expansivity. This represents the change in thermodynamic state of the membrane system that is produced by isothermal expansion. We recognize that the membrane system includes the closed membrane structure plus the interfacial phases. The effects of interfacial interactions are implicit in natural heat of expansion for the force-free state,

$$T \cdot \left(\frac{\partial \tilde{S}}{\partial \alpha} \right)_{T,\beta} = T \cdot K \cdot \left(\frac{\partial \alpha}{\partial T} \right)_{\pi_0} - T \cdot \frac{d\gamma}{dT}$$

where the thermal area expansivity is for constant surface pressure. The heat of expansion of the interfaces is Kelvin's heat of expansion for a free surface. We will show in an example at the end of this section that the heat of expansion of the interfaces may contribute significantly to the natural heat of expansion of the membrane.

When we matched cross derivatives of entropy coefficients, Equations 4.11.26, we eliminated the temperature coefficients of the thermodynamic state functions. The temperature coefficients are determined at constant element area and shape and define the specific heat, C_α, at constant surface area,

$$C_\alpha \equiv T \left(\frac{\partial \tilde{S}}{\partial T} \right)_{\alpha,\beta} = T \left(\frac{\partial \tilde{E}}{\partial T} \right)_{\alpha,\beta} \qquad (4.11.32)$$

In general, it is very difficult, sometimes impossible, to measure the specific heat at constant geometry. Normally, the specific heat is measured under the condition of constant applied force, i.e., constant isotropic tension or constant surface pressure. Specific heats at constant intensive force are derived from enthalpic potentials, e.g.,

$$\tilde{H} \equiv \tilde{E} - \bar{T} \cdot \alpha$$

or

$$\tilde{H}' \equiv \tilde{E} + \pi \cdot \alpha$$

with the following prescriptions:

$$C_{\bar{T}} \equiv \left(\frac{\partial \tilde{H}}{\partial T} \right)_{\bar{T},\beta}$$

$$C_\pi \equiv \left(\frac{\partial \tilde{H}'}{\partial T} \right)_\pi \qquad (4.11.33)$$

The specific heat at constant tension is the reversible heat exchange per degree per unit membrane area that would be measured in a calorimeter. In a calorimeter, the membrane would presumably be free of external force, e.g., no osmotic stress or other surface tractions. It is not possible to maintain constant area and simultaneously measure the reversible heat exchange per degree. However, the two specific heats are related by

$$(C_{\bar{T}} - C_\alpha) = T \cdot K \cdot \left(\frac{\partial \alpha}{\partial T} \right)_{\bar{T}}^{2} \qquad (4.11.34)$$

Consequently, we can calculate the specific heat at constant area from that at constant tension with the thermoelastic properties. We can also compare Equation (4.11.34) with the internal equation of state through

$$(C_\pi - C_\alpha) = T \cdot K \cdot \left(\frac{\partial \alpha}{\partial T}\right)_\pi^2 \tag{4.11.35}$$

The second equation of state, which represents solid elastic behavior, is used with Equations 4.11.27 and 4.11.28 to determine the internal energy and heat of extension. Here, we need the thermoelastic change in the shear resultant at constant element area and shape,

$$\left(\frac{\partial T_s}{\partial T}\right)_{\alpha,\beta}$$

This partial derivative is simply given by

$$\left(\frac{\partial T_s}{\partial T}\right)_{\alpha,\beta} = \left(\frac{\partial \mu}{\partial T}\right)_{\alpha,\beta} \frac{(\tilde{\lambda}^2 - \tilde{\lambda}^{-2})}{2(1+\alpha)}$$

Thus, the reversible heat of extension and internal energy change at constant temperature are expressed as

$$\left(\frac{\partial \tilde{E}}{\partial \tilde{\lambda}}\right)_{T,\alpha} = \left[\mu - T \cdot \left(\frac{\partial \mu}{\partial T}\right)_{\beta,\alpha}\right] \cdot (\tilde{\lambda} - \tilde{\lambda}^{-3})$$

$$T \cdot \left(\frac{\partial \tilde{S}}{\partial \tilde{\lambda}}\right)_{T,\alpha} = -T \cdot \left(\frac{\partial \mu}{\partial T}\right)_{\beta,\alpha} \cdot (\tilde{\lambda} - \tilde{\lambda}^{-3}) \tag{4.11.36}$$

or in terms of β as

$$\left(\frac{\partial \tilde{E}}{\partial \beta}\right)_{T,\alpha} = \mu - T \cdot \left(\frac{\partial \mu}{\partial T}\right)_{\beta,\alpha}$$

$$T \cdot \left(\frac{\partial \tilde{S}}{\partial \beta}\right)_{T,\alpha} = -T \cdot \left(\frac{\partial \mu}{\partial T}\right)_{\beta,\alpha} \tag{4.11.37}$$

In the following examples, we will examine the expected thermoelastic behavior of a phospholipid bilayer and a two-dimensional, elastomeric network. The first example will demonstrate the relationship between the internal equation of state for surface pressure and the mechanochemical state equation for isotropic tension. An important chemical feature, the interfacial interaction of hydrocarbon and water, will be made apparent. The second example will show the entropic character of an elastomeric material and the difficulty in measuring the temperature dependence of the elastomer shear modulus.

Example 1

Consider the example of a phospholipid bilayer introduced in Examples 1 and 2 of Section 4.6. In the absence of experimental evidence on the temperature dependence of elastic properties of a phospholipid layer, we must again use the abstraction of the bilayer as two monolayers with the equation of state given by

$$\pi(\tilde{A} - \tilde{A}_e) = c\,T$$

where $\tilde{A}$ is the area per molecule (in a single layer). $\tilde{A}_e$ is the excluded area per molecule. The isothermal area compressibility modulus and thermal area expansivity are

$$K \equiv -\tilde{A}\left(\frac{\partial \pi}{\partial \tilde{A}}\right)_T = \frac{\tilde{A}\pi}{(\tilde{A} - \tilde{A}_e)}$$

$$\left(\frac{\partial \alpha}{\partial T}\right)_\pi \equiv \frac{1}{\tilde{A}}\left(\frac{\partial \tilde{A}}{\partial T}\right)_\pi = \frac{(\tilde{A} - \tilde{A}_e)}{T\tilde{A}}$$

The change in surface pressure with temperature at constant area is given by

$$\left(\frac{\partial \pi}{\partial T}\right)_\alpha = \frac{\pi}{T}$$

Above the transition temperature, a value $\tilde{A}_e$ equal to 38 Å² per molecule would approximate a lecithin bilayer surface pressure vs. area curve as discussed in Example 2 of Section 4.6. At an area per molecule of 68 Å², which corresponds to a bilayer surface pressure of 70 dyn/cm, the thermal area expansivity at constant pressure and the change in surface pressure with temperature at constant area would be

$$\left(\frac{\partial \alpha}{\partial T}\right)_\pi \simeq 1.5 \times 10^{-3}/^\circ C$$

$$\left(\frac{\partial \pi}{\partial T}\right)_\alpha \simeq 0.23\ \text{dyn/cm/}^\circ C$$

These small numbers indicate the difficulty in measuring the temperature dependence of a monolayer equation of state with the Langmuir trough technique (see Davies and Rideal, 1961).[12]

Now, we will investigate the thermoelastic effects which would be observed in a bilayer membrane experiment. In order to estimate these changes, the temperature dependence of the interfacial free energy density, γ, must be evaluated. As discussed in Example 1 of Section 4.6, the interfacial free energy density of hydrophobic interaction for interfacial regions (the interstitial spaces between polar head groups) is approximately that of simple hydrocarbon-water interfaces. Consequently, we can approximate the temperature dependence by using twice the value of the temperature dependence of surface tension at oil-water interfaces, $d\gamma_o/dT$,

$$\frac{d\gamma}{dT} \simeq 2\,\frac{d\gamma_o}{dT}$$

Measured values for $d\gamma_o/dT$ depend on hydrocarbon chain length, but are typically about -0.1 dyn/cm/°C (see Adam, *The Physics and Chemistry of Surfaces*, 1968).[1] Thus,

$$\frac{d\gamma}{dT} \simeq -0.2 \ \text{erg/cm}^2/^{\circ}\text{C}$$

We recall the following relations:

$$\left(\frac{\partial \alpha}{\partial T}\right)_{\overline{T}} = \left(\frac{\partial \alpha}{\partial T}\right)_{\pi} - \frac{1}{K}\frac{d\gamma}{dT}$$

and

$$\left(\frac{\partial \overline{T}}{\partial T}\right)_{\alpha} = \frac{d\gamma}{dT} - \left(\frac{\partial \pi}{\partial T}\right)_{\alpha}$$

which are the thermal area expansivity and thermoelastic change in tension at constant area as measured in an experiment. With the values for these parameters that were introduced above, we obtain estimates of the thermoelastic effects,

$$\left(\frac{\partial \alpha}{\partial T}\right)_{\overline{T}} \simeq 1.5 \times 10^{-3} \ -(-0.2)/(160) = 2.7 \times 10^{-3}/^{\circ}\text{C}$$

$$\left(\frac{\partial \overline{T}}{\partial T}\right)_{\alpha} \simeq -0.23 - 0.2 = -0.43 \ \text{dyn/cm}/^{\circ}\text{C}$$

It is apparent that thermoelastic observations will be strongly influenced by the temperature dependence of the interfacial interaction. The surface pressure change might even be masked by changes in γ. There is direct evidence that such effects are indeed important. X-ray diffraction measurements of the temperature dependence of the thickness of phospholipid bilayers (arranged in stacked, hydrated layers) have been performed by Drs. Costello and Gulik-Krzywicki (personal communication).[10] Since the thermal volume expansivity, $(1/V)(\partial V/\partial T)_p$, is on the order of $10^{-4}/^{\circ}\text{C}$, the fractional change in thickness is approximately equal and opposite to the fractional change in area produced by temperature change. The stacked bilayers are essentially free of membrane tension. Therefore, we can obtain an estimate of the thermal area expansivity for the natural state, free of external forces. The recent work of Drs. Costello and Gulik-Krzywicki gives values of 2 to $2.7 \times 10^{-3}/^{\circ}\text{C}$ for $(\partial \alpha/\partial T)_{\overline{T}}$ of egg lecithin bilayers and lipids extracted from sarcoplasmic reticulum. These values for thermal area expansivity agree with the estimate we made above, which includes interfacial effects.

We can also estimate the difference in specific heats that would be expected from experiments. The difference in specific heats for a bilayer idealized as two monolayers is obtained from the equation of state,

$$(C_{\pi} - C_{\alpha}) = \frac{\pi}{T} \cdot \frac{(\widetilde{A} - \widetilde{A}_e)}{\widetilde{A}}$$

If we incorporate the temperature dependence of the hydrophobic effect, the difference in specific heats is predicted to be

$$(C_{\overline{T}} - C_{\alpha}) = \frac{(\widetilde{A} - \widetilde{A}_e)}{\widetilde{A}} \cdot \frac{\pi}{T}\left[1 - \frac{T}{\pi}\frac{d\gamma}{dT}\right]^2$$

With the values used previously, we find that

$$(C_{\overline{T}} - C_\alpha) = 3.45 \cdot (C_\pi - C_\alpha) = 0.36 \;\; erg/cm^2 \; /^\circ C$$

The threefold change indicates the importance of the temperature dependence of the hydrophobic effect.

Example 2

Consider a two-dimensional, elastomeric surface network. In Section 4.7, the entropy density was shown to be approximated by

$$(\widetilde{S} - \widetilde{S}_0)_{T,\alpha} = -\rho_0 \left(\frac{N_A k}{M_w} \right) \left\langle \frac{r_0^2}{\sigma_0^2} \right\rangle \beta$$

where ρ_0 is the surface mass density of polymer chains; N_A is Avogadro's number; M_w is the molecular weight of each chain; $<r_0^2/\sigma_0^2>$ is the ensemble average that characterizes the variation in molecular subchain length in the total network, normalized by the free variation of an individual chain.
Therefore,

$$\left(\frac{\partial S}{\partial \beta} \right)_{T,\alpha} = - \left(\frac{\rho_0 N_A k}{M_w} \right) \left\langle \frac{r_0^2}{\sigma_0^2} \right\rangle$$

and the surface shear modulus is given by

$$\mu = \left[\left(\frac{\rho_0 N_A k}{M_w} \right) \left\langle \frac{r_0^2}{\sigma_0^2} \right\rangle \right] T$$

which increases in direct proportion to temperature. Consequently,

$$\left(\frac{\partial \widetilde{E}}{\partial \beta} \right)_{T,\alpha} = 0$$

The free energy density at constant temperature and surface area is totally entropic as the development in Section 4.7 prescribed. The thermoelastic behavior that would be observed in a mechanical experiment is represented by

$$\left(\frac{\partial \widetilde{\lambda}}{\partial T} \right)_{\alpha, T_s} = - \frac{1}{2} \frac{(\widetilde{\lambda}^2 - \widetilde{\lambda}^{-2})}{(\widetilde{\lambda} + \widetilde{\lambda}^{-3})} \left(\frac{1}{T} \right)$$

Since biological materials are only viable at temperatures in the range of 272 to 320 K, it is apparent that the fractional changes in extension ratio would be less than a fraction of a percent per degree.

4.12 Internal Dissipation, Viscosity, Viscoelasticity, Relaxation, and Viscoplasticity

The development of elastic material behavior of a membrane was based entirely on reversible thermodynamics, i.e., conservative free energy potentials. For membrane deformations that are produced at a slow rate without permanent material alteration due to the membrane forces, elastic constitutive relations are reasonable representations of the material behavior of the membrane. On the other hand, as the rate of deformation increases, thermodynamically irreversible processes become evident, which result from internal friction and heat dissipation within the membrane. Irreversible processes create nonconservative forces that depend on the rate of deformation as well as the instantaneous deformation state. If we consider a solid membrane structure, then the static membrane forces will approximate the elastic, conservative level, which characterizes thermodynamic equilibrium. However, the time-dependent response of membrane deformation to applied forces will be limited by the rate of viscous dissipation in the material. In the absence of applied forces, the solid material returns to its initial shape. This is viscoelastic solid behavior and does not involve irrecoverable (plastic) material deformation. This is in contrast to plastic flow that often results from prolonged exposure to applied forces; here, the material changes are permanent. If the magnitude of the applied forces are sufficiently large, the material will immediately begin plastic flow. The forces applied to the material produce work per unit time or mechanical power that is dissipated as heat in the material. For a perfect liquid, there is no shear resultant threshold or yield level; flow commences in response to any shear force resultant or stress. The constitutive relations between the nonconservative or frictional force resultants and rates of deformation are characterized by coefficients of viscosity. Viscosity represents the rate at which entropy is generated in the material. The result is dissipation of mechanical work. We will use a first order, irreversible thermodynamic approach to demonstrate the modes of entropy production and heat generation that are produced by different intensive rates of deformation. Entropy production is assumed to increase in proportion to the square of the rate of deformation divided by temperature. This approach will yield the independent coefficients of viscosity.

When we discuss viscosity and irreversible thermodynamics, we are again treating the membrane as a two-dimensional continuum. Consequently, viscosity coefficients are to be determined directly by mechanical experiments which involve macroscopic regions of the surface and, thus, are continuum properties. Viscosity is a measure of the fluidity of a material as the elastic shear modulus is a measure of its rigidity. However, fluidity and ridigity are subjective terms which must be carefully handled. They are not substitutes for the viscosity of a liquid or the shear modulus of a solid. In membrane biology, such qualitative descriptors are often used to characterize membrane structure; these images can be illusory. Measurements of kinetic or thermal motions of molecules in the membrane are frequently used to quantify membrane fluidity. In classical kinetic theory, diffusive motions of molecules may be related to the coefficients of viscosity of the continuum, but for a complex structure such as membrane, the relationship is difficult to establish.

Dissipative mechanisms that limit the rate of continuous deformation of the membrane are inherently complicated. As irreversible processes, they cannot be represented by a definitive thermodynamic equation — only an inequality, as we discussed before. Consequently, a phenomenological approach must be used to represent these processes. Ultimately, the processes must be defined empirically by experiment. We will briefly outline the phenomenological methods of irreversible thermodynamics. Then, we will use the results to develop first order, constitutive relations for three metaphorical regimes of material behavior: solid, semisolid transition, and plastic or liquid. These material regimes characterize different molecular processes and will be represented by different viscosity coefficients.

The first law of thermodynamics relates the differential change in total material energy, dE, to the incremental exchange of heat, δ Q, and work done on the material, δW,

$$dE = \delta Q + \delta W \tag{4.12.1}$$

Because the total energy is conserved, a cyclic process results in

$$\oint \delta Q = -\oint \delta W \tag{4.12.2}$$

This demonstrates that the irreversible heat loss for an isothermal, cyclic process is provided by additional work in the cycle. The added work is called "hysteresis." If we assume that the irreversible process is totally entropic (in other words, only heat is generated by nonzero rates of deformation), we may write the first law for either a reversible or irreversible process at constant temperature as

$$(dE)_T = (\delta Q)_T + (\delta W)_T$$

where

$$(dE)_T = (dQ)_T + (dW)_T \qquad \text{(reversible)} \tag{4.12.3}$$

From the difference, we may define the irrecoverable work, δW_T^{IRR}, done on the system,

$$(\delta Q - dQ)_T + (\delta W - dW)_T = 0$$

This mechanical energy is lost as heat transfer to the environment. Thus,

$$\delta W_T^{IRR} \equiv (\delta W - dW)_T$$

$$\delta Q_T^{IRR} = -\delta W_T^{IRR} = (\delta Q - dQ)_T \tag{4.12.4}$$

For an isothermal process, exact differentials specify the reversible, elastic material behavior previously examined, and Equation 4.12.4 represents the internal heat dissipation and mechanical losses. The loss of mechanical power is proportional to the rate of internal entropy production which is eventually exchanged with the environment as heat. (See Katchalsky and Curran, Entropy Change in Irreversible Processes, in *Nonequilibrium Thermodynamics in Biophysics,* 1967).[48]

The mechanical power is specified by the work per unit time performed by the material force resultants. The mechanical power density is the scalar product of the material force resultants times the rate of deformation matrix. Therefore, the time rate of change of work per unit initial area of a membrane element is given by

$$\left(\frac{\partial \widetilde{W}}{\partial t} \right)_T = (T_{ij} V_{ij})(\lambda_1 \lambda_2) \tag{4.12.5}$$

The mechanical power dissipated in the material as heat is proportional to the time rate of generation of the material entropy density. The rate of entropy production is measured per unit of instantaneous area and is an Eulerian variable, $\dot{s}$; thus,

$$(\lambda_1 \lambda_2)\,\dot{s} = -\frac{1}{T}\left(\frac{\partial \widetilde{Q}}{\partial t} \right)_T^{IRR} = \frac{1}{T}\left(\frac{\partial \widetilde{W}}{\partial t} \right)_T^{IRR} \tag{4.12.6}$$

The heat exchange is opposite in sign to the rate of entropy production because it is lost to the environment. The mechanical power loss is produced by the nonconservative or frictional part of the force resultants. If we consider only the nonconservative force resultants, then Equations 4.12.5 and 4.12.6 combine to give

$$\frac{1}{(\lambda_1 \lambda_2)} \left(\frac{\partial \widetilde{W}}{\partial t} \right)_T^{IRR} = T_{ij}^{v} \, V_{ij} = T \, \dot{s} \qquad (4.12.7)$$

where T_{ij}^{v} is the inelastic or viscous force resultant matrix. Equation 4.12.7 must be a positive definite quantity, $\dot{s} \geqslant 0$, according to the original statement by Clausius.

In 1931, Onsager postulated a phenomenological equation that relates entropy production, $\dot{s}$, to the scalar product of flows, J_p, and their conjugate forces, $\Delta\Phi_p$.[61] The thermodynamic context of force is the gradient of chemical potential, which includes contributions from mechanical work, particle activities, etc. Flows, on the other hand, are the fluxes of configurational entropy, given by time rates of material movement, e.g., particle flux or rate of deformation. The first order approximation to irreversible processes is provided by

$$\dot{s} = \sum_p J_p \cdot \Delta\Phi_p$$

In the phenomenological approach, the flows are linearly related to the conjugate forces by the coupling coefficients, L_{pq},

$$J_p \equiv \sum_q L_{pq} \cdot \Delta\Phi_q$$

Likewise, an inverse relationship exists for the forces in terms of the flows,

$$\Delta\Phi_p = \sum_q \overline{L}_{pq} \, J_q$$

Consequently, the entropy production is quadratic in the flows,

$$\dot{s} = \sum_p \left(\sum_q \overline{L}_{pq} \, J_q \right) J_p \equiv \overline{L}_{pq} \, J_q \, J_p \qquad (4.12.8)$$

if we use implicit summation of indices.

For a continuum material, the flows are the time rates of deformation. These describe the intensive time rate of change of distances between material points. A membrane which is isotropic in its surface is characterized by two independent rates of deformation (see Section 2.5): the time rate of change of area dilation or condensation of membrane; the time rate of exchange of material extension or shear deformation at constant area. From Section 2.5, these are

$$J_1 \equiv \tfrac{1}{2} \, V_{kk} \, \delta_{ij}$$

$$J_2 \equiv \widetilde{V}_{ij} \qquad (4.12.9)$$

Therefore, the conjugate forces may be expressed as

$$\Delta\Phi_1 = \overline{L}_{11} \, (\tfrac{1}{2} \, V_{kk} \, \delta_{ij}) + \overline{L}_{12} \, \widetilde{V}_{ij}$$

$$\Delta\Phi_2 = \overline{L}_{21} \, (\tfrac{1}{2} \, V_{kk} \, \delta_{ij}) + \overline{L}_{22} \, \widetilde{V}_{ij} \qquad (4.12.10)$$

With Equations 4.12.9 and 4.12.10, Equation 4.12.8 for the entropy production is given by the quadratic sum,

$$\dot{s} = \overline{L}_{11} \, \tfrac{1}{2} \, (V_{kk})^2 + \overline{L}_{22} \, \widetilde{V}_{ij} \, \widetilde{V}_{ji} \qquad (4.12.11)$$

since the identity, $\delta_{ij} \, \widetilde{V}_{ij} \equiv 0$, eliminates cross relations.

Consequently, the viscous or nonconservative force resultant matrix can now be obtained from the phenomenological rate of entropy production, Equation 4.12.11, and the mechanical power loss, Equation 4.12.7. Since $V_{ij} \equiv \tfrac{1}{2} \, V_{kk} \, \delta_{ij} + \widetilde{V}_{ij}$, Equations 4.12.7 and 4.12.11 combine to give

$$T^v_{ij} = \kappa V_{kk} \, \delta_{ij} + 2\eta \, \widetilde{V}_{ij} \qquad (4.12.12)$$

where

$$\kappa \equiv \tfrac{1}{2} \, T \overline{L}_{11}$$

$$\eta \equiv \tfrac{1}{2} \, T \overline{L}_{22}$$

Thus, the phenomenological coefficients produce specific coefficients of viscosity. κ is the viscous proportionality factor for energy dissipation produced by rate of area dilation or condensation. η is the viscous coefficient for energy dissipation produced by rate of shear deformation. This is commonly referred to as "surface viscosity" for surface shear at $\pm 45°$ to the principal extension axis. Both coefficients have units of dyn-sec/cm or poise-cm (surface poise).

In terms of the fractional change in area, α, and the extension ratio at constant area, $\tilde{\lambda}$, the isotropic and shear resultant components of Equation 4.12.12 can be written

$$\overline{T}^v = \frac{\kappa}{(1 + \alpha)} \left(\frac{\partial \alpha}{\partial t}\right) \qquad (4.12.13)$$

and

$$T^v_s = 2\eta \left(\frac{\partial \, \ell n \, \tilde{\lambda}}{\partial t}\right) \qquad (4.12.14)$$

Equation 4.12.14 defines the shear behavior of a two-dimensional liquid membrane such as a phospholipid bilayer (above the order-disorder transition temperature for the hydrocarbon chains).

By contrast to the liquid membrane, the encapsulating membranes of biological cells are often solid materials with complex structure. The first order constitutive behavior for a viscoelastic solid can be obtained by parallel superposition of elastic and viscous force resultants as originally developed by Kelvin,

$$T_{ij} = T^e_{ij} + T^v_{ij} \qquad (4.12.15)$$

where $T_{ij}{}^e$ is the elastic force resultant matrix previously derived from the free energy density in Section 4.5. The superposition in Equation 4.12.15 and the hyperelastic constitutive relations provide the first order constitutive relations for a viscoelastic membrane solid. The isotropic tension is

$$\overline{T} = K\alpha + \frac{\kappa}{(1 + \alpha)} \left(\frac{\partial \alpha}{\partial t} \right) \tag{4.12.16}$$

The maximum shear resultant is

$$T_s = \frac{\mu}{2(1 + \alpha)} (\widetilde{\lambda}^2 - \widetilde{\lambda}^{-2}) + 2\eta_e \left(\frac{\partial \ell n \, \widetilde{\lambda}}{\partial t} \right) \tag{4.12.17}$$

where we use η_e to represent the coefficient of viscosity for the solid regime of material behavior. In tensor form, the constitutive relation is

$$T_{ij} = \overline{T} \, \delta_{ij} + \frac{2\mu}{(1 + \alpha)^2} \, \hat{\epsilon}_{ij} + 2\eta_e \widetilde{V}_{ij} \tag{4.12.18}$$

where the strain matrix, $\hat{\epsilon}_{ij}$, is defined by Equation 4.5.9.

From Equations 4.12.16 and 4.12.17, we observe that there are two dynamic equations for deformation recovery after a sudden removal of applied forces, i.e., $\overline{T} = 0$ and $T_s = 0$. They are, respectively,

$$[\alpha(1 + \alpha)]^{-1} \frac{d\alpha}{dt} = -\frac{K}{\kappa}$$

and

$$4(1 + \alpha) \, [\widetilde{\lambda}(\widetilde{\lambda}^2 - \widetilde{\lambda}^{-2})]^{-1} \frac{d\widetilde{\lambda}}{dt} = -\frac{\mu}{\eta_e}$$

Therefore, there is a characteristic time scale for area recovery, κ/K, and also one for extensional recovery, η_e/μ. For small area compressibility, the first equation will involve small fractional area changes and can be linearized to give

$$\alpha = \alpha_0 \, \exp \, [-tK/\kappa]$$

where α_0 is the area deformation at the time when $\overline{T}$ is set to zero. The extensional recovery is nonlinear but can be integrated to give

$$\left(\frac{\widetilde{\lambda}^2 - 1}{\widetilde{\lambda}^2 + 1} \right) = \left(\frac{\widetilde{\lambda}_0^2 - 1}{\widetilde{\lambda}_0^2 + 1} \right) \exp \, [-t\mu/\eta_e]$$

where $\widetilde{\lambda}_0$ is the maximum extension ratio when T_s is set to zero at $t = 0$ (we have assumed that surface area is constant during extension recovery).

Viscoelastic relations characterize the membrane as a solid material, whereas the simple viscous relations represent a liquid substance. The solid-like or liquid-like behavior exhibited by a material depends on the magnitude and duration of the forces supported by the material. The semisolid transition between solid and liquid behavior is modeled by viscoplastic constitutive relations. These relations describe the rate of irrecoverable or permanent deformation of the material. As such, they represent liquid-like behavior, in contrast to viscoelastic solid behavior where dissipation limits only the rate of recoverable deformations. The constitutive relations for the semisolid transition to liquid behavior differ from the simple liquid relations because some solid character is retained. The principal aspects of semisolid behavior are represented by two types of relations: (1) creep and (2) perfect plastic (i.e., Bingham material). We will discuss first order equations for these two types of behavior. The distinguishing feature of solid- or liquid-like behavior is the material response to shear. Hence, we will assume that the material is an incompressible surface and investigate the constitutive relations for the shear resultant.

Simple creep behavior is characterized by two observations: (1) the rate of shear deformation is proportional to the shear resultant when the material forces are constant or time independent, and (2) the material responds elastically when the forces change instantaneously. This behavior can be modeled by the serial summation of recoverable (elastic) and irrecoverable (plastic) rates of deformation as originally developed by Maxwell. The total rate of deformation is the sum of the rates of deformation in the elastic and viscous elements by definition,

$$V_{ij} \equiv V_{ij}^e + V_{ij}^v \tag{4.12.19}$$

The serial coupling of the elastic and viscous elements implies that the force resultants are equal,

$$T_{ij} = T_{ij}^e = T_{ij}^v \tag{4.12.20}$$

The first order constitutive relations for the elastic and viscous elements are obtained from Equations 4.5.11 and 4.12.12, respectively,

$$T_{ij}^e = \bar{T}^e \, \delta_{ij} + 2\mu \, \hat{\epsilon}_{ij}^e \tag{4.12.21}$$

$$T_{ij}^v = \bar{T}^v \, \delta_{ij} + 2\eta_c \, \tilde{V}_{ij}^v \tag{4.12.22}$$

for the two-dimensionally incompressible membrane. Because the membrane material is assumed to be two-dimensionally incompressible, only the deviatoric parts of the force resultant matrices and rate of deformation matrices need to be considered, i.e.,

$$V_{ij} = \tilde{V}_{ij} = \tilde{V}_{ij}^e + \tilde{V}_{ij}^v$$

and

$$\tilde{T}_{ij} = \tilde{T}_{ij}^e = \tilde{T}_{ij}^v$$

The rate of deformation for the viscous element is that of a simple liquid,

$$\tilde{V}_{ij}^v = \tilde{T}_{ij}^v/2\eta_c = \tilde{T}_{ij}/2\eta_c \tag{4.12.23}$$

where we use η_c to represent the coefficient of viscosity for relaxation and creep in the semisolid regime. The rate of deformation for the elastic element is not so simply obtained. First, we recall the relationship between rate of deformation and the time derivative of the Lagrangian strain matrix,

$$\tilde{V}^e_{ij} = \frac{\partial \epsilon^e_{pq}}{\partial t} \left(\frac{\partial a_p}{\partial x_i} \frac{\partial a_q}{\partial x_j} \right)^e \tag{4.12.24}$$

Second, if we assume that the principal axes of the deformation are independent of time, then we can write the time derivative of the Lagrangian strain matrix as

$$\frac{\partial \epsilon^e_{pq}}{\partial t} = \left(\frac{\partial \epsilon_{pq}}{\partial \tilde{\lambda}} \right)^e \frac{\partial \tilde{\lambda}^e}{\partial t} \tag{4.12.25}$$

for the two-dimensionally incompressible membrane. Since the shear resultant, T_s, is only a function of extension ratio for the elastic element, Equation 4.12.25 can be expressed in terms of the time derivative of the shear resultant,

$$\frac{\partial \epsilon^e_{pq}}{\partial t} = \left(\frac{\partial \epsilon_{pq}}{\partial \tilde{\lambda}} \right)^e \left(\frac{\partial \tilde{\lambda}}{\partial T_s} \right)^e \frac{\partial T_s}{\partial t} \tag{4.12.26}$$

where Equation 4.12.20 has been employed. Therefore, the constitutive relation for material creep and relaxation behavior is given by the combination of Equations 4.12.19, 4.12.23, 4.12.24, and 4.12.26 as

$$\tilde{V}_{ij} = \left[\left(\frac{\partial \epsilon_{pq}}{\partial \tilde{\lambda}} \right)^e \left(\frac{\partial a_p}{\partial x_i} \frac{\partial a_q}{\partial x_j} \right)^e \right] \left(\frac{\partial \tilde{\lambda}}{\partial T_s} \right)^e \frac{\partial T_s}{\partial t} + \frac{\tilde{T}_{ij}}{2\eta_c} \tag{4.12.27}$$

In general, this relation is a nonlinear differential equation. The components of Equation 4.12.27 are determined by the maximum rate of shear deformation, since $\tilde{V}_1 = V_s$ and $\tilde{V}_2 = -V_s$. Hence, Equation 4.12.27 is a dynamic relation between the extension ratio and the maximum shear resultant,

$$V_s = \frac{\partial (\ln \tilde{\lambda})}{\partial t} = \left(\frac{\partial \ln \tilde{\lambda}}{\partial T_s} \right)^e \frac{\partial T_s}{\partial t} + \frac{T_s}{2\eta_c} \tag{4.12.28}$$

The coefficient $(\partial \ln \tilde{\lambda} / \partial T_s)^e$, is an inverse elastic relation that may depend on the extension ratio. For example, a hyperelastic material with a constant elastic shear modulus, μ, would give

$$\left(\frac{\partial \ln \tilde{\lambda}}{\partial T_s} \right)^e = \left[\mu (\tilde{\lambda}^2 + \tilde{\lambda}^{-2})^e \right]^{-1}$$

For small deformations, i.e., $\tilde{\lambda} \simeq 1$, the linear equation of Maxwell is obtained for Equation 4.12.28,

$$\frac{\partial \tilde{\lambda}}{\partial t} = \frac{1}{2\mu} \frac{\partial T_s}{\partial t} + \frac{T_s}{2\eta_c}$$

However, in general, the coefficient for the time derivative of the shear resultant is a function of the shear resultant,

$$\left(\frac{\partial \ln \tilde{\lambda}}{\partial T_s}\right)^e = f(T_s)$$

This is the instantaneous response provided by the elastic element. The constitutive relation for membrane creep behavior is given to first order by the nonlinear differential equation,

$$\frac{\partial(\ln \tilde{\lambda})}{\partial t} = f(T_s)\ \frac{\partial T_s}{\partial t} + \frac{T_s}{2\eta_c} \tag{4.12.29}$$

An example of the nonlinearity can be shown by considering the hyperelastic relation with constant coefficient, μ. In this case, the constitutive behavior is derived from

$$T_s^e = \frac{\mu}{2}\ (\tilde{\lambda}^2 - \tilde{\lambda}^{-2})^e$$

$$f(T_s) = \left(\frac{\partial \ln \tilde{\lambda}}{\partial T_s}\right)^e \cong \frac{1}{2T_s}\left[\frac{1}{\sqrt{1 + u^2}}\right]$$

where $u \equiv \mu/T_s$. For large extension ratios at constant area, the nonlinear constitutive relation is approximated by

$$\frac{\partial(\ln \tilde{\lambda})}{\partial t} \simeq \tfrac{1}{2}\ \frac{\partial(\ln T_s)}{\partial t} + \frac{T_s}{2\eta_c}$$

For material that is held at a fixed extension, the shear resultant would relax with time according to

$$\frac{T_s}{T_s^0} \simeq \left[1 + t \cdot \left(\frac{\eta_c}{T_s^0}\right)\right]^{-1}$$

where T_s^o is the initial shear resultant. On the other hand, for a fixed level of the shear resultant, T_s, the extension ratio increases exponentially,

$$\tilde{\lambda} = \tilde{\lambda}^e \cdot \exp\left[t \cdot (T_s/2\eta_c)\right]$$

where $\tilde{\lambda}^e$ is the initial extension ratio that is given by the instantaneous elastic response. This equation embodies the basic premise of the Maxwell model,

$$\tilde{\lambda} = \tilde{\lambda}^e \cdot \tilde{\lambda}^p$$

that is, the extension ratio of the material is given by the product of recoverable (elastic) and irrecoverable (plastic) extension ratios, since

$$\frac{\partial \ln \tilde{\lambda}}{\partial t} = \frac{\partial \ln \tilde{\lambda}^e}{\partial t} + \frac{\partial \ln \tilde{\lambda}^p}{\partial t}$$

We see that the rate of permanent structural alteration of the material increases with the level of applied force. The rate of dissipation is characterized by the viscosity coefficient, η_c, and represents relaxation and reorganization of molecular structure. Thus, the characteristic period of time within which the material can be considered solid is on the order of η_c/μ. As such, this coefficient of viscosity will greatly exceed the viscosity that represents frictional resistance in the solid regime of recoverable deformations.

The other type of semisolid membrane behavior that we will discuss is that of a viscoplastic material, called a Bingham plastic. This theory of viscoplasticity was introduced by Bingham in 1922 and generalized by Hohenemser and Prager 10 years later (see Prager, 1961).[69] Two material properties are used to model the behavior of a plastic body: a yield stress and viscosity. Adapting this theory to two-dimensional materials, we write the following constitutive definitions and relations (Evans and Hochmuth, 1976):[19] (1) $T_s < \hat{T}_s$: viscoelastic behavior; (2) $T_s > \hat{T}_s$: plastic flow with the rate of deformation given by

$$2\eta_p \, \tilde{V}_{ij} = F \cdot \tilde{T}_{ij} \tag{4.12.30}$$

The two material constants are the yield shear, $\hat{T}_s$, and viscosity, η_p, for plastic flow. The yield function, F, depends on the ratio of the yield shear to the maximum shear resultant,

$$F = f(\hat{T}_s/T_s)$$

The simple relation for a Bingham plastic is

$$F = 1 - \hat{T}_s/T_s \tag{4.12.31}$$

which is only defined for $T_s \geq \hat{T}_s$. When the maximum shear resultant is less than the yield shear, the membrane exhibits solid-like behavior. On the other hand, when the yield shear is exceeded, plastic flow begins. The irrecoverable deformation is given by Equations 4.12.30 and 4.12.31 as

$$V_s = \frac{\partial(\ell n \, \tilde{\lambda})}{\partial t} = \frac{1}{2\eta_p} \, (T_s - \hat{T}_s) \tag{4.12.32}$$

for the rate of shear deformation when $T_s > \hat{T}_s$. The coefficient of viscosity, η_p, represents dissipation in the material when the structure flows like liquid.

For a constant level of shear resultant in excess of the yield shear, the irrecoverable deformation is given by a plastic extension ratio, λ^p, relative to the initial dimension at time, $t = 0$; the extension increases exponentially,

$$\lambda^p = \exp \left[t \cdot (T_s - \hat{T}_s)/2\eta_p \right]$$

Two types of constitutive relations for semisolid behavior are given in equations 4.12.29 and 4.12.32. Certainly, they are not unique, but they illustrate two different aspects common to transitions from solid to liquid behavior. In Equation 4.12.29, the distinction between solid and liquid behavior is determined by the duration of the material force resultant. If the material sustains forces for periods of time less than the order of η_c/μ, it behaves essentially as a solid and vice versa. With Equation 4.12.32, the distinction between solid and liquid behavior is determined by the magnitude of the material force resultant. If the material sustains forces which are smaller than the yield value, it does so again essentially as a solid and vice versa. Both semisolid constitutive relations, Equations 4.12.29 and 4.12.32, are defined by the extension ratio, shear resultant, and time derivatives that are measured at a fixed material location. Thus, they are Lagrangian representations. As we have shown, the semisolid regime exhibits liquid-like flow behavior, e.g., the exponential growth of plastic extension. Consequently, many situations are more easily analyzed with the introduction of Eulerian or spatial variables. The advantage is that Eulerian variables are functions of fixed spatial coordinates.

In general, the Eulerian representation of the constitutive relations simply requires the appropriate representation for the rate of deformation matrix in Equations 4.12.28 and 4.12.32. Since we assume that the membrane surface is incompressible, the rate of deformation for extensional flow (i.e., with the orientation of the principal extension axes fixed in space) is given by the gradient of the velocity in the direction of extension, say x_1,

$$V_s = \frac{dv_1}{dx_1}$$

For the specific case where the shear resultant is constant, the constitutive relations, Equations 4.12.28 and 4.12.32, predict that the velocity increases linearly in the direction of flow. For instance, the viscoplastic constitutive relation, Equation 4.12.32, would be simply written as

$$\frac{dv_1}{dx_1} = \frac{1}{2\eta_p}\,(T_s - \hat{T}_s) \tag{4.12.33}$$

On the other hand, the semisolid relation for relaxation and creep involves a time derivative that is local to a material point. We can relate the time derivative at a material location to the time derivative that is observed from a fixed position in space. This relationship includes the time rate of change of a function, f, that occurs explicitly in the material, plus the apparent temporal change that is viewed from a point fixed in space as the material flows past because of spatial variations of the function, f. The latter is called the convective derivative. Together, these terms give the time rate of change of the function that is observed in the spatial reference frame. We use df/dt to represent the time derivative of the function as measured at a fixed point in space. In general,

$$\frac{df}{dt} = -v_i\,\frac{\partial f}{\partial x_i} + \frac{\partial f}{\partial t}$$

where we retain the time derivative, $\partial(\ \ )/\partial t$, as the rate of change local to the material point. If we use this relation in Equation 4.12.29, we can specify an Eulerian representation for the extensional flow of material in semi-solid relaxation,

$$\frac{dv_1}{dx_1} = f(T_s)\cdot\left[\frac{dT_s}{dt} + v_1\,\frac{dT_s}{dx_1}\right] + \frac{T_s}{2\eta_c} \tag{4.12.34}$$

where t and x_1 are independent functions. This relation is especially useful if the flow of material is steady as viewed in space, i.e., $dT_s/dt \equiv 0$. In this case, the spatial variation of the shear resultant is related to the rate of deformation,

$$\frac{dv_1}{dx_1} = f(T_s) \cdot \left(v_1 \frac{dT_s}{dx_1} \right) + \frac{T_s}{2\eta_c}$$

We will demonstrate applications of the Eulerian representations, Equations 4.12.33 and 4.12.34, in Section V for axisymmetric surfaces, where the rate of deformation is expressed by curvilinear variables, and the velocity along the meridian is

$$\frac{dv_1}{dx_1} \equiv \frac{dv}{ds} = -\frac{v}{r}\frac{dr}{ds}$$

The brief examples which follow illustrate the magnitudes of the parameters that characterize time dependent material behavior for red blood cell membrane and lipid bilayers. In Section V, we will consider the red cell membrane behavior in detail.

Example 1

Electrocompression of lipid bilayers (White, 1970 and 1974; Requena et al., 1975; Evans and Simon, 1975)[23,74,94,94a] produces a small reduction in bilayer thickness that is accommodated by a small increase in area per molecule because the bilayer is essentially incompressible in volume. (The volumetric compliance is about 100 times smaller than the area compliance.) The experiment involves the application of fast voltage changes to the bilayer (time intervals of 10^{-3} to 10^{-6} sec). It is of interest to estimate the viscoelastic response time for area changes. The uniform expansion of the bilayer, which corresponds to change in thickness, produces shear of the hydrocarbon chains in the membrane interior. Therefore, a reasonable approximation to the area viscosity, κ, is the shear viscosity of the hydrocarbon interior, $\tilde{\eta}_{HC}$, times the bilayer thickness, δ_m,

$$\kappa \simeq \tilde{\eta}_{HC} \cdot \delta_m$$

Equation 4.12.16 gives the time scale for area recovery, t_a,

$$t_\alpha = \frac{\kappa}{K} \simeq \frac{\tilde{\eta}_{HC} \cdot \delta_m}{K}$$

The viscosity of the hydrocarbon interior has been measured to be on the order of 1 poise (dyn-sec/cm²). For a thickness of 3 to 5×10^{-7} cm and an area elastic modulus of 10^2 dyn/cm, the area recovery time would be on the order of $t_a \sim 10^{-9}$ to 10^{-8} sec. This is considerably less than the minimum time intervals for the electrocompression experiments.

Example 2

As was first observed by Hoeber and Hochmuth (1970)[46] and subsequently by Waugh and Evans (1976),[92] the time constant for extension recovery in shear deformation is of the order 10^{-2} to 10^{-1} sec for mammalian red cell membranes. From Equation 4.12.17, it is possible to estimate the surface viscosity for the membrane surface in the solid material regime. The time constant and the elastic shear modulus, μ, combine to give the estimate,

$$\eta_e \approx t_{\tilde{\lambda}} \cdot \mu$$

where $t_{\tilde{\lambda}}$ is the recovery time. The membrane elastic shear modulus is on the order of 10^{-2} dyn/cm. Hence,

$$\eta_e \approx 10^{-4} - 10^{-3} \text{ dyn-sec/cm} \quad \text{(poise-cm)}$$

This value is about one to two orders of magnitude greater than the surface viscosity of a lipid bilayer of the same thickness above the order-disorder phase transition of the hydrocarbon chains as deduced from lateral diffusion constants of marker particles (Evans and Hochmuth, 1978).[22] This suggests that the additional structure in the red membrane contributes significantly to dissipation. In Section V, we will discuss recent recovery experiments in detail.

Example 3

When the red cell membrane is subjected to a shear resultant, $\hat{T}_s$, greater than 2 to 8×10^{-2} dyn/cm, the membrane commences plastic flow. This yield shear, T_s, was calculated by Evans and Hochmuth (1976)[19] from data originally taken by Hochmuth et al. (1973),[44] on microfilaments of red cell membrane. The microfilaments are produced by fluid shear deformation of glass-attached red cells. The microfilament or membrane tether is pulled from the cell membrane at the attachment location. The rate of deformation for membrane material which forms the microfilament is on the order of 1 sec^{-1}. Therefore, for shear resultants in excess of the yield shear by about 10^{-2} dyn/cm, the surface viscosity which represents the dissipation in plastic flow is estimated to be $\eta_p \sim 10^{-2}$ dyn-cm/sec (poise-cm). The detailed analysis is presented by Evans and Hochmuth (1976).[19] In addition to this semisolid behavior, Evans and LaCelle (1975)[22a] reported observations of creep in red cell membranes. The permanent deformation was produced by aspiration of membrane regions with a small micropipet for periods of time on the order of 5 to 20 min. The characteristic time limit for the domain of solid behavior is on the order of 10 min. From this data, the surface viscosity that represents the creep behavior would be on the order of 10 dyn-sec or surface poise. Thus, we see that there are large differences in coefficients of viscosity which represent dissipative processes in the different regimes of material behavior. We will give examples of these processes in Section V.

SECTION V

5.1 Biological Membrane Experiments

In Sections II and III, we introduced the principles and concepts of intensive deformation, rate of deformation, and force resultants that characterize a two-dimensional continuum such as a single molecular layer or thin membrane structure. Deformation and rate of deformation were derived from differential geometry, and force distribution, from mechanical equilibrium. In Section IV, we used thermodynamic methods to develop constitutive relations between these intensive variables, i.e., we obtained force resultants as functions of deformation and rate of deformation. These constitutive relations define material properties, such as elastic and viscous coefficients, that must be determined experimentally. Often, we must examine whether or not the material properties can be considered as constants, independent of force resultants, deformation, or rate of deformation. The outcome of an experiment may lead to more complicated relations. In this section, we will illustrate the application of the constitutive relations and methods of analysis to some selected membrane mechanical experiments. Clearly, the specific choices of experimental examples reflect our personal interests and are not intended to give the impression that all biological membranes exhibit the same behavior. However, the examples will represent the majority of material characteristics developed in Section IV. In addition, these types of experiments are of contemporary interest to many membrane bioscientists. We believe that the concepts will be useful in analysis of future experiments as new and novel procedures are developed.

Our method of approach will be to describe essential features of the experiment and then to set up a model for the material deformation and force equilibrium with a discussion of assumptions. Finally, we will introduce the appropriate constitutive relations in the analysis and correlate the analytical results within the experimental observation. This approach will provide values for intrinsic membrane elastic and viscous coefficients. The examples will be grouped by material behavior, i.e., elastic (reversible) and dissipative (irreversible). The particular topics in the elastic group include

1.   Area dilation vs. isotropic tension in human red blood cell membrane and sea urchin egg membrane-cortex
2.   Extensional (shear) deformation vs. the deviatoric (maximum shear) resultant in red blood cell membranes, with results for cells from various vertebrate animals
3.   Curvature change determined by bending and shear rigidity in osmotic swelling of human red cells
4.   Thermoelasticity and the temperature dependence of the area compressibility modulus and shear modulus of human red cell membrane
5.   Thermoelasticity and area compressibility of phospholipid multibilayer and water systems

The topics in the dissipative group are

1.   Time dependent extensional recovery and response (viscoelastic) of human red cell membrane
2.   Permanent plastic extension by relaxation and flow (viscoplastic) of human red cell membrane

In several analyses, numerical solutions are necessary because of algebraic complications. In these cases, we will describe the method of solution and give the numerical result obtained by digital computation.

As a preface to the presentation of mechanical experiments on biological membranes, it is important to note two of the major difficulties that confront the experimentalist. First of all, the size of many cells and regions of interest on the cell membrane surface are microscopic with dimensions that are on the order of 10^{-4} cm (1 μm). Dimensions at this level are difficult to observe accurately with the optical system because of diffraction. Hence, we can only reliably detect the changes in major dimensions, and precise measurement of the deformation field may be impossible. The second difficulty is that the forces required to deform (or even destroy) the membrane may be 10^{-6} dyn (10^{-9} g) or less for a region 10^{-4} cm across. Accurate measurement and control of such small forces is extremely difficult. With these difficulties in mind, we recognize that often it is only possible to obtain the first order behavior of the membrane material, and we expect that subtle but perhaps important higher order effects may not be apparent from the correlation of analysis with experimental observation.

5.2 Elastic Area Dilation Produced by Isotropic Tension

Conceptually, the simplest experiment to perform on a membrane is area dilation. This is accomplished by selecting a membrane-encapsulated system that is spherical or nearly spherical, e.g., an osmotically preswollen red blood cell, a large (single-walled) artificial lipid vesicle, or a sea urchin egg. If the interior volume of a spherical capsule is fixed, any mechanical experiment will produce area dilation where the dominant force resultant in the surface is isotropic tension. Examples of three such experiments are illustrated schematically in Figure 5.1: (1) micropipet aspiration, (2) compression between two flat surfaces, and (3) deflection of the surface by a rigid spherical particle. Figure 5.2 presents graphs of the fractional change in area of the surface for each type of experiment, with the assumption that the area dilation is uniform over the entire surface. In brief review, we recall that the micropipet aspiration technique was originally developed by Mitchison and Swann in 1954 for the study of sea urchin eggs.[61] Subsequently, Rand and Burton utilized the method with human red blood cells (1964).[71] They were followed by a host of researchers (e.g., LaCelle, 1969 and 1970;[49,50] Leblond, 1972; Waugh and Evans, 1976; Evans et al., 1976; Evans and Waugh, 1977),[24,27,49,50,52,92] who have been applying this technique to membrane mechanical studies with significant improvements in methods of data acquisition and analysis. The second type of area dilation experiment also has a long history. Cole first used compression of sea urchin eggs between flat surfaces in the 1930s.[9] Since then, Yoneda and Hiramoto have extensively used the method. The third experiment was developed by Hiramoto, but quantitative data on the force applied to the magnetic particle that displaces the surface are not available. Nonetheless, this experiment is ingenious and perhaps has potential applications to membrane systems other than sea urchin eggs. We will discuss each technique separately with examples of experimental data.

Consider first the micropipet aspiration of spherical or nearly spherical vesicles. This procedure produces isotropic tension in the surface. As the transmembrane pressure differential is increased, the surface approaches a spherical shape. Therefore, the isotropic tension becomes much greater than the maximum shear resultant and dominates the equations of equilibrium for forces supported by the membrane. Equation 3.5.9 can be used to demonstrate this feature for the direction normal to the surface. We decompose the principal tensions into isotropic and deviatoric parts. Thus,

$$\overline{T}\left(\frac{1}{R_m} + \frac{1}{R_\phi}\right) + T_s\left(\frac{1}{R_m}\diagup\frac{1}{R_\phi}\right)^{0} = \Delta P$$

(5.2.1)

as the principal curvatures become equal. Thus, the deviatoric part cannot contribute to the normal force equilibrium. The second equation of equilibrium (Equation 3.5.9) demonstrates that the isotropic tension must also be uniform over the spherical region, i.e.,

$$\frac{d\overline{T}}{ds} \to 0$$

(5.2.2)

where s is the curvilinear distance along the surface meridian. This is easily observed experimentally with osmotically preswollen red blood cells. The nearly spherical red cell is aspirated into the micropipet with suction pressures on the order of 10^3 dyn/cm^2 until the outer portion of the cell surface is made spherical. From this point on, large pressures on the order of 10^5 dyn/cm^2 are required to produce a very small displacement of the aspirated projection. Figure 5.3 shows videotape images of a single, preswollen red cell for two micropipet suction pressures on the order just described. Details of the experimental method are to be found in Evans et al. (1976)[27] and Evans and Waugh (1977).[25]

The micropipet suction pressure may be related to the membrane tension by satisfying the equation of equilibrium (5.2.1) both on the hemispherical cap in the pipet and on the spherical surface of the cell outside the pipet. It must be assumed that no friction

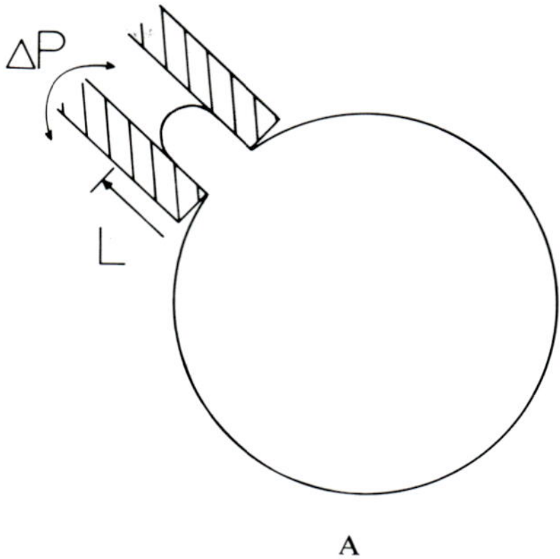

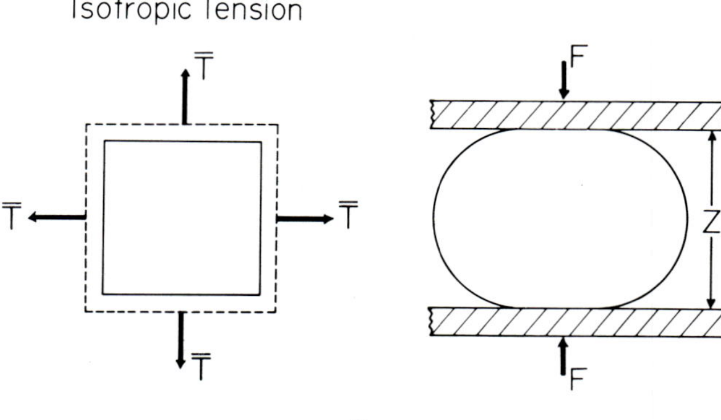

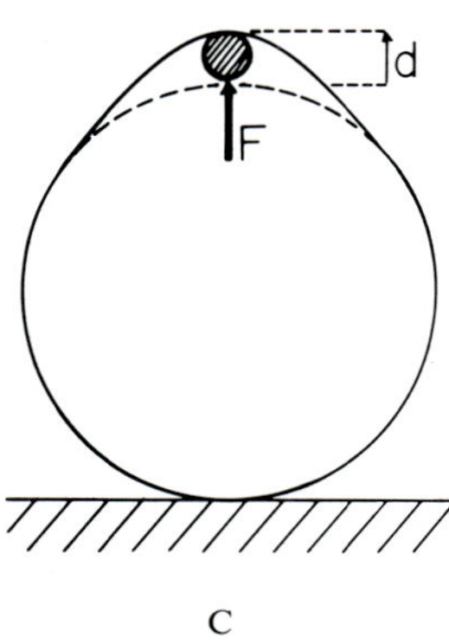

FIGURE 5.1. Three examples of mechanical experiments in which the dominant force resultant in the surface is isotropic tension. If the membrane encapsulated system is spherical or nearly spherical and the interior volume is fixed, then area dilation will be the primary deformation produced by mechanical experiments. The examples schematically illustrate (A) micropipet aspiration, (B) compression between two flat surfaces, and (C) deflection of the surface by a rigid spherical particle. In (A), ΔP is the pressure difference between the pipet interior and the outside medium and L is the length of the aspirated projection of the membrane. In (B), F is the force on each plate where the plates are separated from each other by a distance, z. In (C), F is the force which displaces the magnetic particle and the surface by a distance, d.

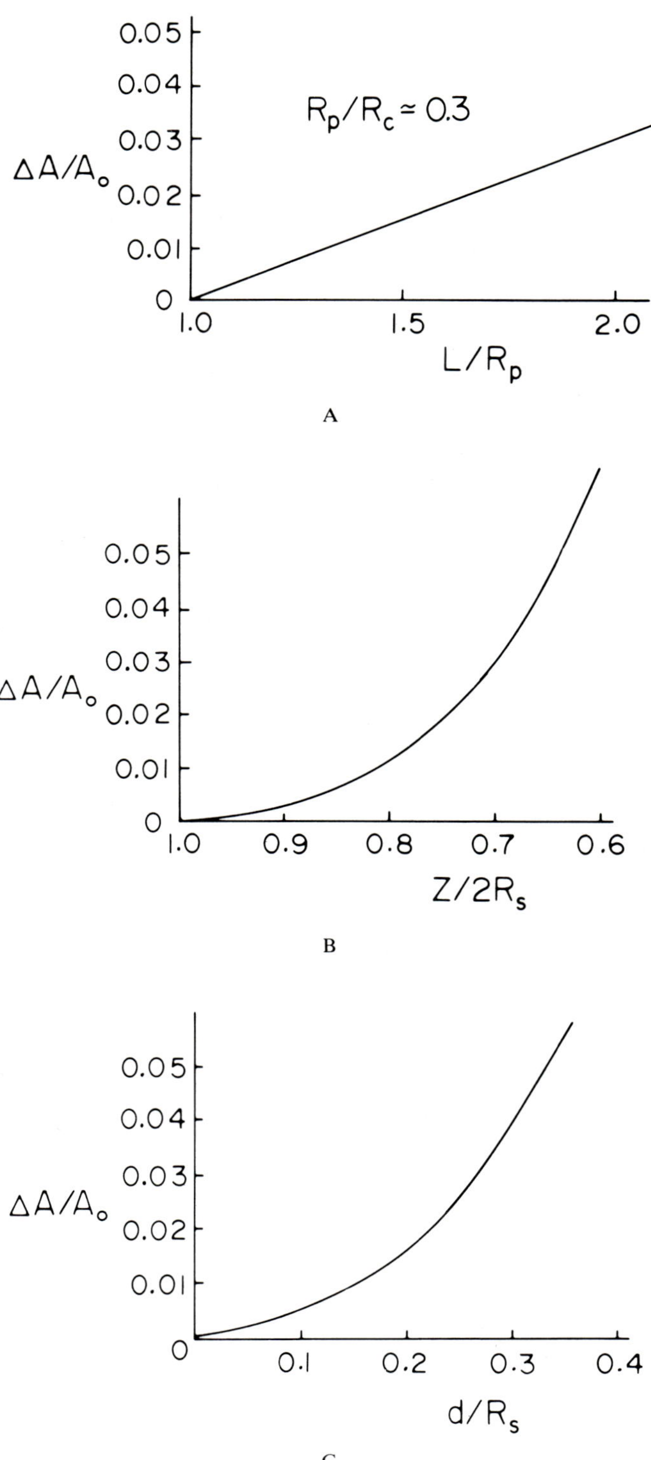

FIGURE 5.2. Plots of the fractional change in area, $\Delta A/A_0$ of the vesicular surfaces for the three experiments illustrated in Figure 5.1. We assume that the area dilation is uniform over the entire surface. (A) R_p is the pipet radius and R_c is the radius of the cell surface exterior to the pipet. (B) and (C) R_s is the radius of the spherical surface before deformation. Other symbols are defined as for Figure 5.1.

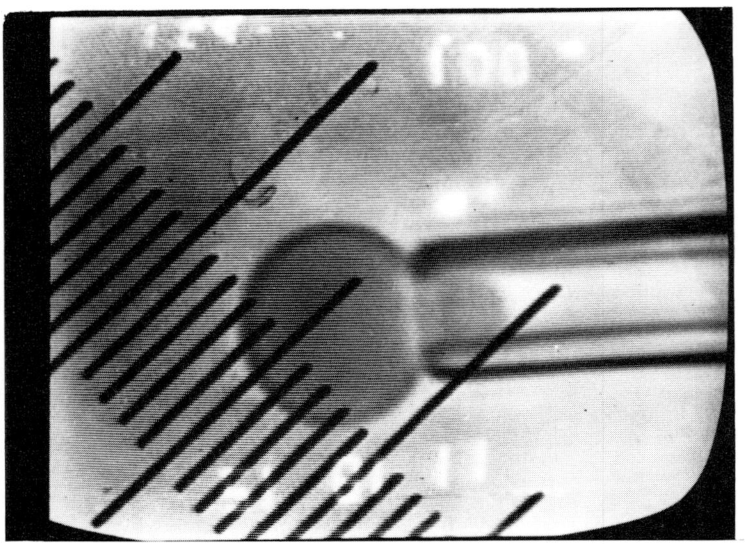

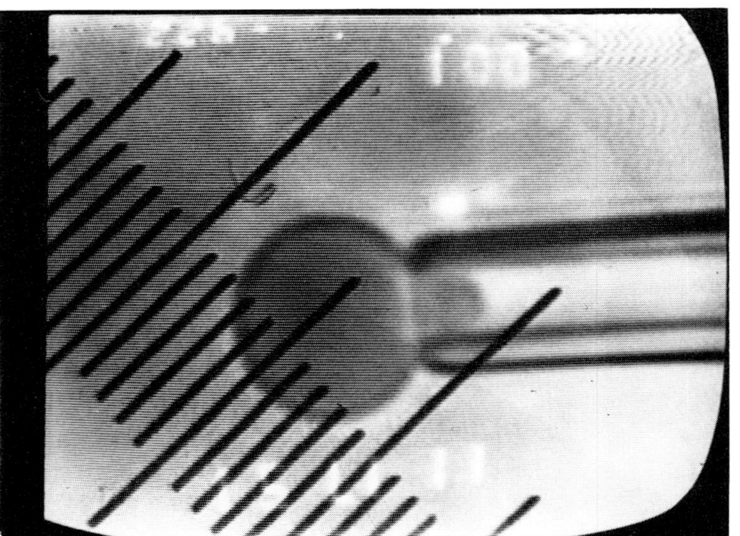

FIGURE 5.3.   Videotape images of a single, preswollen, red blood cell. The micro-pipet suction pressures that are applied to the cell are on the order of 10^4 dyn/cm^2 (above) and 10^5 dyn/cm^2 (below). The micrometer scale that appears in the photographs is calibrated at 0.9 μm per division. Equation 5.2.3 may be used to calculate the membrane force resultant from the suction pressure. The results are shown in Figure 5.4. For this order of magnitude change in suction pressure, the cell projection only increases about a half micron in length. Larger suction pressures lyse the red cell. (Photographs from Evans, E. A. and Hochmuth, R. M., *Current Topics in Membranes and Transport,* Vol. 10, Kleinzeller, A. and Bronner, F., Eds., Academic Press, New York, 1978, 1. With permission.)

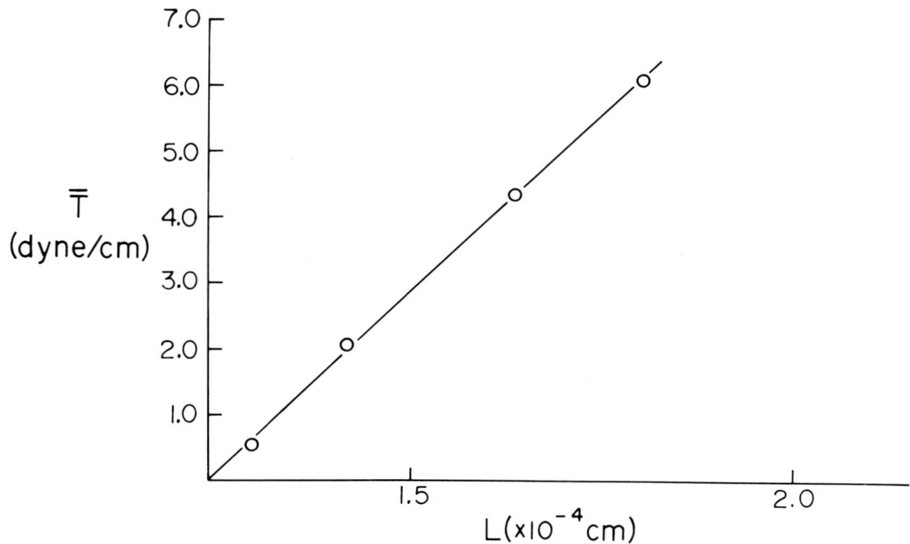

FIGURE 5.4. Isotropic tension, $\overline{T}$, is calculated from Equation 5.2.3 vs. length of the projection, L, for a single cell experiment. This data may be combined with Equation 5.2.4 to obtain the relationship between isotropic tension and the fractional change in membrane surface area (see Figure 5.5).

occurs at the pipet wall and that negligible flow of water occurs between the pipet wall and cell surface from outside to inside the pipet. These assumptions have been challenged experimentally and appear to be valid (Waugh, 1977).[91] The result of the simple analysis is that the negative pressure is proportional to the isotropic tension,

$$\Delta P = 2\overline{T} \left(\frac{1}{R_p} - \frac{1}{R_c} \right)$$

(5.2.3)

where R_p and R_c are the radii of the micropipet and outer portion of the cell, respectively. Movement of the lead edge of the cell projection in the pipet is small, and essentially no detectable change in outer cell dimension occurs. Figure 5.4 shows the isotropic tension as calculated with Equation 5.2.3 and the suction pressure vs. length of the projection which is observed in a typical cell experiment.

Displacement of the cell projection into the pipet is actually due to both an increase in membrane surface area and a decrease in cell volume. The movement is reversible and therefore represents conservative, thermodynamic processes. It has been demonstrated that the reversible volume change is due to water fluxes across the cell membrane (Evans and Waugh, 1977)[25] and that the water movement can be essentially eliminated by artificially "loading" the red cells with high salt concentrations before swelling. Consequently, we can treat the cell volume as constant and simply relate the small change in surface area, ΔA, of the cell to the movement, ΔL, of the lead edge of the cell projection. This is approximated by the Taylor series expansion to first order as

$$\Delta A \simeq 2\pi \left[R_p \cdot \Delta L \cdot \left(1 - \frac{R_p}{R_c} \right) \right]$$

(5.2.4)

provided that the projection length, L, is initially greater than or equal to the pipet radius.

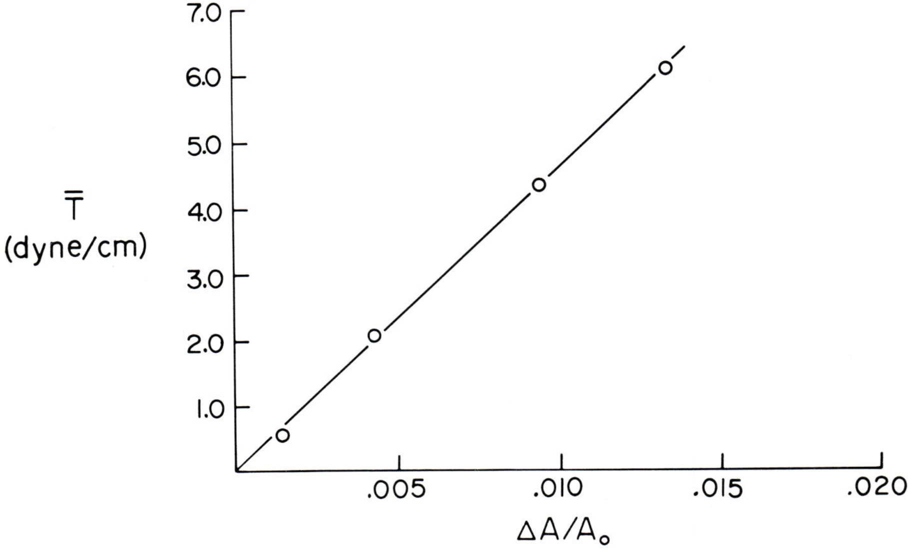

FIGURE 5.5.   The relationship of isotropic tension to fractional increase in area as calculated from the data in Figure 5.4. The data exhibit linear behavior that can be represented by the first order constitutive equation, Equation 4.4.12. With this equation and the results above, we obtain the area compressibility modulus, K, at constant temperature from the slope of the regression line. (From Evans, E. A. and Waugh, R., *Biophys. J.*, 20, 307, 1977. With permission.)

If we combine the length vs. isotropic tension data (Figure 5.4) with Equation 5.2.4, the relationship of isotropic tension to fractional change in area is obtained (Figure 5.5). The data exhibit linear behavior that can be represented by the first order constitutive equation (4.4.12) for isotropic tension vs. fractional change in area, α,

$$\bar{T} = K \cdot \alpha \qquad (5.2.5)$$

where $\alpha = \Delta A/A_0$. (Note that for the red cell membrane only small area changes on the order of 2 to 4% can be produced above which lysis or rupture of the cell membrane occurs.) From Equation 5.2.5 and the data of Figure 5.5, we directly obtain the elastic area compressibility modulus, K, at constant temperature. The value for this intrinsic membrane property is found to be about 450 dyn/cm at 25°C. Later on, we will show that the surface elastic shear modulus, μ, is four to five orders of magnitude smaller than the area modulus.

Our next discussion deals with the compression of sea urchin eggs between two flat surfaces. Unlike the mammalian red blood cell which normally has a liquid cytoplasm, the sea urchin egg interior is more complicated. Indeed, its membrane is closely associated with a cortical layer or shell which probably contributes to its mechanical rigidity. It is still appropriate to model the constitutive behavior of the complicated membrane-cortex by the anisotropic (two-dimensional) surface elastic relations that were developed in Section IV. We make the assumption that the membrane-cortex is responsible for the elastic solid behavior of the unfertilized sea urchin egg when it is exposed to external forces over long time periods (e.g., greater than 1 min in duration). [For such time periods, it appears that the endoplasm is essentially viscoplastic and cannot support stresses other than hydrostatic pressure (Hiramoto, 1970).][42] Since the thickness of the membrane and cortex of the egg are less than 10% of the radius of curvature of the egg, we will treat the force equilibrium with the equations of thin shell theory. In other words, the forces applied to the egg are primarily balanced by the force resultants that act in the plane of the shell, i.e., by tensions. The bending moments and other stress resultants produce second order contributions, but may be important in situations where curvature changes are appreciable.

We will assume that the material composition of the membrane-cortex is isotropic in the surface plane. In other words, the material properties do not depend on the choice of surface directions. However, implicit in this assumption is the provision for material properties that may be different in the direction normal to the surface, i.e., anisotropic in the third dimension. For instance, the membrane-cortex may be a laminated structure like an onion or layered fabric which is isotropic in a plane but anisotropic with respect to the direction normal to that plane. Most biological membrane systems exhibit such stratified complexity in their ultrastructure.

Even though the concept of stress as a continuous function of thickness through the membrane-cortex shell is invalid, it has been used to represent the material constitutive behavior by Hiramoto (1963).[40] However, there are important differences between a three-dimensionally isotropic material and an anisotropic membrane material which is only isotropic in its surface plane. We will demonstrate the differences by converting the linear elastic behavior of a three-dimensionally (bulk) isotropic material into surface elastic relations. For the three-dimensionally isotropic membrane, the principal stresses that act normal to the edges of the material element would be given by the force resultants (tensions) divided by the shell thickness, t_m,

$$\sigma_1 = T_1 / t_m$$

$$\sigma_2 = T_2 / t_m$$

For small deformations of a bulk isotropic material, the first order elastic constitutive relation is expressed in terms of the Young's elastic modulus, E, and the Poisson's ratio, v (see Landau and Lifshitz, *Theory of Elasticity*).[51] The constitutive equations would be

$$\sigma_1 = \frac{\overline{T}_0}{t_m} + \frac{E}{(1 - v^2)} (\epsilon_1 + v\epsilon_2)$$

$$\sigma_2 = \frac{\overline{T}_0}{t_m} + \frac{E}{(1 - v^2)} (\epsilon_2 + v\epsilon_1)$$

where ϵ_1 and ϵ_2 are the surface strains along the principal directions defined by subscripts, 1 and 2. Based on these relations, the surface elastic properties of the membrane-cortex can be obtained by analogy to the constitutive relations (4.4.12), i.e.,

$$K = \frac{t_m E}{2(1 - v)}$$

$$\mu = \frac{t_m E}{2(1 + v)}$$

for the corresponding bulk, isotropic material. [We have taken advantage of the small strain approximation that gives $\alpha \cong \epsilon_1 + \epsilon_2$ and $\tilde{\epsilon} \cong (\epsilon_1 - \epsilon_2)/2$.] Most materials are essentially volumetrically incompressible, which implies that Poisson's ratio is 0.5. In this case, the elastic constants are in the ratio of three to one, i.e.,

$$K \equiv 3\mu$$

Thus, a major difficulty is apparent: the surface elastic moduli are not mutually independent as they would be for an anisotropic membrane material. In general, the only way to evaluate elastic constitutive relations is to compare the static material deformations to the distribution of tensions.

If the shape of the membrane surface and the magnitudes of applied forces are known, the distribution of tensions is specified. Presumably, the initial shape of the surface is known *a priori* (e.g., a sphere); therefore, the material deformation or strain distribution can be calculated from geometry alone. Consequently, the relationship between force resultants, tensions T_1 and T_2, and the deformation is determined. This yields the material constitutive relation. On the other hand, if we know what the material elastic behavior is and if we are given the applied forces, we can predict the static shape.

In the sea urchin egg experiment, for example, Yoneda (1964)[96] demonstrated that the surface which was not in contact with the compression plates exhibited constant total curvature, i.e.,

$$\left(\frac{1}{R_m} + \frac{1}{R_\phi} \right) = \text{constant}$$

for the toroidal rim of the cell. Thus, in the compression experiment, the isotropic tension, $\overline{T}$, is significantly greater than the shear resultant contribution, T_s. Furthermore, the isotropic tension must be essentially independent of position along the contour,

$$\frac{d\overline{T}}{ds} \simeq 0$$

as with the micropipet experiment discussed previously.

If the interior of the cell is incompressible, i.e., the cell has a constant volume, the surface area must increase as the cell is compressed. Figure 5.6 shows the predicted cross section of the cell. The contour is calculated by a digital computer for the case where the sum of the principal curvatures is equal to a constant and the volume is equal to that of the sphere. The following transformation is useful in performing such computations (refer to Figure 3.5 and Equations 3.5.10 in Section III):

$$\frac{1}{R_\phi} = \frac{\sin \theta}{r} = \frac{u}{r}$$

$$\frac{1}{R_m} = \frac{d\theta}{ds} = \frac{du}{dr}$$

where

$$u \equiv \sin \theta$$

and $-\cos\theta = dr/ds$ from differential geometry. Therefore, the total curvature is a constant given by

$$\frac{1}{R_m} + \frac{1}{R_\phi} = \frac{du}{dr} + \frac{u}{r} = c \tag{5.2.6}$$

This relation can be integrated directly,

$$r \cdot u = \frac{c}{2} (r^2 - R_i^2)$$

Since the conditions on the contour of the sea urchin egg surface are that

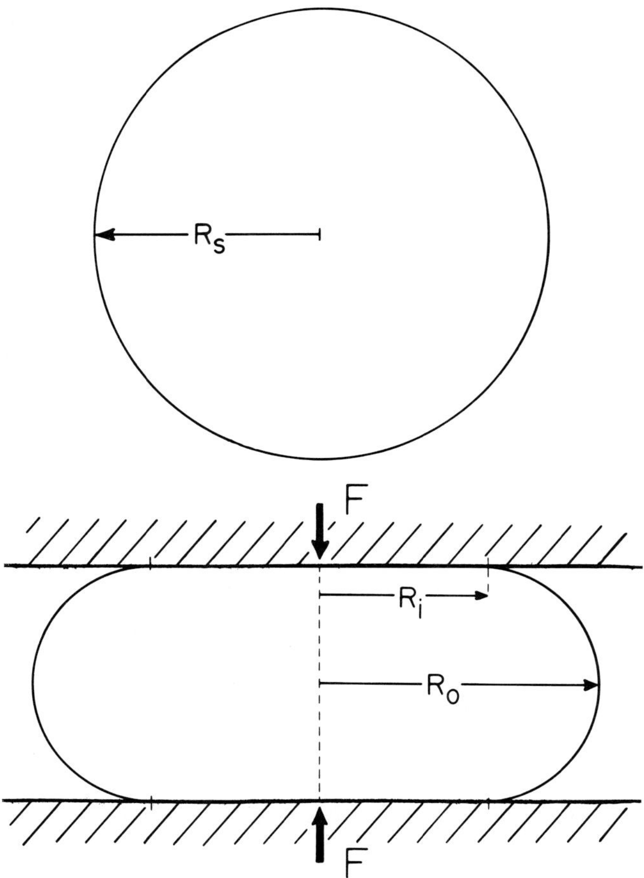

FIGURE 5.6. Cross sections of a spherical membrane capsule and the shape that results from compression between flat, parallel plates at constant volume. This compression can only be accomplished if the surface area increases.

$$u = 1 \text{ at } r = R_0 \quad \text{(the equator)}$$

and

$$u = 0 \text{ at } r = R_i \quad \text{(the radius of the contact region)}$$

hence,

$$\frac{c}{2} = R_0 / (R_0^2 - R_i^2)$$

Consequently, the compressed shape of the originally spherical egg can be determined with the constraint that the interior contents maintain constant volume (Figure 5.6). Figure 5.2B is a plot of the fractional increase in area of the sphere at constant volume as a function of the distance between the compression plates.

Next, we calculate the total force, F, on each plate from the axial force balance. This is determined by the hydrostatic pressure, ΔP, that acts on the contact surface area, πR_i^2,

$$F = \Delta P \circ \pi R_i^2$$

The balance of forces at the equator of the toroidal rim is given by

$$2\pi R_0 \cdot \overline{T} = \Delta P \cdot \pi (R_0^2 - R_i^2)$$

in terms of the equatorial radius, R_0, and the isotropic tension, $\overline{T}$. Therefore, the hydrostatic pressure difference may be eliminated. The result is the relation between the intensive force resultant and the extensive force,

$$F = \frac{2\pi \cdot R_0 \cdot R_i^2 \cdot \overline{T}}{(R_0^2 - R_i^2)} \tag{5.2.7}$$

Equation 5.2.7 can be rearranged into dimensionless form with the initial sphere radius, R_s,

$$\frac{F}{\overline{T} \circ R_s} = \frac{2\pi R_0 \cdot R_i^2}{R_s (R_0^2 - R_i^2)} = f\left(\frac{z}{2 R_s}\right)$$

This is a simple function of the deformed geometry of the surface that is uniquely determined by the distance between the plates, z. If the constitutive relation for the isotropic tension is independent of area change, the dimensionless force is given solely by the function, f,

$$\frac{F}{\overline{T}_0 \cdot R_s} = f\left(\frac{z}{2 R_s}\right) \tag{5.2.8}$$

Equation 5.2.8 is the result expected for immiscible drops of liquid. The surface may exhibit isotropic tension proportional to change in area; then the total force is given by

$$\frac{F}{K \circ R_s} = f\left(\frac{z}{2 R_s}\right) \cdot \alpha \tag{5.2.9}$$

since $\overline{T} = K\alpha$, e.g., like a red cell membrane or single-walled, phospholipid vesicle. For a surface which has both an initial isotropic tension and an area elastic behavior, the total force is expressed by the sum,

$$F = R_s \cdot \left[\overline{T}_0 + K \cdot \alpha\right] \cdot f\left(\frac{z}{2 R_s}\right) \tag{5.2.10}$$

Figure 5.7 shows the results calculated by computer for Equations 5.2.8 and 5.2.9. Also shown in Figure 5.7 is the correlation of Equation 5.2.10 with the published data of Hiramoto (1963).[41] This correlation gives values for an initial isotropic tension of 0.12 dyn/cm and an area elastic modulus of 0.23 dyn/cm for an unfertilized sea urchin egg, e.g.,

$$\overline{T}_0 \simeq 0.12 \text{ dyn/cm}$$

$$K \simeq 0.23 \text{ dyn/cm}$$

Hiramoto analyzed the sea urchin egg experiments with the constitutive relation for a bulk isotropic, elastic material with a Young's modulus, E, and Poisson's ratio of 0.5. His results gave

$$\overline{T}_0 = 0.03 \text{ dyn/cm}$$

$$t_m \cdot E = 0.68 \text{ dyn/cm}$$

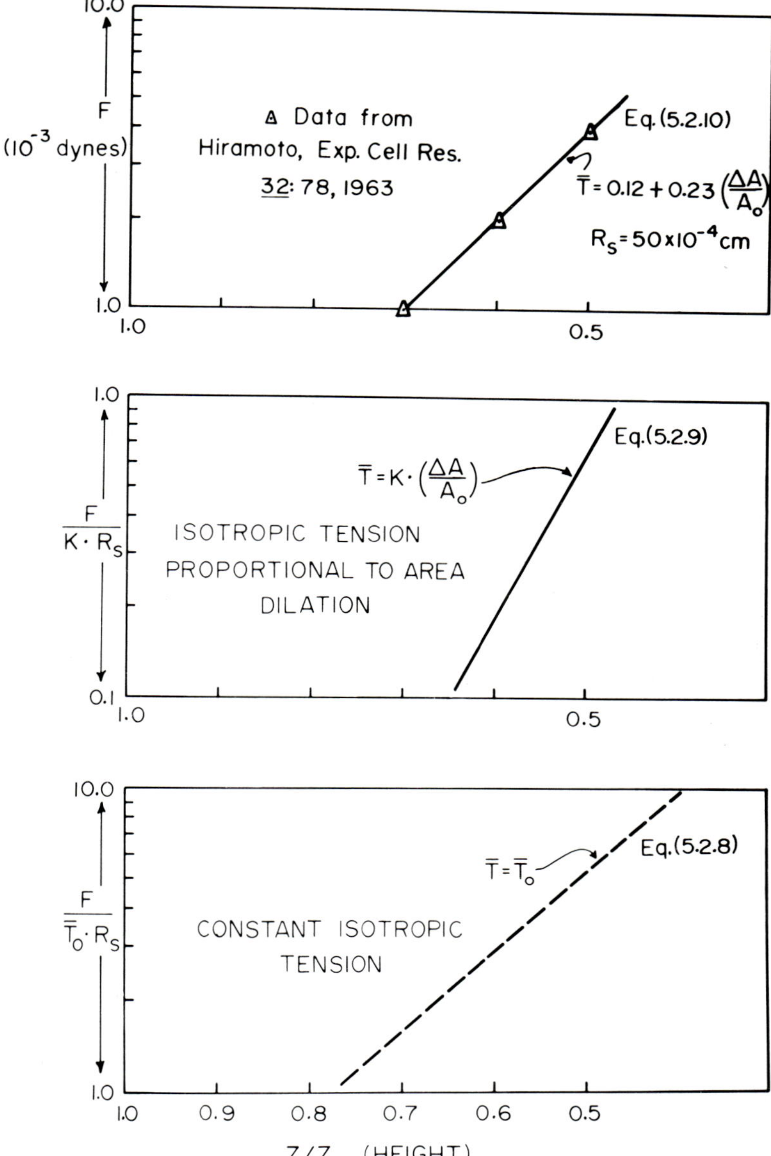

FIGURE 5.7. Theoretical results for compression force vs. the distance between plates as calculated by digital computation. The upper curve is the correlation with data from Hiramoto's work (1963).[40] The middle plot is the semilog behavior that is predicted for elastic membranes such as phospholipid vesicles. The lower plot is the semilog behavior that is predicted for a constant isotropic tension, e.g., a liquid drop interface.

where the thickness, t_m, of the material is implicit in the material parameters. Hiramoto derived these values from his experiment by calculating the principal force resultants (tensions), T_m and T_o, at the equator. The ratio of the principal tensions at the equator is given by

$$\frac{T_\phi}{T_m} = \frac{2}{(1 - \tilde{A})} - \frac{R_0}{R_m}$$

$$(5.2.11)$$

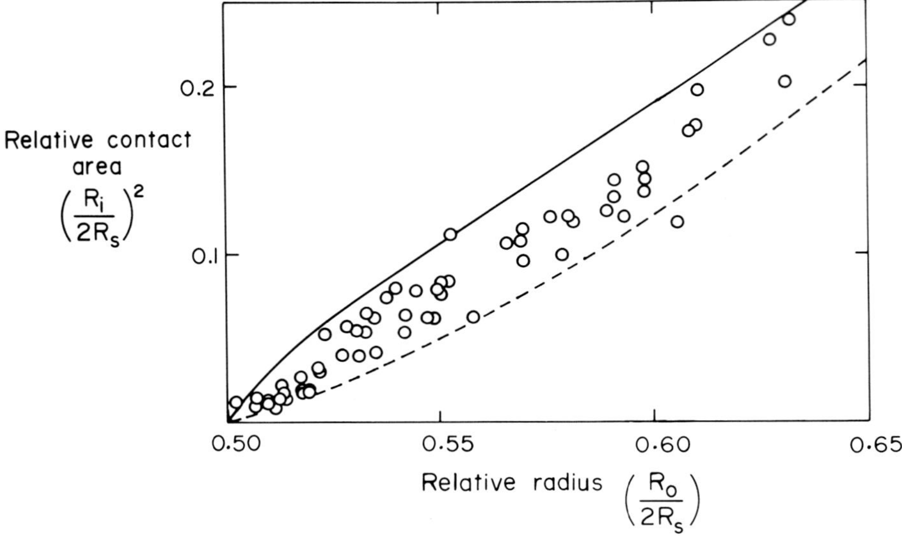

FIGURE 5.8.   Experimental and theoretical values for the area of contact of membrane with the compression plates that were obtained: from photographic measurements (solid line); by observing Brownian motion of small marker particles in the extracellular fluid (open circles); and from the computer calculation of the surface contour for uniform total curvature and constant cell volume (dashed line). (The photographic and Brownian motion data are from Yoneda, M., *J. Exp. Biol.*, 41, 893, 1964.)

where T_m is the meridional tension and T_ϕ is the latitudinal tension; $\tilde{A}$ is the ratio of the contact area, πR_i^2, to the area of the equatorial plane, πR_0^2. The equatorial radius is R_0, and the radius of curvature of the meridian at the equator is R_m. The above relationship is independent of the material constitutive behavior; it is determined by force equilibrium. According to Hiramoto, the tension ratio is the order of 2:1 or 3:1 over the majority of the experimental range even for small applied forces. The observations of Yoneda, on the other hand, demonstrated that the force resultants appear to be isotropic in the surface plane (because the sum of the curvatures is constant). The discrepancy seems to result from an overestimation of the area in contact with the compression plate, πR_i^2. An overestimate produces an apparent difference between the principal tensions in accordance with Equation 5.2.11. Data taken from Yoneda (1964)[96] provide the contact area as measured photographically through the optical system (solid line) and as measured with the use of Brownian motion of small marker particles in the extracellular fluid (open circles). The latter were taken as the region where no Brownian motion was observed. The dashed line in Figure 5.8 shows the contact area predicted from the computer calculation of surface contour with isotropic tension. This forms a lower bound for the Brownian motion data. Here, the ratio of principal tensions is identically one, i.e.,

$$T_m = T_\phi = \overline{T}$$

If any material is associated with the cell surface that is not structurally active or if the location of the action of the force resultants is not coincident with the outer surface, then the results shown in Figure 5.8 are to be expected. Indeed, in most membrane systems, such effects are anticipated from the ultrastructure or microanatomical data, i.e., the surface rigidity is provided by subsurface material. Clearly, it is very difficult to evaluate elastic constitutive relations unless detailed information exists about material deformations. In this case, the measurement of contact area is uncer-

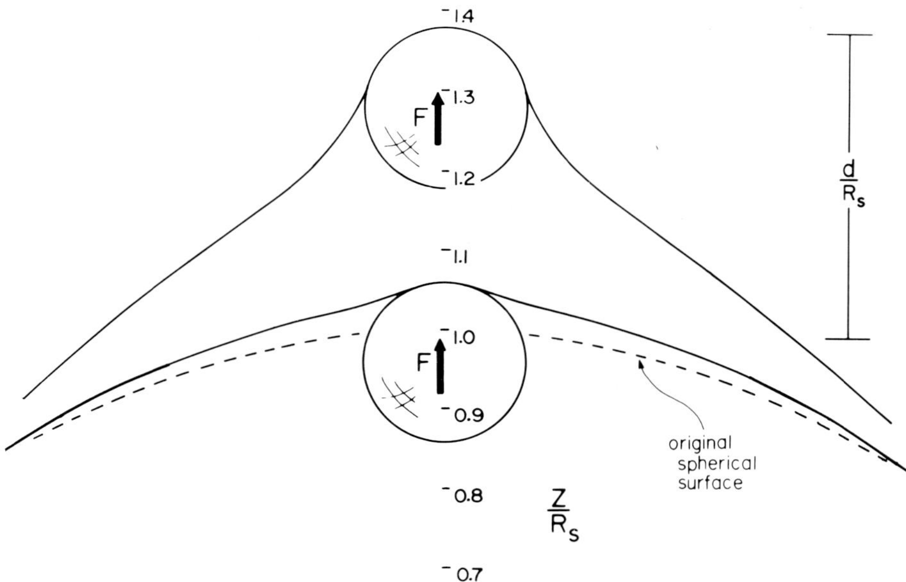

FIGURE 5.9. An illustration of surface contours for two different particle force levels. This figure is based on computer results that model the experiment performed by Hiramoto on sea urchin egg membrane-cortex (no force data have been published). The calculations correspond to a ratio of particle radius to sphere radius of 0.1 and uniform isotropic tension.

tain. Therefore, only the cell contour can provide information about the distribution of tensions.

Hiramoto performed another interesting and clever experiment on sea urchin egg membrane-cortex. This experiment involved the impingement of a solid particle against the egg cortex; the force was produced by a magnetic field. No data have been published on the particle forces. However, we will provide some computer results for surface contours and particle force vs. displacements. We assume that the cell is not free to move because of attachment to a rigid support. Again, we assume in advance that the force resultants or tensions in the membrane-cortex are dominated by isotropic tension because of the small displacement of the spherical surface. Figure 5.9 is an illustration of surface contours for two different particle force levels, where the ratio of particle radius to sphere radius is 0.1. Figure 5.10 is a plot of the magnitude of the dimensionless force vs. displacement that is expected for a constant isotropic tension (e.g., a liquid drop). Figure 5.11 is a plot of dimensionless force vs. displacement that is expected for an area elastic contribution (e.g., a phospholipid vesicle). The results in Figures 5.10 and 5.11 may be combined to correlate experimental data for force vs. displacement if the elastic surface has an initial isotropic tension, $\bar{T}_0$. Deviations of the surface contour from that shown in Figure 5.9 in the vicinity of the particle would provide critical information for determining the contributions of shear elasticity (given by the shear modulus, μ) and bending moments. For instance, in the region close to the spherical particle, shear rigidity would resist the "necking" of the surface below the particle. Perhaps such an investigation of surface geometry in this experiment could provide additional elastic moduli. However, because of the relatively large shear deformations next to the particle, force relaxation and plastic flow of material may occur. These will be observed as permanent surface deformations.

To conclude this section on elastic area changes produced by isotropic tension, we will comment on the large difference between the elastic area compressibility modulus

FIGURE 5.10.   A plot of dimensionless force vs. particle displacement calculated for a constant isotropic tension such as would be present in a liquid drop surface.

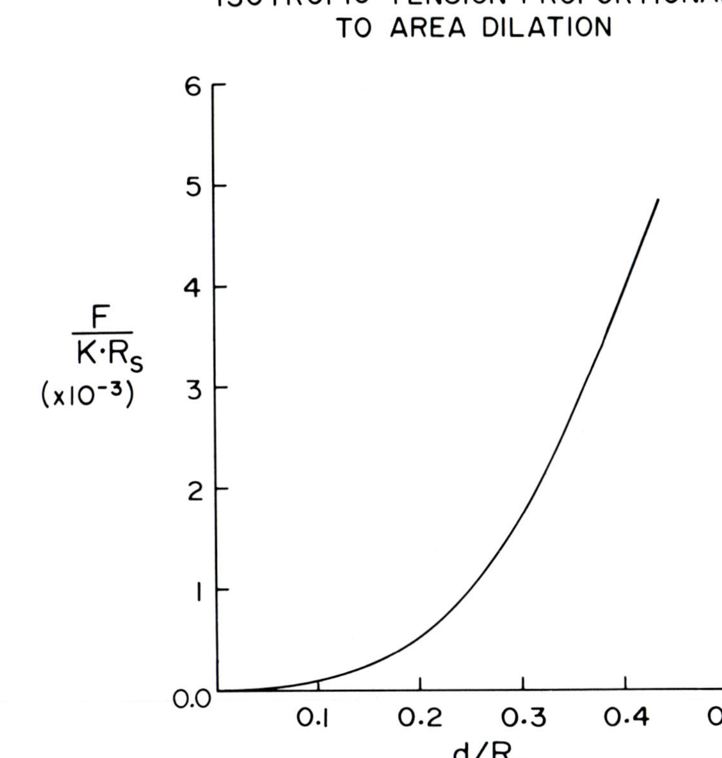

FIGURE 5.11.   A plot of dimensionless force vs. particle displacement calculated for an area elastic contribution such as would be expected for a phospholipid vesicle.

of red cell membranes, $\sim 10^2$dyn/cm, and that estimated for the sea urchin egg surface, $\sim 10^{-1}$dyn/cm. The large value for the red cell membrane is on the same order as that expected for lipid bilayer systems (Evans and Waugh, 1977),[24] where the lipid amphiphiles form a highly condensed, two-dimensional liquid, which is held together primarily by hydrophobic forces. If the surface is smooth like the red cell, then area changes of the spherical shape will produce changes in area per molecule in the surface. Thus, there is a large area compressibility modulus. On the other hand, a ruffled surface or a surface studded with submicroscopic villi, like that of the sea urchin egg, can accommodate major changes in surface area on a scale which is much larger that the surface villi. The surface is expanded by unwrinkling. The expansion of this carpet-like material would not produce intensive changes in surface composition and would not necessarily require large membrane tensions. In addition, the constant isotropic tension present in the surface might be a force resultant which is associated with the maintenance of the microvillar projections.

5.3 Elastic Extensional Deformation Produced by Membrane Shear

As discussed in Sections III and IV, membrane shear forces are associated with the difference or deviation between principal membrane force resultants (tensions). In simple uniaxial tension, the maximum shear resultant in the membrane surface is one half of the tension; the other half equals the isotropic tension. Uniaxial tension produces extensional deformation along one axis by membrane shear. As shown previously, general deformations include both shear deformation and dilation or condensation of surface area. Consequently, in order to investigate the constitutive relation between membrane shear deformation and surface shear force resultant, we require experiments that can produce deviatoric force resultants and membrane extension in a manner similar to uniaxial extension.

In the previous section, we outlined the experimental evidence for the isotropic elastic behavior of membranes. We observed that swollen red cell membranes and lipid bilayer vesicles greatly resist area changes and that isotropic tensions on the order of dynes per centimeter produce only a percent change in surface area. In contrast to this behavior, membrane force resultants or tensions that are one hundred times smaller can produce extensional deformations of several hundred percent in flaccid (unswollen) red cells. For example, Hochmuth et al. (1973)[44] used the flow of extracellular fluid to deform red cell disks that were point-attached to a glass substrate. The forces on the exposed cell surface were on the order of 10^{-6} dyn. For dimensions on the order of 1 μm, this gives force resultants of 10^{-2} dyn/cm. Figure 5.12 shows a single cell that has been deformed by shear flow of the extracellular fluid. The cell deformation is elastic for the extensions shown here. Even though the overall cell extension is only 1.4 (40% greater than the original cell diameter), the local membrane extension is very large in the vicinity of the attachment location. Since the membrane tensions are four orders of magnitude smaller than the area compressibility modulus, we can consider that the local surface area of a membrane region remains constant. In other words, the extension, λ_1, of a material element of the membrane is accompanied by a constriction, λ_2, of the element which is inversely proportional to the extension, $\lambda_2 \equiv 1/\lambda_1$. To illustrate the nonuniform extension of membrane surface, we consider the extension of a circular disk with a single-point attachment as a model for the uniaxial extension of the red cell. Figure 5.13 shows a typical case with the force along any vertical ''cut'' in the membrane disk equal to the traction or shear force of a fluid flowing over the portion of the disk downstream of the cut (free-body force balance). The constitutive behavior which is assumed for the disk is that given by Equation 4.4.12 for the deviatoric or shear resultant. Figure 5.14 shows the nonuniform extension of material from

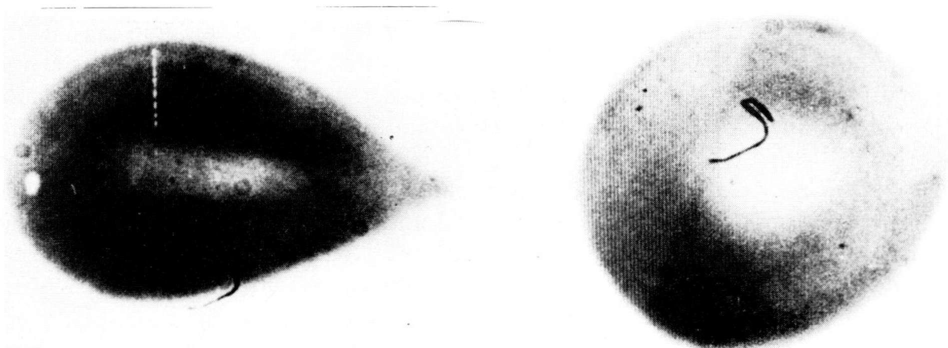

FIGURE 5.12.   Photographs of videotape images of a point-attached red blood cell. On the right is an undeformed, biconcave red blood cell; on the left is the same cell deformed by shear flow of extracellular fluid. The diameter of the undeformed cell is about 8 μm. Cell deformation is elastic for the extensions shown here. (From Evans, E. A. and Hochmuth, R. M., *J. Membr. Biol.*, 30, 351, 1977. With permission.)

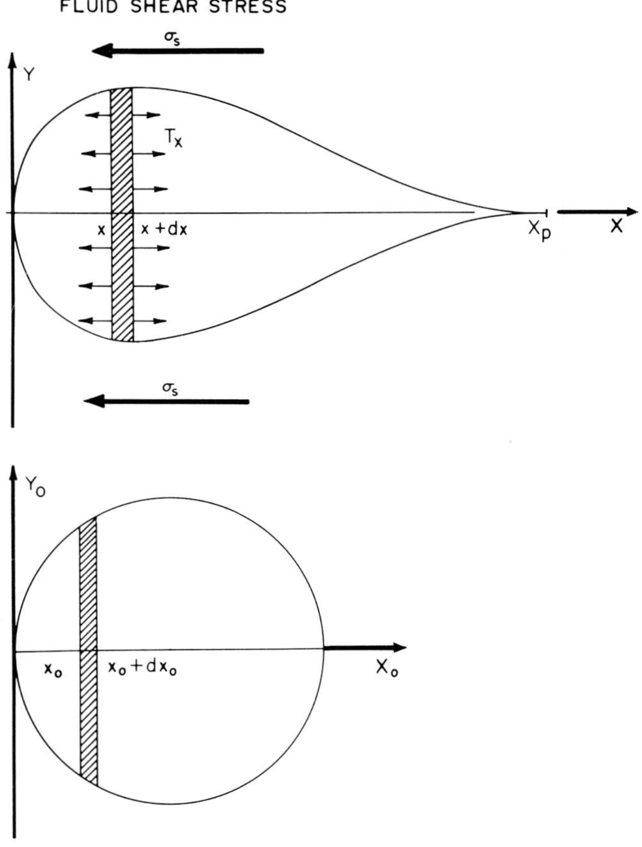

FIGURE 5.13.   Schematic illustration of the uniaxial extension at constant area of a membrane element defined by $(x_0, x_0 + dx_0)$ into the element $(x, x + dx)$ under the action of a uniform fluid shear stress, σ_s. x_p is the location of the attachment point for the disk model. (From Evans, E.A., *Biophys. J.*, 13, 926, 1973. With permission.)

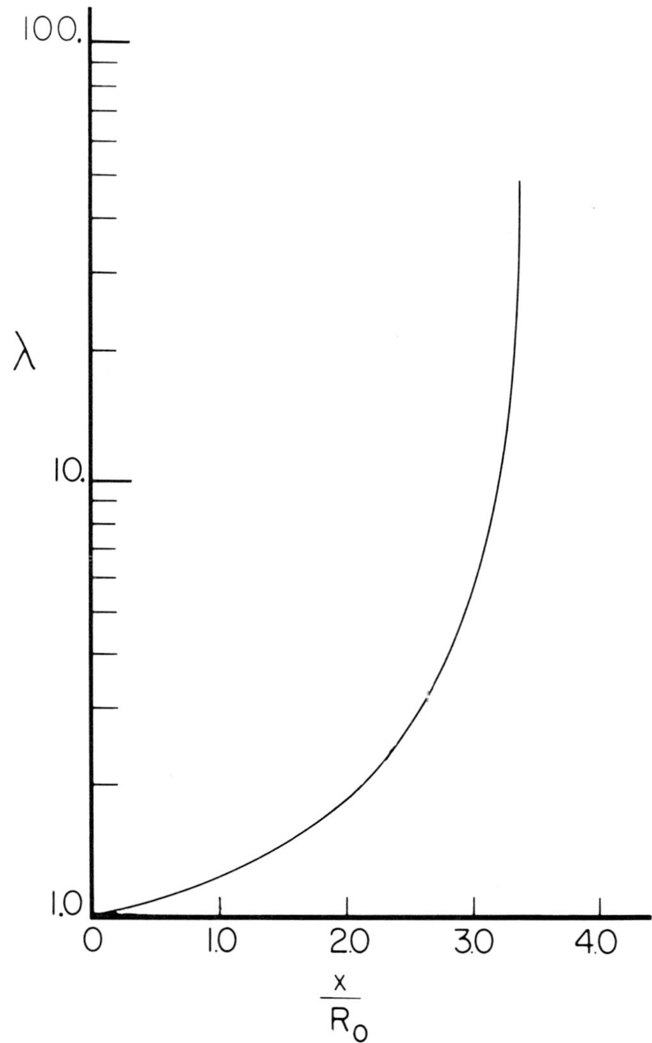

FIGURE 5.14.   The principal extension ratio, λ, plotted against the coordinate, x, of the deformed disk, divided by the initial radius, R_o. The extension ratio increases asymptotically in the region of the point attachment. The overall extension of the disk is 1.67 in this example. (From Evans, E. A., *Biophys. J.*, 13, 941, 1973. With permission.)

the rear of the disk to the point of attachment (details for the results in Figures 5.13 and 5.14 are given by Evans, 1973).[16] Hochmuth's experiment demonstrates the hyperelastic behavior of a red cell membrane in extension when the surface area has not been forced to increase. This hyperelasticity permits the flaccid red cell to easily flow through small capillaries in the microcirculation and through other small apertures, for instance, in the spleen and bone marrow. On the other hand, passage of a spherical lipid bilayer vesicle would require a channel diameter at least as large as the vesicle diameter. Otherwise, the sphere would have to increase its surface area in order to make it into the channel.

With the disk as a model for the elastic deformation of the point-attached red cell in a shear flow, it is possible to calculate the membrane elastic shear modulus, μ, by comparing the theoretical deformation of a disk to the actual deformation of a red

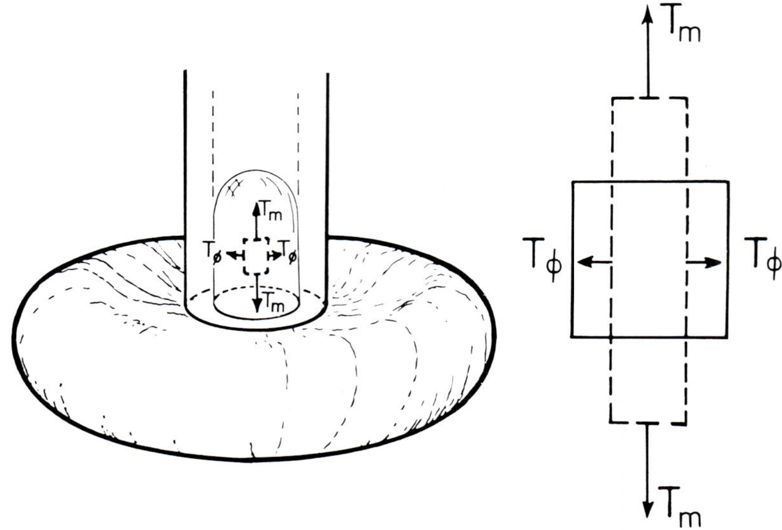

FIGURE 5.15.   Illustration of membrane extension and deviation between force resultants produced by micropipet aspiration of a flaccid red blood cell. The principal force resultants (tensions) can be decomposed into isotropic and deviatoric (shear) contributions.

cell. The data of Hochmuth et al. (1973)[44] give a value on the order of 10^{-2} dyn/cm for the surface shear modulus. However, there are several uncertainties which limit the accuracy of such a comparison: the red cell membrane is a curved surface, the extension ratio in the surface increases tremendously near the attachment location (indeed, plastic deformation eventually occurs at these locations), and it is difficult to accurately determine the force on the cell surface which is due to the fluid shear flow. A particular method of micropipet aspiration appears to be preferable, although it lacks the simplicity and convenience of the flow channel experiment. Various techniques of micropipet aspiration have been used to assess whole red cell "deformability" (e.g., Rand and Burton, 1964; LaCelle, 1969 and 1970).[49,50,71] The method that we will describe was originally used by Leblond (1973).[52] Unlike the other micropipet techniques, this procedure can be analyzed to determine the intensive state of deformation in the membrane (Evans, 1973).[16]

The micropipet technique that we will consider is illustrated schematically in Figure 5.15. We will demonstrate (Evans, 1973)[16] that the membrane tensions produced by a small caliber pipet are concentrated in and near the pipet mouth when the pipet is applied to a relatively flat region of the cell membrane. We will show that these tensions drop off inversely as the square of the distance from the pipet center, normalized by the pipet radius. Thus, the pipet experiment shown here investigates the intrinsic membrane material behavior local to the pipet tip, independent of the extrinsic cell shape. Aspiration of the cell to such an extent that the membrane tensions at the cell rim or periphery cannot be neglected results in curvature changes at the cell rim. When the cell begins to change curvature at the rim, the pressure in the cell cytoplasm is no longer negligible, and isotropic tension in the membrane begins to dominate shear. For such excessive aspiration of the cell, the experiment will not provide a valid measurement of membrane shear elasticity.

The analysis of the micropipet aspiration experiment involves two independent procedures (as does any mechanical experiment): first, we must determine the membrane force resultants which act to balance the applied forces, i.e., mechanical equilibrium;

second, we analyze the deformation of the membrane surface to determine the extension ratios of the material as a function of location. The results establish the relationship of membrane shear force resultant to membrane shear deformation. In turn, the experimental results can be correlated with the elastic constitutive relation for membrane shear, Equation 4.4.12, to evaluate the membrane surface elastic shear modulus, μ, and any possible dependence of the modulus on deformation.

We begin with the balance of forces (mechanical equilibrium) for the membrane. In the analysis, we will make two critical assumptions: (1) the membrane is pushed tightly against the glass wall of the pipet without wall friction; (2) only membrane force resultants determine the mechanical equilibrium with negligible contribution from bending moments. These assumptions can be evaluated from experimental results. The membrane is normally held against the pipet inner wall by the hydrostatic pressure inside the cell which is in excess of the effect of the circumferential tension. When hydrostatic pressure is not high enough, the membrane will pull away from the wall, and the projection will neck down. Necking behavior does occur when the outside portion of the cell becomes spherical as the isotropic tension increases (see Evans, 1973, for discussion).[16] Frictional interaction with the pipet inner wall will produce hysteresis between the curve for aspiration (loading) and that for release (unloading) of the cell. Recently, Waugh (1977)[91] has shown that frictional effects are on the order of only 10%. Therefore, the first assumption can be taken as valid. The second assumption which concerns bending moments is more difficult to evaluate. Elastic energy storage that results from membrane curvature changes or bending will obviously be largest at the entrance to the pipet (where the membrane bends around the edge of the entrance) and over the spheroidal cap of the aspirated cell projection. Since this energy storage occurs during the aspiration of the initial, small spheroidal bump (the order of a pipet radius), it is expected that further aspiration will not appreciably change this free energy. This is because the formation of the cylindrical portion of the projection requires curvature elastic energy which increases linearly with the projection length. A linear increase in energy implies that the applied force or pipet suction pressure would not increase with the length of the projection. Therefore, the curvature or bending elastic effects will be an initial bias in the force required to aspirate the membrane projection into the pipet, independent of lengths greater than a pipet radius. Experimentally, it has not been possible to detect any bias. Thus, the effect is less than the experimental uncertainties inherent to observation of the initial aspiration of a small membrane bump. Also, in Example 3 of Section 4.9, we estimated that the bending or curvature elastic energy contribution would only be about 20% for the bump formation and would decrease rapidly with further aspiration and extension of the membrane surface. From an intuitive viewpoint, if the bending resistance of the membrane were large, then the membrane region just outside the entrance of the pipet would perceptibly bend away from the tip of the pipet; this feature is not observed. Hence, our second assumption appears to be reasonable for the large extensions produced by micropipet aspiration.

The two assumptions permit us to treat the principal force resultants (tensions) as continuous functions throughout the projection region and outer membrane surface. The meridional tension, T_m (that acts along a direction tangent to the cylindrical generator of the projection surface and which continues along the radial direction in the membrane region exterior to the pipet entrance), is constant along the cylindrical section and is given by the force balance,

$$2\pi \cdot R_p \cdot T_m = \Delta P \cdot \pi R_p^2 \tag{5.3.1}$$

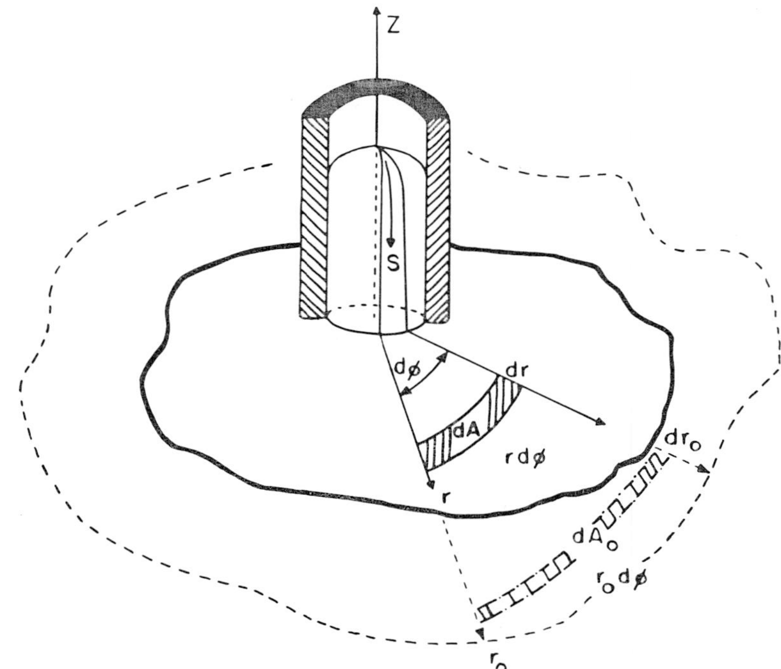

FIGURE 5.16.   A diagram of the membrane region local to the pipet entrance for the experiment illustrated in Figure 5.15. We assume that the membrane surface outside the pipet entrance is essentially flat and use cylindrical coordinates (z, r, ϕ) to describe the surface geometry (also, curvilinear coordinate, s, is used to describe distance along the meridian of the cell surface). The displacement and constant area deformation of a segment of an annular ring is shown by the curved element.

(The principal membrane tensions are illustrated schematically in Figure 5.15.) The pipet radius is given by R_p. The pipet suction pressure is given by ΔP (where it is implicitly assumed that the hydrostatic pressure inside the cell is equal to the pressure in the suspending medium outside the cell. This is valid since the cell remains flaccid far away from the pipet tip). Equation 5.3.1 provides the boundary condition for the meridional or radial tension at the edge of membrane region outside the pipet entrance. The membrane region local to the pipet is diagrammed in Figure 5.16. If we assume that the membrane surface outside the pipet entrance is essentially flat, then the equation of equilibrium in the surface is given by Equation 3.5.9,

$$\frac{d}{dr}(rT_m) - T_\phi = 0 \qquad (r \geq R_p) \tag{5.3.2}$$

where the principal force resultant or tension, $T_\circ$, acts circumferentially and tangent to a latitude circle of the cylinder inside the pipet or azimuthal circle of the outside surface plane (Figure 5.15). Equation 5.3.2 can be rewritten as

$$\frac{dT_m}{dr} = -\frac{2T_s}{r} \qquad (r \geq R_p) \tag{5.3.3}$$

Here, the deviatoric or maximum shear force resultant, T_s, is introduced with the definition that

$$T_s \equiv \frac{T_m - T_\phi}{2}$$

We conceptualize the exterior membrane surface as much larger than the pipet dimension and integrate Equation 5.3.3 to obtain

$$T_m \bigg|_{R_p} = - \int_{R_p}^{\infty} dT_m = \int_{R_p}^{\infty} \frac{2T_s}{r} \, dr \qquad (5.3.4)$$

since the force resultants are assumed to be negligible at distances far from the pipet (i.e., at the periphery of the cell; this is consistent with the flaccid, essentially unchanged shape of the equatorial region of the red cell throughout the experiment). From Equations 5.3.1 and 5.3.4, we obtain a relation between the intrinsic membrane shear force resultant and the externally applied pipet suction pressure,

$$\Delta P = \frac{4}{R_p} \int_{R_p}^{\infty} \frac{T_s}{r} \, dr \qquad (5.3.5)$$

For a reversible elastic process, the shear resultant depends only on the material deformation, as in Equation 4.4.12. Next, we will show that the material extension decreases inversely as the square of the radial distance from the pipet entrance. Therefore, the shear resultant also decreases rapidly away from the pipet as is indicated by the experimental observations.

Now, we must determine the state of deformation in the membrane surface as a function of the length of the cell projection, which is aspirated into the pipet. Then, with the material extension ratios as a function of radial position and with a postulated constitutive relation (e.g., Equation 4.4.12), we can integrate Equation 5.3.5 to predict the suction pressure that is required to aspirate a membrane.projection of length, L, given the pipet radius and material elastic shear modulus, μ. Correlation of the prediction and observed relation between pressure and length provides the value for the elastic shear modulus and critical evaluation of the postulated constitutive behavior.

The mechanical area dilation of swollen cells has demonstrated that the red cell membrane greatly resists changes in surface area (as we discussed in the previous section). Therefore, we assume that the surface is incompressible in the two dimensions which characterize the membrane surface, i.e.,

$$\lambda_1 \cdot \lambda_2 \equiv 1$$

We need only to consider a single principal extension ratio, say, along the meridional direction, λ_m; the other extension ratio is given by the reciprocal, $\lambda_\phi = \lambda_m^{-1}$. Incompressibility or the constant area requirement implies that each element of area of the membrane surface remains constant in magnitude when deformed even though its shape has changed. Figure 5.16 illustrates the displacement and constant area deformation of a segment of annular ring that occurs in pipet aspiration. In order to evaluate the material deformation, we must distinguish between the undeformed and deformed coordinate systems. The undeformed coordinate system is defined by the radial position, r_0, along a surface meridian from the symmetry axis of the pipet and the azimuthal angle, ϕ, which is the polar angle in the initially flat surface (Figure 5.16). The deformed coordinate system is repesented by the curvilinear distance, s, along the meridian that begins at the pole of the spheroidal cap and which continues along the cylindrical generator and eventually becomes the radial coordinate, r, in the outer membrane surface. The azimuthal angle is obviously the same because of symmetry. Elemental areas of undeformed and deformed states of the membrane must be equal,

$$r_0 \, dr_0 \, d\phi = r ds \, d\phi$$

or

$$r_0 \, dr_0 = r \, ds \qquad (5.3.6)$$

The principal extension ratio along the meridional direction, λ_m, is given by (see Section II),

$$\lambda_m = \frac{ds}{dr_0}$$

and, therefore, by the radius ratio,

$$\lambda_m = \frac{r_0}{r} \qquad (5.3.7)$$

for the same material locations. The total areas of membrane from the origin ($r = r_0 = 0$) must be equal for both states and are obtained by integration of Equation 5.3.6 over the membrane surface,

$$2\pi \int_0^{r_0} r_0 \, dr_0 = 2\pi \int_0^{r} r \, ds$$

The total surface area can be broken down into contributions of the spheroidal cap, A_{cap}; the cylindrical section, A_{cyl}; and the outside annulus, $\pi(r^2 - R_p^2)$, for any location outside the pipet. This gives

$$r_0^2 = \frac{A_{cap} + A_{cyl}}{\pi} + (r^2 - R_p^2) \qquad (r \geq R_p)$$

Hence, the local extension ratio is determined from the relation,

$$\lambda_m^2 = \frac{A_{cap} + A_{cyl}}{\pi r^2} + \left(1 - \frac{R_p^2}{r^2}\right) \qquad (r \geq R_p) \qquad (5.3.8)$$

if we divide by the square of the radial coordinate. From Equation 5.3.8, it is apparent that the square of the material extension ratio drops off as $1/r^2$ from the pipet entrance. The remaining step is to give the geometric relations for the area of the spheroidal cap and cylindrical section. The exact shape of the spheroidal cap can be established only by satisfying mechanical equilibrium for the membrane surface in this region exposed to a hydrostatic pressure difference. This is not possible in closed form, but can be done on a digital computer. However, this region contributes minimally to the maximum degree of extension as the project increases. Therefore, we use spherical surface segments to approximate the cap geometry as the projection length goes from $0 \leq L \leq R_p$ and then treat the cap as a hemisphere for projection lengths greater than a pipet radius. With this approach, Equation 5.3.8 is given by

$$\lambda_m^2 = 1 + \left(\frac{R_p}{r}\right)^2 \left(\frac{2L}{R_p} - 1\right) \qquad (L \geq R_p) \qquad (5.3.9)$$

which is valid for projection lengths greater than one pipet radius. From Equation 5.3.9, we see that the maximum extension ratio in the membrane occurs at the pipet entrance and is equal to

$$\hat{\lambda} = \sqrt{\frac{2L}{R_p}}$$

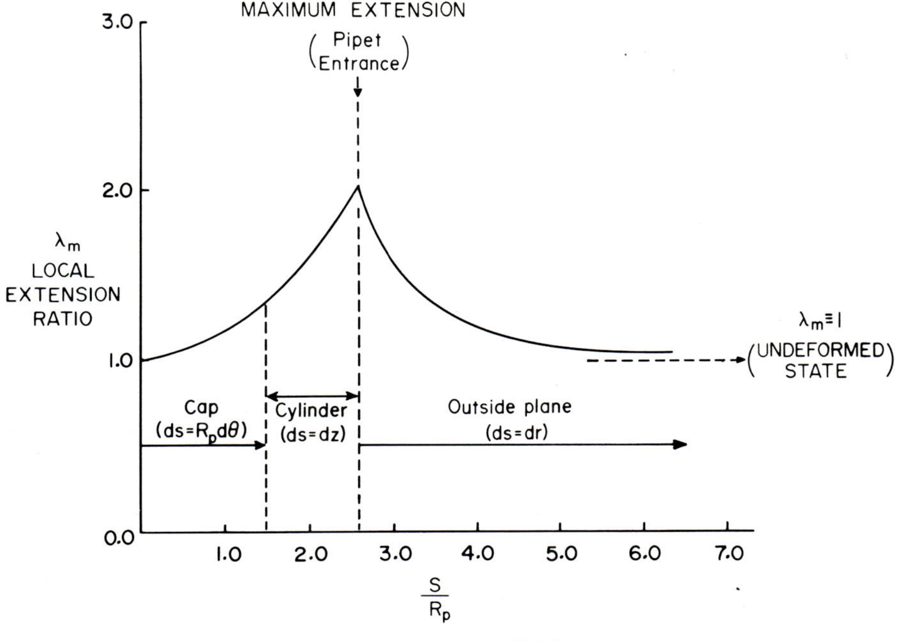

FIGURE 5.17.   Nonuniform extension produced by micropipet aspiration of a surface at constant area. The extension ratio is plotted as a function of curvilinear distance, s, along the meridian from the pole of the spheroidal cap to the outer membrane surface.

For an aspirated length equal to two pipet radii, Figure 5.17 shows the extension ratio as a function of curvilinear distance, s, along the meridian from the pole of the spheroidal cap to the outer membrane surface.

If we postulate that the membrane obeys the first-order elastic constitutive equation, (4.4.12), in shear,

$$T_s = \frac{\mu}{2}\,(\lambda_m^2 - \lambda_m^{-2})$$

then we can use Equation 5.3.9 in the integral relation, Equation 5.3.5, to obtain an anticipated relation between the pipet suction pressure and the aspirated projection length,

$$\Delta P = \frac{2\mu}{R_p} \int_{R_p}^{\infty} (\lambda_m^2 - \lambda_m^{-2})\,\frac{dr}{r}$$

or

$$\Delta P = \left(\frac{\mu}{R_p}\right)\left[\left(\frac{2L}{R_p} - 1\right) + \ln\left(\frac{2L}{R_p}\right)\right]\quad (L \geq R_p)$$

$$(5.3.10)$$

Since the logarithm is a very weak function, we note that the predicted relation between pressure and aspirated projection length is essentially linear even though the elastic constitutive relation is nonlinear or quadratic in the extension ratio. Experimental results for a single red cell experiment are shown in Figure 5.18A alongside a photograph (Figure 5.18C) of the video image of the micropipet aspiration of a flaccid red cell disk. Note that the experimental data is linear. Consequently, we can use Equation

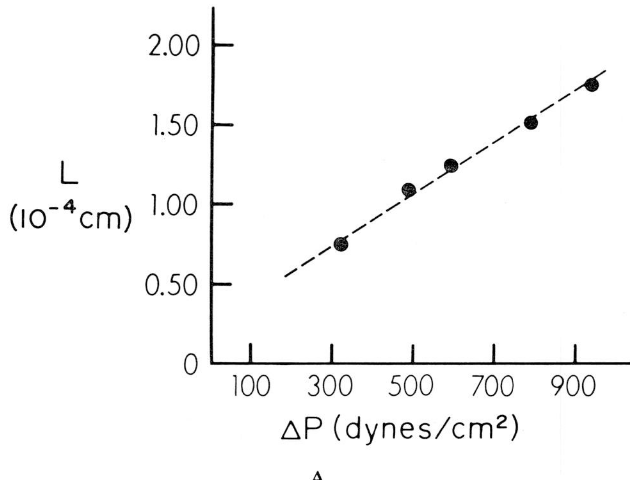

FIGURE 5.18(A) Data for the cell projection length, L, vs. the suction pressure, ΔP, for micropipet aspiration of a flaccid red cell. (B) Prediction of the length of the cell projection vs. the suction pressure for a 2-D hyperelastic material at constant area. Data points correspond to those in (A) as correlated by a specific value for the shear modulus, μ. (C) Photograph of the aspiration of a flaccid cell as used to measure the membrane elastic shear modulus. (From Waugh, R. E., Temperature Dependence of the Elastic Properties of Red Cell Membrane, Ph.D. dissertation, Duke University, Durham, N.C., 1977. With permission.).

5.3.10 to correlate the material constitutive relation with these results. The length vs. pressure relation can be plotted in a dimensionless form of Equation 5.3.10, as shown in Figure 5.18B, and the data can be correlated with this curve to obtain the membrane surface elastic shear modulus, μ. The value for the shear modulus is on the order of 10^{-2} dyn/cm as was the value obtained from the fluid shear deformation of glass-attached red cells by Hochmuth et al. (1973).[44] Recently, Waugh (1977)[91] has carefully investigated the effects of hysteresis (e.g., caused by glass friction). His results are shown in Figure 5.19, which indicates that the effect is small. The confidence limits for the measured shear moduli are also shown in Figure 5.19. Waugh's measurements give a value of 6.6×10^{-3} dyn/cm at 25°C with an 18% standard deviation for 30 samples. The frictional effects were measured to be on the order of 10%, which is better than the statistical resolution.

There are some other notable aspects of the elastic extension experiments that have been performed on red cell membranes. First, the same value of elastic shear modulus was obtained from two different mechanical experiments, which demonstrates that the intrinsic material behavior can be studied independent of the extrinsic geometry of the cell shape. Next, the red cell membrane behaves in a hyperelastic manner for extension ratios of up to 3:1, i.e., it exhibits first order, nonlinear behavior given by Equation 4.4.12, with a constant shear modulus. This is unusual, since most materials, with the exception of rubber-like elastomers, possess such elastic behavior only for small deformations. Indeed, we saw that the area dilation properties were entirely different, i.e., small deformations ($\sim$1%) gave isotropic tensions that are four to five orders of magnitude greater than the shear resultant. Even though the shear modulus of the red cell membrane is small in comparison to the area compressibility modulus, the membrane has sufficient rigidity to provide for maintenance of cell shapes that are not spherical.

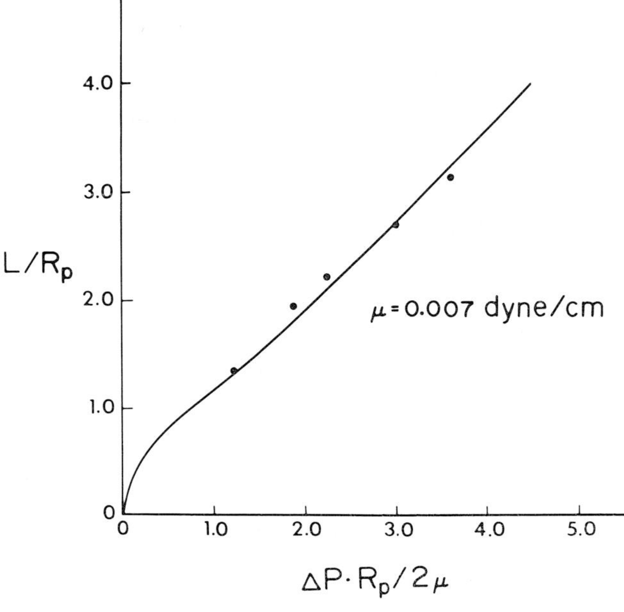

FIGURE 5.18 B

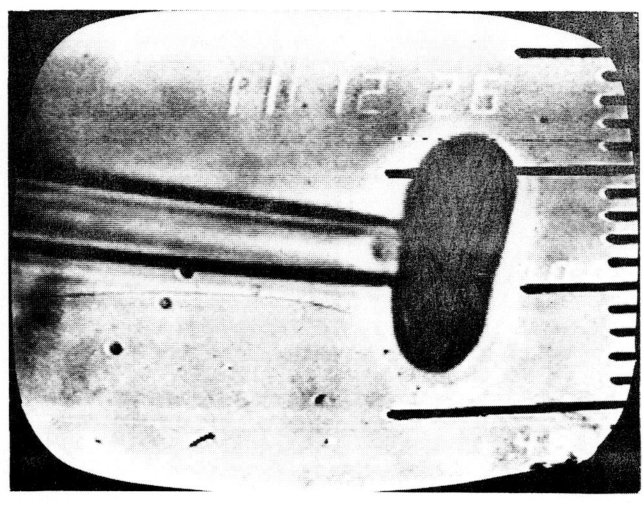

FIGURE 5.18C

On the other hand, when the shear rigidity of the red cell membrane is destroyed (e.g., by temperatures above 45 to 50°C), the cell surface breaks up into spherical remnants of various sizes (Ponder, 1971).[68] The sphere is a stable shape for lipid vesicles as well.

Red cells of vertebrate animals other than mammals retain a ''dead'' nucleus and are generally larger than mammalian (nonnucleated) red cells. Because the aspiration of membrane surface with a small micropipet is localized, it has been possible to investigate the elastic shear moduli of nucleated cells. (Figure 5.20 shows some examples of nucleated cell experiments.) These membrane surfaces exhibit the same hyperelastic material behavior as human red cells, but the maximum extension ratios that can be

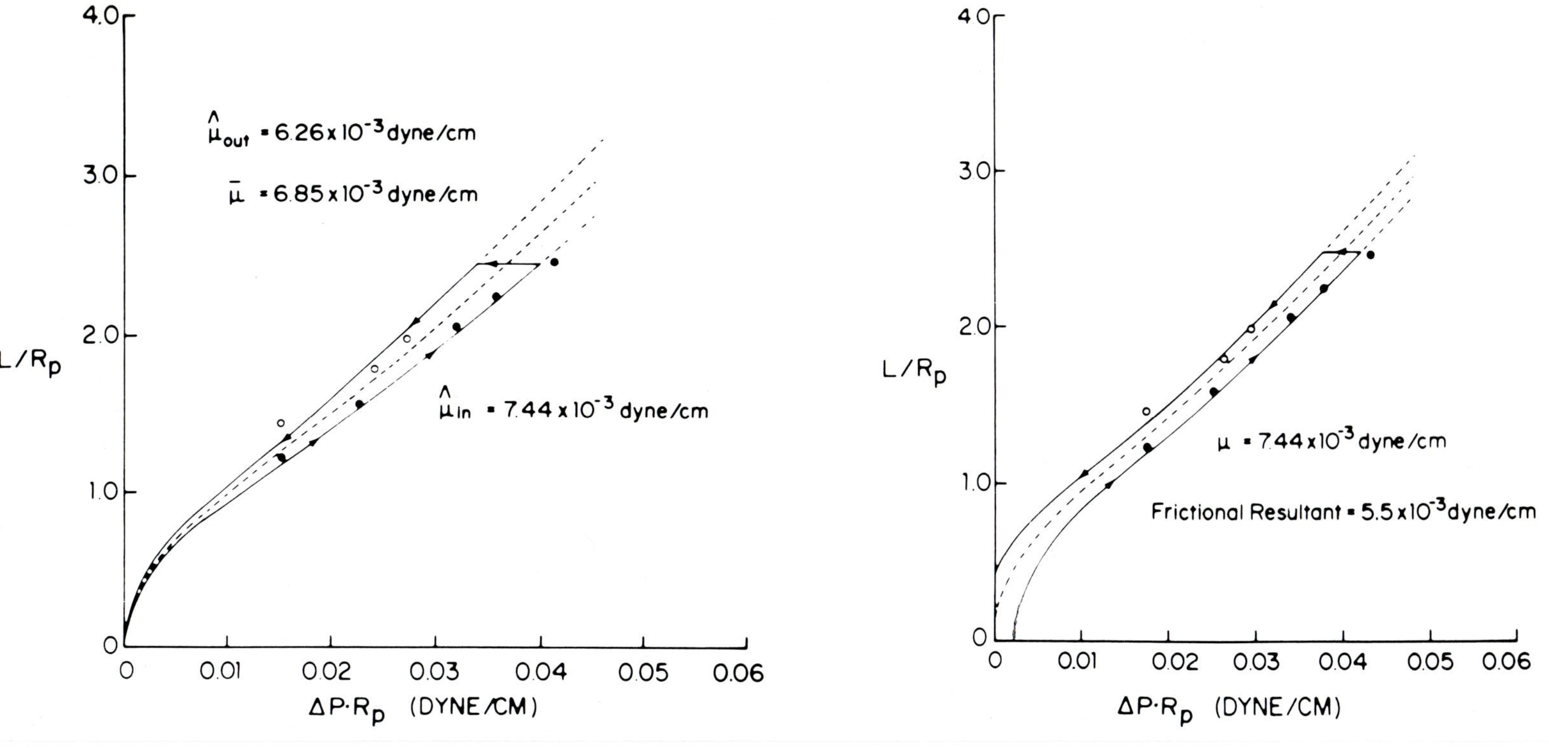

FIGURE 5.19.   (A) Correlated behavior of cell aspiration and release if there is a length dependent frictional force. (B) Correlated behavior of the same cell aspiration and release if there is a constant frictional force, such as a friction at the pipet entrance. In each figure, the aspiration phase is the lower curve and the release phase is the continuation as the upper curve. (From Waugh, R. E., Temperature Dependence of the Elastic Properties of Red Cell Membrane, Ph.D. dissertation, Duke University, Durham, N.C., 1977. With permission.)

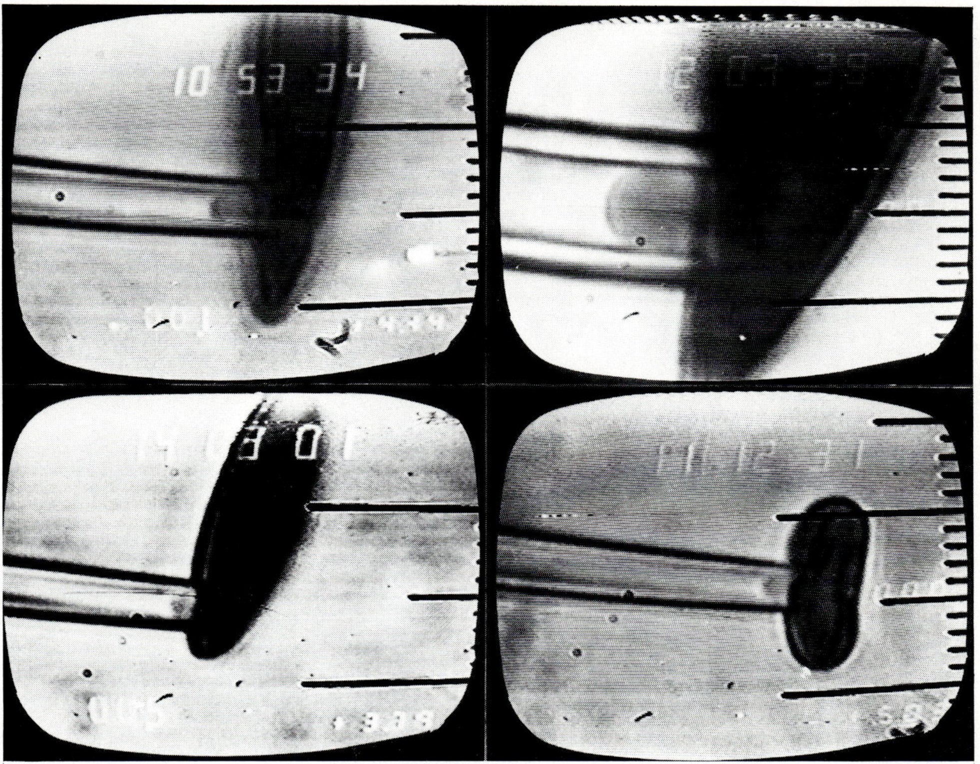

FIGURE 5.20. Photographs of videotape images of micropipet experiments using nucleated red blood cells. These membranes exhibit the same hyperelastic material behavior, but the maximum extension ratios that can be obtained are lower because the presence of the nucleus limits the available surface. Cells are from frog (upper left), turkey (lower left), *Amphiuma* (upper right), and opossum (lower right). (From Waugh, R. and Evans, E. A., *Microvasc. Res.*, 12, 291, 1976. With permission.)

attained are lower because the presence of the nucleus limits the available surface. Figure 5.21 gives data for some specific single cell experiments taken from Waugh and Evans (1976).[92] The elastic shear moduli are as much as 50 times larger than mammalian red cell membrane moduli for some animal cells, e.g., reptiles. On the other hand, the mammalian red cell membranes that have been tested (opossum, human, cat) all have essentially the same magnitude of elastic shear modulus.

5.4 Elastic Curvature Changes: Bending vs. Shear Rigidity in Osmotic Swelling of Flaccid Cells

In the previous two sections, we discussed measurements of elastic constitutive relations between in-plane force resultants and membrane surface deformations. In analyzing these experiments, we neglected curvature or bending elastic effects because their contributions to the total work of deformation appeared to be small. For instance, we showed in Section 4.9 that bending energy is expected to be much smaller than shear elastic energy in micropipet experiments, since the shear deformation is large. However, there are situations where the bending or curvature elastic energy is a major component of the reversible work required to deform the cell membrane. We will consider a specific example, which is the osmotic swelling of a red blood cell from a flaccid biconcave disk to a spheroid. In red cell swelling, large regions of the cell membrane undergo very small extensional deformations, but significant changes in surface cur-

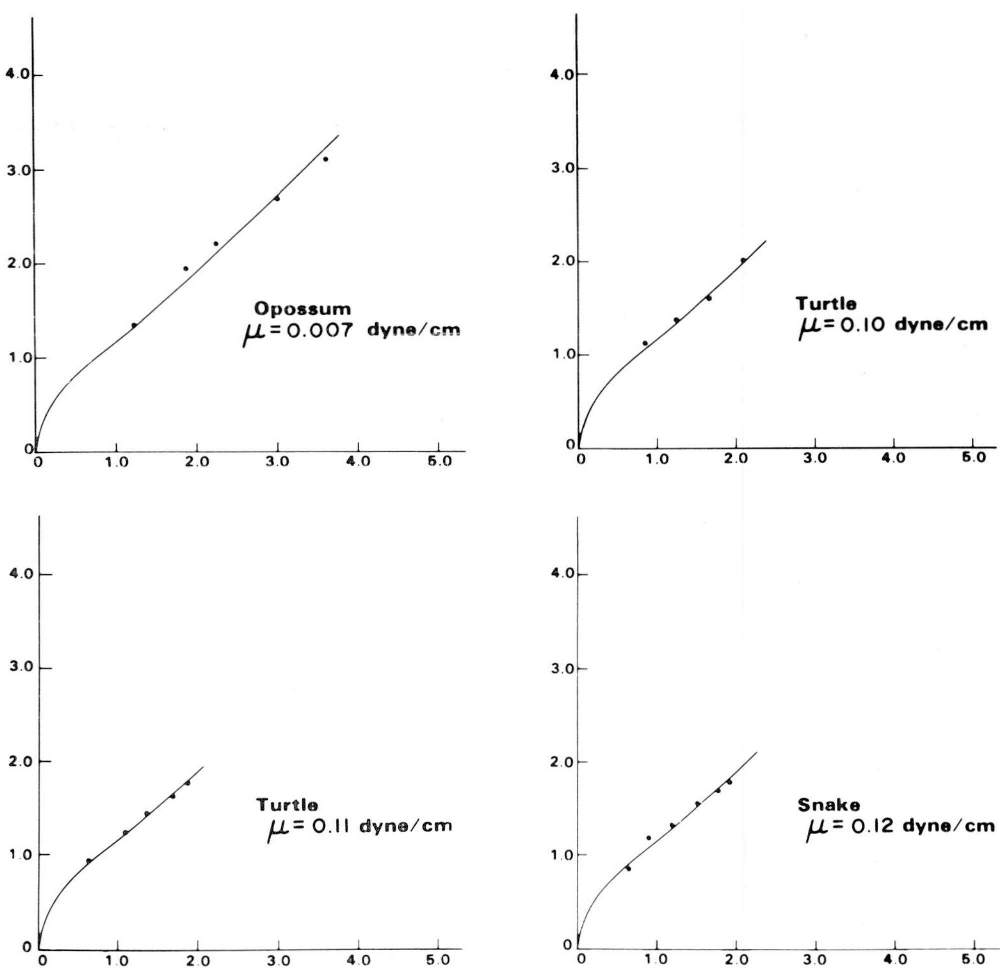

FIGURE 5.21.   Correlation between experimental data and predicted behavior for a single individual cell from each animal type. The solid line is the theoretical prediction of Evans (1973),[16] and dots represent data from Waugh and Evans (1976).[92] (From Waugh, R. and Evans, E. A., *Microvasc. Res.*, 12, 291, 1976. With permission.)

vature occur in these regions. The central regions of the red cell disk are deformed into polar regions of the spheroid with very little membrane extension (Evans, 1973).[15] Large extension occurs primarily over the peripheral or equatorial belt of the biconcave disk. Thus, curvature or bending elasticity is essential in the determination of equilibrium shapes for the intermediate swollen states of the red cell.

Observations of osmotic sphering of red blood cells have been extensively described in the past (e.g., Ponder, 1971).[68] The first attempt at a detailed computation of the membrane force resultants and strains in the sphered red blood cell was made by Fung and Tong (1968).[31] However, their analysis treated only the initial discoid and the final spherical shapes, but not the intermediate shapes. Also, they chose elastic constitutive relations for the membrane which were not realistic because the properties were based on the assumption of an isotropic material in three dimensions. With a more realistic set of material properties and a more detailed analysis, Zarda (1974)[99] and Zarda et al. (1977)[100] have used a finite-element method to compute red cell shapes for all stages of osmotic swelling. They have assumed that the unstressed shape of the cell is the typical biconcave discocyte, e.g., measured by Evans and Fung (1972),[18] shown in Figure 5.22A, or taken from Ponder (1971)[68] by Fung and Tong (1968),[31] shown in Figure 5.22B. Qualitatively, the differences in computed behavior starting with either of these two shapes were shown to be insignificant. The choice of initial shape in the calculation procedure was an important assumption, i.e., the assertion that the unstressed shape is the normal biconcave discocyte. Alternative assumptions could have been chosen, e.g., the unstressed shape is a sphere or even a flat circular disk. None of the computations and experiments made to date has resolved the issue. We will discuss this point further at the end of this section.

The analysis given by Zarda (1974)[99] and Zarda et al. (1977)[100] starts from a variational principle which essentially minimizes a functional, ψ, of the mechanical power,

$$\delta\psi(v_i, \overline{T}) = 0 \qquad (5.4.1)$$

where v_i is the local velocity of the membrane and $\overline{T}$ is isotropic tension in the membrane. The functional is expressed as

$$\psi(v_i, \overline{T}) = c \int_A (v_k v_k)\, dA - 2 \int_A (\sigma_k v_k)\, dA$$

$$+ 2 \int_{A_0} \left(\frac{\partial \tilde{F}_s}{\partial t}\right)_T dA_0 + \int_{A_0} \left(\frac{\partial \tilde{F}_c}{\partial t}\right) dA_0$$

$$+ 2 \int_A \frac{\overline{T}}{(1 + \alpha)} \left(\frac{\partial \alpha}{\partial t}\right) dA \qquad (5.4.2)$$

where dA_0 and dA are the initial and instantaneous areas of membrane elements. The variation of the first integral on the right hand side of Equation 5.4.2 creates a dissipative term that facilitates the computation. This term is similar to, but not an exact expression for, viscous dissipation in the membrane. The term, σ_i, is the surface traction, i.e., applied force per unit area exerted on the membrane; $(\tilde{F}_s)_T$ and $(\tilde{F}_c)_T$ are the membrane elastic free energy or "strain" energy densities due to shear force resultants (deviatoric tension) and bending moments, respectively. It can be shown that the variational equation, (5.4.1), is equivalent to the proper equations of mechanical equilibrium of the axisymmetric membrane (Section 3.5). Zarda et al. (1977)[100] express the principal membrane tensions by the following elastic constitutive relations:

$$T_1 = \frac{1}{\lambda_2} \left(\frac{\partial \tilde{F}_s}{\partial \lambda_1}\right)_T + \overline{T}$$

$$T_2 = \frac{1}{\lambda_1} \left(\frac{\partial \tilde{F}_s}{\partial \lambda_2}\right)_T + \overline{\overline{T}} \qquad (5.4.3)$$

which are the forms derived by Skalak et al. (1973)[84] using the nonlinear elastic development of Green and Adkins (1970). (Note: the term "strain energy density" is employed by these authors for the Helmholtz free energy density.) Equations 5.4.3 are equivalent alternatives to the elastic constitutive relations that we developed in Section 4.4, which are designed to provide a thermodynamic perspective that can be linked to membrane chemistry. If we introduce the deformation variable, β, we obtain

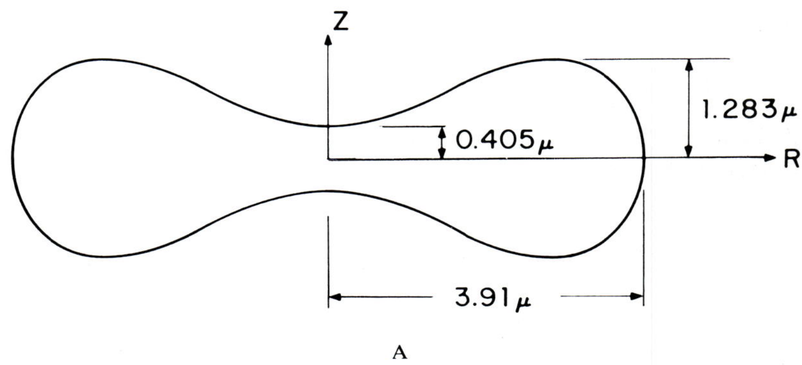

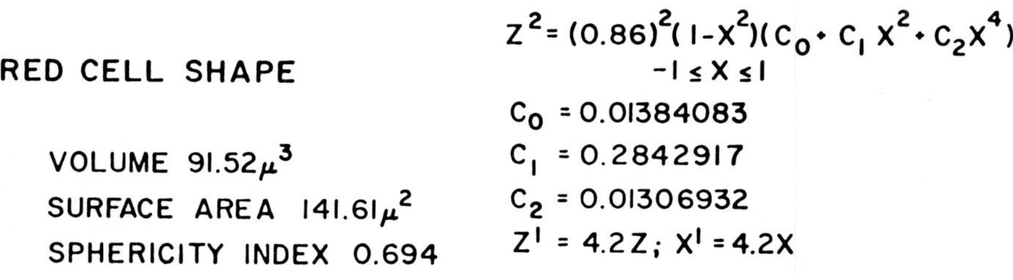

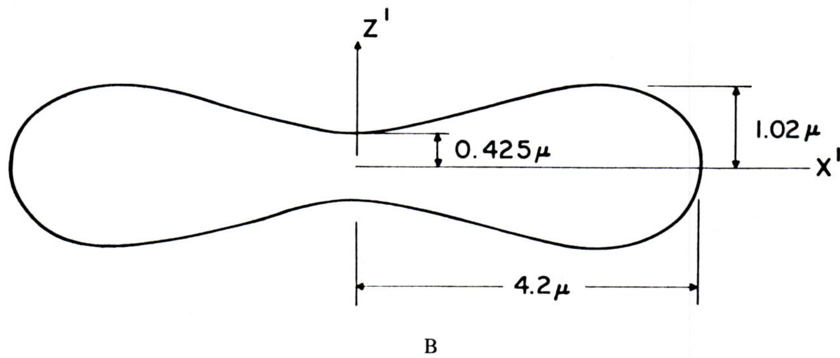

FIGURE 5.22.   (A) Average unstressed shape measured for normal human red blood cells by Evans and Fung (1972).[18] (B) Unstressed shape of the human red blood cell, adapted from Fung and Tong (1968).[31] (From Zarda, P. R., Chien, S., and Skalak, R., *J. Biomech.*, 10, 211, 1977. With permission.)

$$T_s = \frac{1}{\lambda_2} \left(\frac{\partial \tilde{F}_s}{\partial \lambda_1}\right) = \frac{1}{\lambda_2} \left(\frac{\partial \tilde{F}_s}{\partial \beta}\right)_T \left(\frac{\partial \beta}{\partial \lambda_1}\right)$$

$$= \frac{1}{2\lambda_1^2 \lambda_2^2} \left(\frac{\partial \tilde{F}_s}{\partial \beta}\right)_T (\lambda_1^2 - \lambda_2^2) \tag{5.4.4}$$

and

$$-T_s = \frac{1}{\lambda_1} \left(\frac{\partial \tilde{F}_s}{\partial \lambda_2}\right)_T = \frac{1}{\lambda_1} \left(\frac{\partial \tilde{F}_s}{\partial \beta}\right)_T \left(\frac{\partial \beta}{\partial \lambda_2}\right)$$

$$= \frac{-1}{2\lambda_1^2 \lambda_2^2} \left(\frac{\partial \tilde{F}_s}{\partial \beta}\right)_T (\lambda_1^2 - \lambda_2^2)$$

which are the deviatoric or shear resultant components that we derived in Section 4.4. Also, the requirement of constant area may be used if desired,

$$\lambda_1 \lambda_2 = 1, \text{ i.e. } \alpha \equiv 0 \tag{5.4.5}$$

Most of the computations by Zarda et al. (1977)[100] have been performed at constant area in which case the principal membrane force resultants are

$$T_1 = \tilde{\lambda} \left(\frac{\partial \tilde{F}_s}{\partial \tilde{\lambda}}\right)_T + \overline{T}$$

$$T_2 = -\tilde{\lambda} \left(\frac{\partial \tilde{F}_s}{\partial \tilde{\lambda}}\right)_T + \overline{T} \tag{5.4.6}$$

where $\tilde{\lambda} \equiv \lambda_1$ and $\overline{T}$ is an unknown, isotropic tension.

If the internal pressure in the cell is large, the isotropic tension will be large and the area of the membrane may expand elastically. Thus, the elastic free energy density will include area dependent terms which are introduced as a constitutive relation for the isotropic tension in Equations 5.4.2 and 5.4.3. Zarda (1974)[99] and Zarda et al. (1977)[100] have used an elastic free energy density which was originally proposed by Skalak et al. (1973),[84]

$$(\tilde{F})_T = \frac{\overline{\mu}}{2} \left(\beta^2 + \beta - \alpha - \frac{\alpha^2}{2}\right) + \frac{\overline{K}}{2} \left(\alpha + \frac{\alpha^2}{2}\right)^2 \tag{5.4.7}$$

For the case of constant area, Equation 5.4.7 is used with $\alpha \equiv 0$ and Equation 5.4.6 to obtain the constitutive relations. $\overline{K}$ is related to the area modulus, and Zarda et al. (1977)[100] assumed the value to be 100 dyn/cm. The coefficient, $\overline{\mu}$, is related to the shear modulus, which they assumed to be 0.005 dyn/cm. The quadratic powers of the deformation variables, α and β, in the elastic free energy density, Equation 5.4.6, have not yet been measured in mechanical experiments on red cells. Only the first order terms can be determined within the resolution of the experimental techniques. In the analysis of osmotic swelling, only the first order terms are effective because of the small values for the deformation variables that are associated with the surface deformation. Therefore, the constitutive relations derived from Equations 5.4.3, 5.4.4, and 5.4.7 have the same influence as those used in Sections 5.2 and 5.3 for red cell micropipet experiments.

Zarda et al. (1977)[100] expressed the curvature elastic energy or bending energy as

$$(\widetilde{F}_c)_T = \frac{B}{2}\,(C_1^2 + 2\nu_c C_1 C_2 + C_2^2)$$

(5.4.8)

where B is a curvature elastic or bending modulus taken equal to 10^{-12} dyn/cm; ν_c is a material parameter that is assumed to be ½. In Section 4.9, the curvature or bending elastic energy was derived from thermodynamic considerations. The result in Equation 4.9.16 is an isotropic bending which implies a value of $\nu_c = 1$ for the stratified membrane. The curvature or bending measures of strain, C_1 and C_2, are those we used previously and given by Reissner (1949).[73] The components equal the changes in the principal curvatures for deformation from the initial, unstressed position to any other configuration (see Section 4.9).

We will now present a summary of the results from Zarda (1974)[99] and Zarda et al. (1977).[100] These investigators employed a finite element, computational method to satisfy the variational principle which is expressed by Equation 5.4.1. In their calculations, the initial unstressed cell shape was assumed to be either the contour shown in Figure 5.22A or 5.22B. In the finite-element method, the location of the meridional curve was specified by nodal points (25 were used for each quarter of the curve). Polynomials were passed through these points (nodes) in groups of three or five to form the finite elements. The computation kept track of the location of each node as the volume of the cell was increased. The reader is referred to the paper of Zarda et al. (1977)[100] for details of the method.

With the initial shape taken as Figure 5.22B, the equilibrium shapes for the various cell volumes are shown in Figure 5.23 where curve B is the initial shape of one quadrant of Figure 5.22B. Curves C-D are the swelling phase and curve A is computed for a reduced cell volume. The deformation of the cell was calculated with the area constant for the curves A through H in Figure 5.23 because the internal pressures and isotropic tensions involved were estimated to be very small. This feature is shown in Figure 5.24. In the determinaion of such equilibrium shapes, bending moments are essential because force resultants cannot support the load across the membrane when the tangent to the surface is perpendicular to the axis of symmetry (except at $r = 0$), e.g., the location of maximum thickness of the toroidal rim. With bending moments, we see that such locations are acceptable, since they occur in the curves A, B, C, and D. In the last stages of sphering (curves I and J in Figure 5.23), the bending moments were negligible. The pressures build up rapidly as the surface attempts to increase in area. Until the cell volume reached the stage shown in curve H, the increased volume was incorporated by a change in shape without any appreciable increase in area. For the final stages (I and J, Figure 5.23), the elastic free energy of expansion was included, represented by the modulus, $\overline{K}$. The curve J in Figure 5.23 is not realistic because it requires a larger increase in area than red blood cells can sustain without lysis. The curves I and J are interesting theoretical predictions because they are nearly perfect spheres. This is due to the fact that the isotropic tension was much larger than the shear resultant. Also, the area increase is small in these figures because the area compressibility modulus greatly exceeded the surface elastic shear modulus. Similar computations were carried out with a red blood cell-shaped membrane that has the properties of rubber (see Green and Adkins, 1970).[35] The results were qualitatively different, as shown in Figure 5.25. The area of the rubber cell increased substantially at moderate increases of volume and the spherical shape was difficult to reach. The early stages of swelling in Figure 5.25 (dotted curve) were not accurately represented because bending effects were omit-

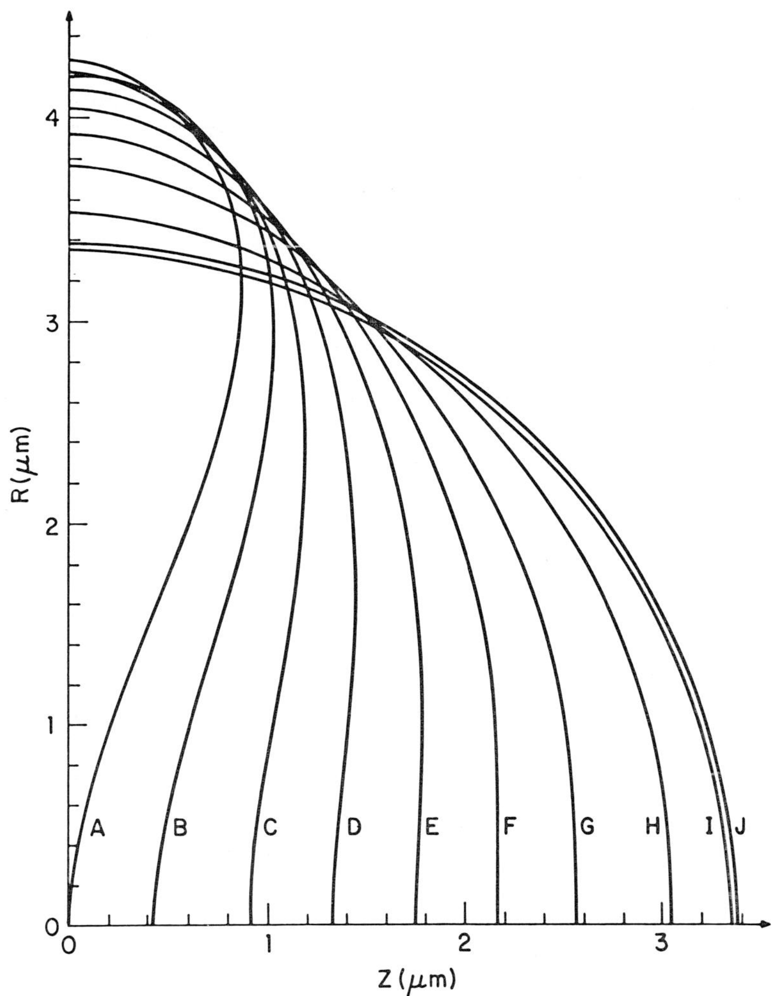

FIGURE 5.23.   Computed shapes of various stages of sphering of a red blood cell. Curve B was taken as the initial unstressed shape from Figure 5.22(B). Curve A was computed for a volume reduced below the normal shape, B. The surface area did not increase until the stage, I, was reached. (From Zarda, P. R., Chien, S., and Skalak, R., *J. Biomech.*, 10, 211, 1977. With permission.)

ted in order to focus on the later stages of expansion. For a rubber sheet, the area modulus and the shear modulus have the same order of magnitude.

The deformations of the red blood cell shown in Figure 5.23 closely resemble the experimental results published by various investigators (Canham and Parkinson, 1970; Evans and Fung, 1972;[18] Evans and Leblond, 1973).[5a,18,22b] For instance, Figure 5.26 presents the average shape of red blood cells measured by Evans and Fung (1972)[18] for an intermediate swollen stage with a volume of 116 μm^3. The initial unstressed shape in this case was that shown in Figure 5.22A. Figure 5.26 also shows the computed shape at a volume of 116 μm^3, starting from the unstressed shape of Figure 5.22A. The agreement of the computed and measured shapes is excellent. This indicates that values assumed for the material properties are reasonable.

In the computations described above, it was assumed that the unstressed shape of the red blood cell was the normal biconcave discocyte such as shown in Figure 5.22A or 5.22B. However, various authors have proposed other abstract forms for the initial

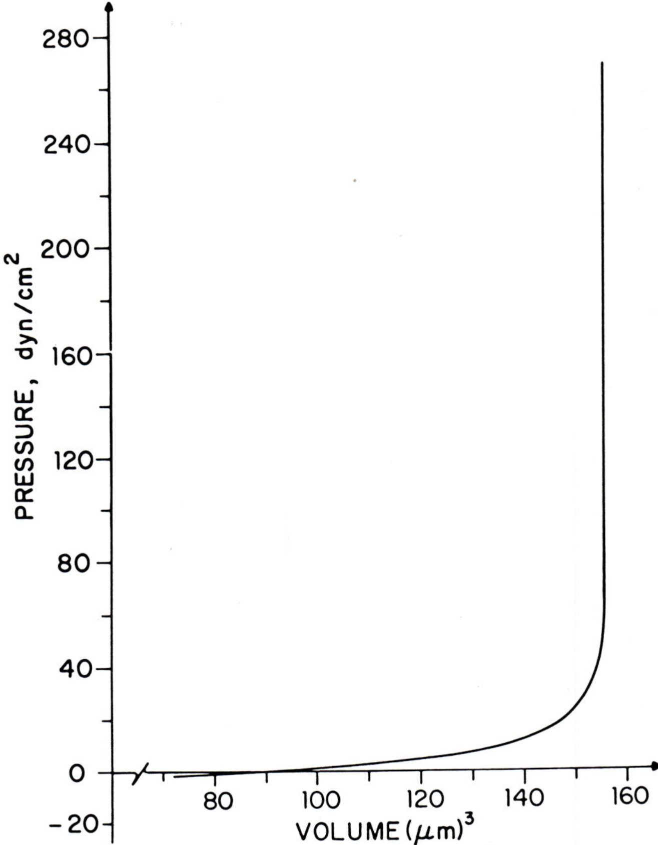

FIGURE 5.24.   Pressure vs. volume computed for the sphering of a
red blood cell shown in Figure 5.23. The pressure increased rapidly
in the final stage of sphering. (From Zarda, P. R., Chien, S., and
Skalak, R., *J. Biomech.*, 10, 211, 1977. With permission.)

membrane shape, e.g., a flat plane (Canham, 1970; Bull and Brailsford, 1975)[4,5] or a
surface of uniform curvature (Helfrich and Dueling, 1975).[39] These authors consider
only bending energy with no surface elastic behavior. The normal biconcave shape was
interpreted as a prestressed shape, which resulted from the minimization of the bending
or curvature elastic energy. The prestress was supplied by supposing the interior of the
cell to be at a slightly negative pressure. The constraints imposed by the fixed volume
and surface area resulted in the normal biconcave shape. In this kind of model (where
the shear elastic energy of the membrane has been neglected in favor of the curvature
or bending energy effects), the dimples may appear at any point of the membrane with
equal ease. Thus, such a cell could easily exhibit the tractor tread type of motion in
very slow shearing flows as reported by Schmid-Schönbein and Wells (1969)[78] and by
Bull and Brailsford (1975).[4] No prestress or residual membrane force resultants can
be detected in the micropipet aspiration of normal red cell disks. The resolution of
these experiments is limited to suction pressures greater than 10 dyn/cm². Therefore,
the maximum value expected for transmembrane pressure differential would be of this
order of magnitude.

The simplest abstract shape that can be assumed for the unstressed shape of the red
blood cell membrane is a sphere. For instance, a radius of 3.35 μm (curve S in Figure
5.27) has the same surface area as the biconcave red blood cell shape in Figure 5.22B.

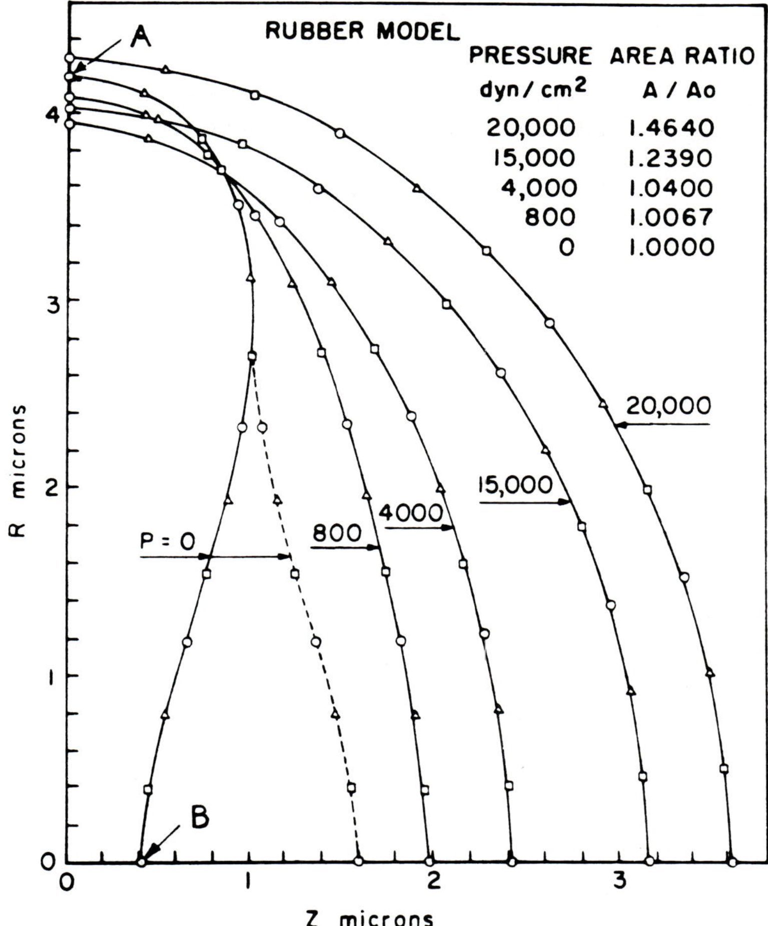

FIGURE 5.25.   Shapes computed for a red blood cell with a membrane 100 A thick having the material properties of latex rubber. The early stages (P = 0) were not accurately represented because bending stiffness was omitted. Note that the later stages are not spherical as in Figure 5.23. (From Zarda, P. R., Chien, S., and Skalak, R., *J. Biomech.*, 10, 211, 1977. With permission.)

In Figure 5.27, the shapes were computed by reducing the volume of the sphere to that of the normal red blood cell volume, while the membrane surface area was held constant. The curves A, B, C, and D in Figure 5.27 correspond to different values of a dimensionless parameter $(\mu \ell^2/B)$ where μ and B are the membrane shear and bending moduli, respectively, and ℓ is a characteristic radius of curvature, taken to be 1 μm (to approximately represent the red blood cell rim). Curve B in Figure 5.27 for $(\mu \ell^2/B) = 1.84$ is very close to the shape of the red blood cell in Figure 5.22B. However, to achieve this shape with $\mu = 0.005$ dyn/cm would require the bending modulus to be $B = 27.2 \times 10^{-12}$ dyn/cm, instead of the values estimated in Section 4.9, and approximately measured, to be 10^{-13} to 10^{-12} dyn/cm. The results in Figure 5.27 show that shapes very much like red blood cells can be produced by deflating an unstressed sphere. The deflated shapes in Figure 5.27 required small negative pressure inside the cell to maintain these computed shapes. The pressures were on the order of 1 to 10 dyn/cm², which could very easily be maintained by weak chemical effects. There appears to be very little hope of measuring such small pressures inside a red blood cell

MEASURED RED CELL SHAPE:

$$Y = 0.5\left[1-X^2\right]^{1/2}(C_0 + C_1 X^2 + C_2 X^4)$$
$$-1 \le X \le 1$$

VOLUME $116\,\mu^3$ $C_0 = 0.552632$

SURFACE AREA $134.1\mu^2$ $C_1 = 1.994737$

$$C_2 = -1.471053$$

$$Z = 3.80Y; \qquad R = 3.80X$$

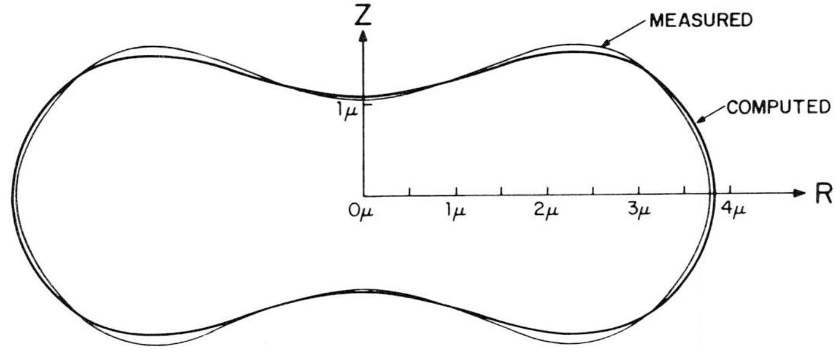

FIGURE 5.26.   Average shape of the human red blood cell at an intermediate state of sphering as measured by Evans and Fung (1972).[18] The computed shape shown has the same volume (116 μm³) as the measured cell. The unstressed shape was that shown in Figure 5.22(A). (From Zarda, P. R., Chien, S., and Skalak, R., *J. Biomech.*, 10, 211, 1977. With permission.)

directly. Hence, a direct verification of the unstressed shape by this means is unlikely.

Although the unstressed shape of the red blood cell is not entirely clear, the relative roles of surface elasticity (area and shear) and curvature elasticity are evident from the computations discussed above. The obvious feature is that the large area compressibility modulus results in a nearly constant surface area of the membrane in every case. Since hemolysis ensues if the increase in area is greater than a few percent, it is a good approximation to assume the area of the membrane is constant in most cases. A spatially variable isotropic tension is used instead of an elastic constitutive relation. Curvature elasticity (bending rigidity) plays an essential role in producing the smooth shapes observed for cross-sections of red blood cells in many situations. Exceptions occur when large extensional deformations are present such that the bending rigidity is negligible. Here, the mechanical equilibrium is primarily maintained by the membrane force resultants (tensions). Bending moments are essential in any case where there is a region of curvature reversal in the membrane such as in the stages of red blood cell swelling before the dimple disappears. The membrane force resultants alone cannot satisfy the equations of equilibrium as pointed out by Fung (1966).[29]

Available evidence from observations of red cell swelling does not provide a definite conclusion as to the relative roles of bending or curvature elasticity and surface shear elasticity for the intermediate stages of swelling. In the first example, Figure 5.23, where the initial shape is a biconcave disk, the curves E-I are for fixed volume and surface area. These remain essentially unchanged if either the bending modulus is set equal to zero or if the shear modulus is set to zero with the bending modulus retained (see Helfrich and Dueling's results, 1975, for the latter).[39] On the other hand, if the unstressed shape of the red blood cell were a sphere, then very distinct differences in the intermediate shapes could be demonstrated which would depend on the relative values of the shear and bending moduli. This is illustrated in Figure 5.28 which shows computed shapes that all have the same surface area and volume and that start from the same unstressed sphere, curve S. The differences between the curves A, B, C, and

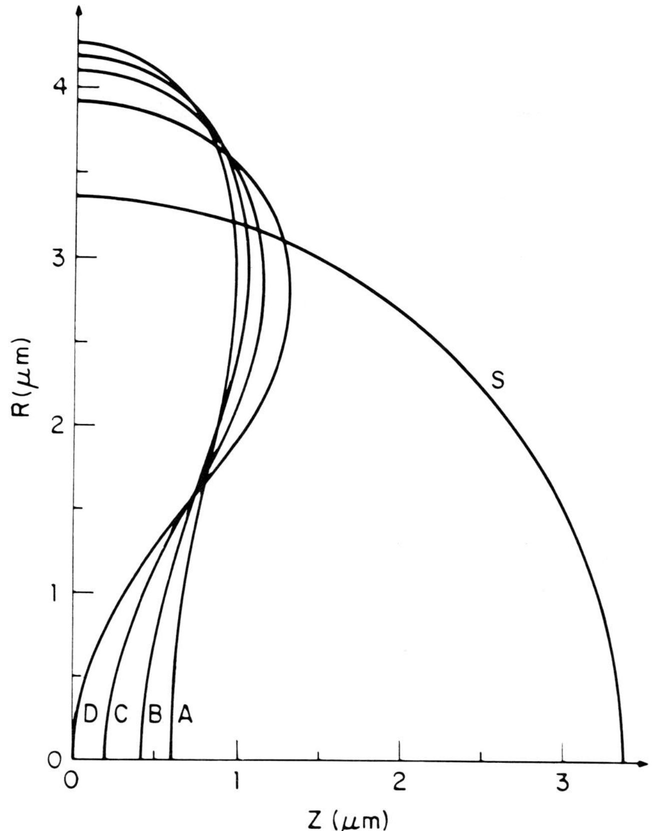

FIGURE 5.27. Equilibrium shapes of red blood cells whose unstressed shape was chosen as the sphere, S. The cells A, B, C, D all enclose the same volume, 91.5 μm³. The sphere and cells A, B, C, D all have the same area, 141.6 μm³. The ratios of shear rigidity to bending stiffness ($\mu\ell^2/B$) for A, B, C, D were 0, 1.84, 4.0, and 8.0, respectively. (From Zarda, P. R., Chien, S., and Skalak, R., *J. Biomech.*, 10, 211, 1977. With permission.)

D correspond to the ratio of shear modulus to bending modulus as expressed by the dimensionless group ($\mu\ell^2/B$). Curve A is the case where no shear rigidity exists and curve D is the opposite case where a large shear rigidity exists, compared to the bending stiffness. If we were to choose from these curves the ones that most resemble the red blood cell shapes at comparable volumes, we would pick A or B. Bull and Brailsford (1975)[4] pointed out that the shapes of curves C and D (Figure 5.28) resemble the forms that result from the deflation of a rubber sphere like a basketball or tennis ball. In rubber-like materials with three-dimensionally isotropic character, the ratio ($\mu\ell^2/B$) is proportional to (ℓ/h)² where h is the wall thickness. Since (ℓ/h) is usually large compared to unity, the role of the shear rigidity is dominant in these cases.

Even though this discussion remains inconclusive and is purely academic with respect to red blood cells, the curves in Figure 5.23 look more like red cell shapes than those in Figure 5.28, and the curvature or bending modulus appears to be small, which would seem to imply that the unstressed (initial) shape of the cell is not a sphere. The concept of an unstressed state as an absolute reference may be untenable and at best abstract because the chemical environment strongly influences the cell shape. Thus, the initial shape must be taken only as a reference geometry in the same way as any chemical

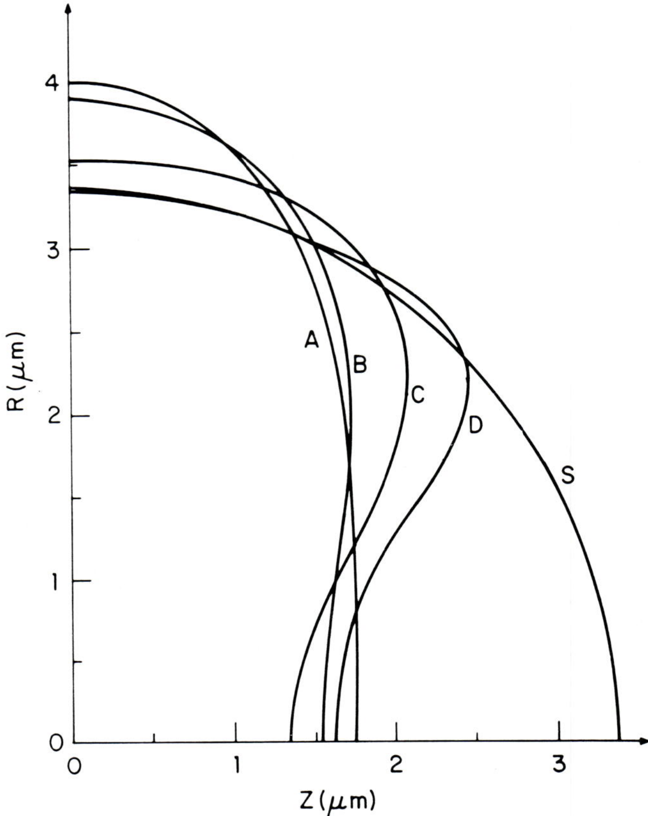

FIGURE 5.28.   Equilibrium shapes of cells whose initial shapes were taken as the unstressed sphere, S. The cells, A, B, C, D all enclose the same volume, 134.6 μm^3. The cells differ in the ratio of shear rigidity to bending stiffness. The dimensionless groups ($\mu \ell^2/B$) for A, B, C, D were 0, 4, 10, and 200, respectively. (From Zarda, P. R., Chien, S., and Skalak, R., *J. Biomech.*, 10, 211, 1977. With permission.)

reference state. Thus, we can measure changes in energy that are produced by deformation relative to the reference geometry. The results are material properties, e.g., surface elastic moduli, bending modulus, etc. However, we recognize that over time the reference geometry may not be stationary and may evolve to a new equilibrium conformation.

5.5 Thermoelasticity and Thermodynamics of Cell Membranes

When we do mechanical work on an elastic substance at constant temperature the displacements of forces supported by the material produce a coherent or deterministic change in the internal energy and entropy states of the material which is reversible. By this we mean that the elastic forces are determined exclusively by the material deformation and that they have specific directions and magnitudes. On the other hand, if we change the temperature of the material, internal energy and entropy states may be changed in a reversible manner without displacements of deterministic forces. This occurs by the processes of heat exchange, i.e., by random exchanges of momentum between the material surface and the adjacent environment or by radiative transfer and molecular excitation. The two reversible processes for alteration of the thermodynamic state functions are related by the combined first and second laws of thermodynamics and the material equations of state, as we developed them in Section 4.11. The interrelation of these processes is evidenced by the reversible heat exchange that is measured in a calorimeter when isothermal work is done on the material and by the thermoelastic forces that are produced in a constrained material when its temperature is changed. It is impossible to construct a calorimeter to measure the reversible heat exchange that occurs during micromechanical experiments on cell membranes, so we turn to the thermoelastic behavior to deduce the changes in material internal energy and entropy states that are produced by elastic deformations at constant temperature. This decomposition provides a direct assessment of the reversible changes in material structure that are produced by deformation, e.g., how much the configurational state of molecular complexes is altered by the deformation and how much the energy of these complexes is changed by the deformation. For a biological membrane structure composed of amphiphilic components, thermoelasticity may be used to study the cohesive force that is provided by the hydrophobic interaction at the membrane interfaces. Also, thermoelasticity, or the reversible heat of expansion, represents the effects of thermal repulsive forces (i.e., surface pressure) internal to the membrane.

Membrane anisotropy results from the natural reluctance of amphiphilic molecules to dissolve in an aqueous phase. As a result of the hydrophobic effect, short-term mechanical experiments do not alter the composition of the membrane and thus provide information about the membrane structure as a closed system. This information complements the results from chemical analysis of the molecular components in open or freely exchanging systems. Comparison of the thermoelastic properties of a composite membrane (such as a red blood cell membrane) to those of pure, single-component lipid bilayers (e.g., large vesicles) gives quantitative information about the chemical state of the composite mixture in relation to the pure system (both of which are assembled as anisotropic structures). If we change the composition of the vesicle, we may determine what effects are produced by various components, e.g., cholesterol, integral membrane proteins, peripheral membrane proteins, etc. These components contribute in different ways to the solid and liquid material behavior of the membrane. For the red cell, current evidence supports the thesis that peripheral proteins (e.g., spectrin) are the origin of the solid elastic properties of the membrane. Also, the lipid components in the outer bilayer are considered to exist in the liquid state. Thus, the bilayer system is a mixture of a lipid-cholesterol solvent with integral protein solutes. In this section, we will analyze the thermoelastic data that have been obtained for red cell membranes. We will compare the thermoelastic properties that are expected for a lipid bilayer system to those measured for the red cell membrane. Then, we will speculate on the relationship of the thermodynamic results to the hypothetical membrane model described above.

First, we consider the surface equation of state which represents both solid and liquid forms of membrane materials such as red cell membrane and lipid bilayer vesicles, respectively. In Section 4.11, we demonstrated that two equivalent forms for the equation of state exist in a closed membrane system: (1) surface pressure as a function of surface area and temperature; (2) isotropic tension as a function of surface area and temperature. In any experiment, we can only measure differential changes in surface pressure. These are equal and opposite in sign to the changes in isotropic tension. The absolute value of surface pressure (which is the time average exchange of momentum between membrane molecules in the plane of the surface) is inaccessible. We can only model it by arbitrary relations such as a form of Van der Waal's equation of state in two dimensions. The two forms for the equation of state have one observable in common, i.e., the isothermal compressibility modulus, K, which is proportional to the change in surface pressure or isotropic tension for a fractional change in surface area at constant temperature. As we discussed in Section 5.2, the area compressibility modulus may be measured by mechanical experiments which cause area dilation. We can also measure the heat of expansion of the surface, which is the product of temperature times the thermoelastic change in isotropic tension. The area compressibility modulus and thermoelastic effect characterize the differential equation of state for the closed surface, i.e., mechanochemical equations of state. From Equations 4.11.6 and 4.11.15 we recall that

$$d\overline{T} = K \, d\alpha + \left(\frac{\partial \overline{T}}{\partial T}\right)_{\alpha} \, dT \tag{5.5.1}$$

and

$$d\pi = -K \, d\alpha + \left[\frac{d\gamma}{dT} - \left(\frac{\partial \overline{T}}{\partial T}\right)_{\alpha}\right] dT$$

The reversible heat of expansion for the membrane is given by the thermoelastic effect, which is the temperature coefficient in the differential equation of state,

$$T \left(\frac{\partial \widetilde{S}}{\partial \alpha}\right)_{T} = -T \left(\frac{\partial \overline{T}}{\partial T}\right)_{\alpha} \tag{5.5.2}$$

We have already discussed the area compressibility measurements, so we will now consider the reversible heat of expansion and the thermoelastic effect. In Section 4.11, we derived the heat of expansion as

$$T \left(\frac{\partial \widetilde{S}}{\partial \alpha}\right)_{T} = T \cdot K \cdot \left(\frac{\partial \alpha}{\partial T}\right)_{\overline{T}} = T \cdot K \cdot \left(\frac{\partial \alpha}{\partial T}\right)_{\pi} - T \frac{d\gamma}{dT} \tag{5.5.3}$$

or

$$T \left(\frac{\partial \widetilde{S}}{\partial \alpha}\right)_{T} = T \left(\frac{\partial \pi}{\partial T}\right)_{\alpha} - T \frac{d\gamma}{dT} \tag{5.5.4}$$

If the surface pressure equation of state is simply proportional to temperature,

$$\pi = T \cdot f(\alpha)$$

then the internal heat of expansion is equal to the surface pressure,

$$T \left(\frac{\partial \pi}{\partial T}\right)_{\alpha} = \pi$$

This feature demonstrates that the heat of expansion is a measure of thermal repulsive forces in the membrane. An example is a Van der Waal's equation of state for a surface gas or noncohesive liquid,

$$\pi = \frac{cT}{(\widetilde{A} - \widetilde{A}_{e})}$$

which was used in examples for Sections 4.6 and 4.11. Even for more complicated equations of state, e.g., with cohesive forces between surface molecules, it is apparent

THERMAL AREA EXPANSIVITY

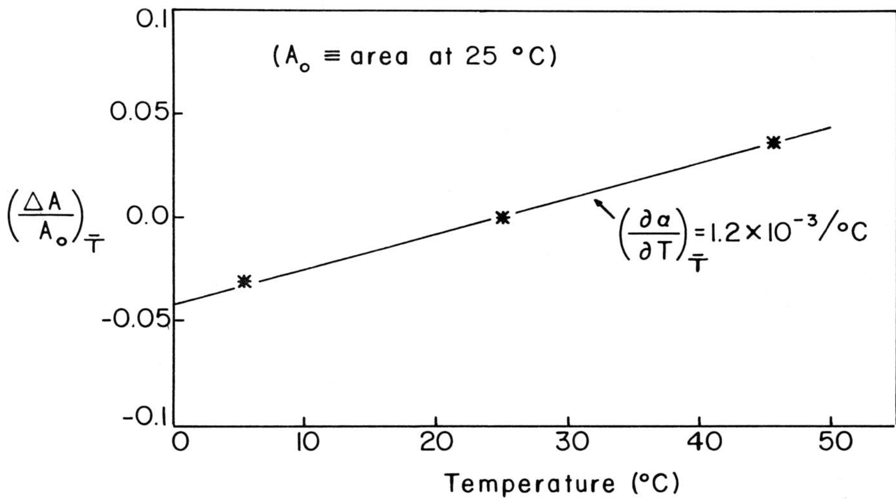

FIGURE 5.29.   Apparent area change of a single red cell relative to its area at 25°C vs. temperature. The apparent area change is calculated with the assumption that the cell volume is constant. Data given here are for a cell swollen in 575 mOsm salt solution. Thus, the correction for cell volume change is less than 10%. The slope of the line is essentially the thermal area expansivity of the red cell membrane.

that the internal heat of expansion represents the thermal repulsive forces in the surface. This is analogous to Van der Waal's original concept.

The heat of expansion for the closed membrane surface is proportional to the thermal area expansivity at constant isotropic tension. With micropipet aspiration of spheroidal red cells or phospholipid vesicles (as described in Section 5.2), the isotropic tension can easily be controlled by the level of suction pressure. For instance, if a preswollen red cell is held with constant pressure, an increase in temperature produces an increase in the length of the aspirated cell projection that is proportional to the area increase of the cell membrane. Using this method, Waugh and Evans (1978)[26] obtain a value for the thermal area expansivity for red cell membrane of $1.2 \times 10^{-3}/°C$ with a standard deviation of ±10%. The surface area appears to increase linearly over the range of 0 to 50°C (Figure 5.29). This value is about half that measured for phospholipid multilayers with X-ray diffraction techniques (Costello and Gulik-Krzywicki;[10] see the next section for a discussion of X-ray data). The heat of expansion is given by the product of temperature, area compressibility modulus, and thermal area expansivity. The area compressibility modulus for the red cell membrane has been found to depend on temperature (Waugh, 1977).[91] The results are shown in Figure 5.30. From these results, we calculate that the heat of expansion decreases from 196 erg/cm² at 0°C to 116 erg/cm² at 50°C. As we will show, the latter value at 50°C is interesting because it is similar to the value anticipated for a phospholipid bilayer. In the range of 45 to 50°C, the red cell membrane loses its shear rigidity and breaks up into a large vesicle plus many smaller vesicles. Presumably, this is the result of the disruption of the adsorbed spectrin layer into aggregated patches on the interior surface (see Elgsaeter and Branton, 1975).[13b]

The range of values, 196 to 116 erg/cm², that are determined for the red cell membrane is the range of natural heats of expansion when the membrane is free of tension. If the membrane is subject to a nonzero tension, the heat of expansion is given by Equation 4.11.31,

ELASTIC AREA COMPRESSIBILITY MODULUS

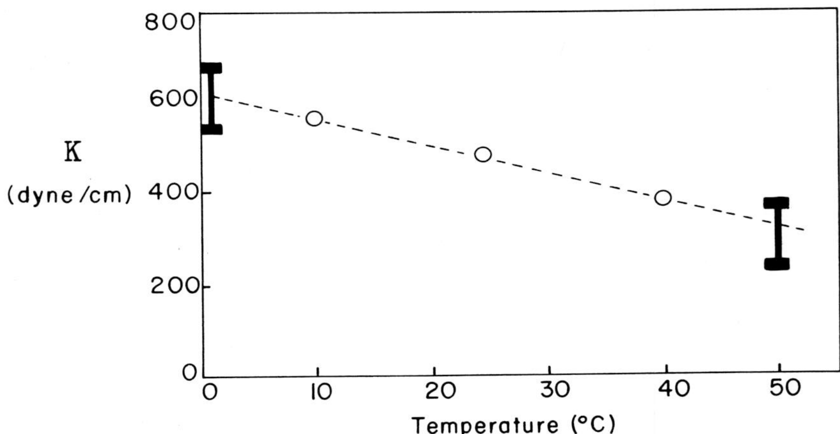

FIGURE 5.30.   The membrane area compressibility modulus measured by micropipet aspiration of red blood cells as a function of temperature. The circles are average moduli for about 20 cells at each temperature. These cells were artificially loaded with high salt concentrations so that cell volume changes were negligible. By comparison, the error brackets are the limits of plus and minus one standard deviation of moduli measured for hundreds of cells with normal salt concentrations, but with the cell volumes corrected for reversible water movements. (From Evans, E. A. and Waugh, R., *Biophys. J.*, 20, 307, 1977. With permission.)

$$T \cdot \left(\frac{\partial \tilde{S}}{\partial \alpha}\right)_T = T \cdot K \cdot \left(\frac{\partial \alpha}{\partial T}\right)_{\overline{T} = 0} - T \left(\frac{d \ln K}{dT}\right) \overline{T}$$

Thus, the temperature gradient of the area compressibility modulus must also be determined. From Figure 5.30, we estimate this to be

$$\left(\frac{d \ln K}{dT}\right) \simeq -0.014/°C$$

With isotropic tensions on the order of 1 dyn/cm, the heat of expansion only differs from the value for the natural state by 4 erg/cm², which is negligible.

Now, we will estimate the heat of expansion for a lecithin bilayer. As in Sections 4.6 and 4.11, we idealize the lipid bilayer behavior as twice the behavior of a lipid monolayer above its phase transition for ordered acyl chains. The bilayer equation of state is chosen as

$$\pi (\tilde{A} - \tilde{A}_e) \simeq (0.7 \times 10^{-15})T$$

which models data for lecithin monolayers at oil-water interfaces (Yue et al., 1975)[98] with $\tilde{A}_e = 38$ Å² per molecule. The surface pressure for the lipid bilayer at constant tension is governed by the effective interfacial tension, γ. Therefore, the change in surface pressure with respect to temperature at constant tension is equal to the change in the interfacial free energy density with temperature.

$$\left(\frac{\partial \pi}{\partial T}\right)_{\overline{T}} \equiv \frac{d\gamma}{dT}$$

TABLE 5.1

Lipid Bilayer Idealized as Two Monolayers (at an Oil-Water Interface)

T (°C)	π(dyn/cm)	$\tilde{A}(\overset{\circ}{A}{}^2)$	K (dyn/cm)	$(\partial\alpha/\partial T)$ (°C^{-1})
0	75	63	190	2.5×10^{-3}
25	70	68	160	2.7×10^{-3}
50	65	73	135	2.9×10^{-3}

With this relation and the equation of state for surface pressure, we can calculate the surface pressure and area per molecule, respectively, for the tension-free bilayer. Table 5.1 demonstrates the estimated values for these variables. The temperature dependence of the surface pressure is assumed to be that of twice the interfacial free energy density for water-hydrocarbon interaction, -0.2 dyn/cm°C. In the next section, we will show that these estimates correlate with values which we will calculate from X-ray diffraction measurements on hydrated multibilayers.

The thermal area expansivity of the lipid bilayer at constant tension (equal to zero) is calculated from Equation 4.11.14,

$$\left(\frac{\partial\alpha}{\partial T}\right)_{\overline{T}=0} = \left(\frac{\partial\alpha}{\partial T}\right)_{\pi} - \frac{1}{K}\frac{d\gamma}{dT} \tag{5.5.5}$$

and the heat of expansion from Equation 5.5.4,

$$T\left(\frac{\partial\tilde{S}}{\partial\alpha}\right)_{T} = \pi - T\frac{d\gamma}{dT} = 130 \text{ erg/cm}^2 \tag{5.5.6}$$

In the spherical state at 50°C, the red cell membrane surface is a two-dimensional liquid mixture. It also appears to have a heat of expansion that is expected for a lipid bilayer. On the other hand, the area compressibility modulus of the red cell membrane at 50°C is more than twice the value expected for a phospholipid bilayer, and the thermal area expansivity is less than half of the expected value for a bilayer membrane. These results indicate that the red cell membrane at 50°C behaves like a mixed phase of compressible lipids and much less compressible components (e.g., cholesterol, integral proteins, etc.). If such a simple concept is reasonable, then the area compressibility modulus would be elevated by an amount proportional to the total area divided by the area occupied by the compressible lipid phase and, likewise, the thermal area expansivity is expected to be lower by the inverse of this quotient. Thus, the heat of expansion would be the same for both the ideal mixture and the lipid phase, which is also consistent with the membrane surface pressure being equal in both cases, i.e., a simple two-phase mixture of compressible and incompressible components.

Such deductions are speculative and often invalid when heterogeneous mixtures are considered. The simple model does provide an abstract basis for comparison to the red cell measurement. For example, if we assume that 45% of the red cell membrane area is occupied by the simple lipid phase at 50°C and the other half by an incompressible component, we can calculate the expected values for the area compressibility modulus, thermal area expansivity, and heat of expansion for the temperature range of 0 to 50°C. Table 5.2 contains these results.

From Table 5.2, it is apparent that this simple model differs from the observed behavior for the red cell membrane as shown by the area compressibility modulus and heat of expansion. The modulus at 0°C is lower than the average value of 600 dyn/

TABLE 5.2

Red Cell Membrane Modeled as 45% Lipid Bilayer Mixture (by
Area at 50°C) with the Remaining Components Incompressible

T(°C)	K(dyn/cm)	$\left(\dfrac{\partial \alpha}{\partial T}\right)$ (°C^{-1})	$T\left(\dfrac{\partial \tilde{s}}{\partial \alpha}\right)_T$ (erg/cm^2)
0	470	1.0×10^{-3}	128
50	300	1.3×10^{-3}	126

cm measured for red cell membrane and the heat of expansion is lower than the calculated value, 196 erg/cm^2, for red cell membrane at 0°C. This is not surprising because the model is a gross oversimplification. A major consideration that was neglected is the effect of the adsorbed spectrin layer underneath the membrane bilayer. Desorption of spectrin with increase in temperature would greatly influence the heat of expansion of the interface. For instance, if we double the temperature gradient of the interfacial free energy density (i.e., to -0.4 dyn/cm°C), then we would match the red cell membrane thermoelastic measurements.

We need to obtain direct measurements of the properties of vesicle membrane systems that contain pure lipids and lipid mixtures with specific components which make up a biological membrane. Alterations of the membrane structure *in situ* by viral insemination and immune reaction, for example, can also be investigated with thermoelastic measurements. The natural heat of expansion is a direct measure of the surface cohesion due to interfacial effects. As such, the heat of expansion may be used to assess the intrinsic resistance to lysis of a cell membrane, which often results from disease or with exposure to degrading chemical agents.

We now consider the surface equation of state that is peculiar to solid membrane materials, i.e., the membrane shear resultant as a function of extension and temperature. Above the phase transition temperature for ordered acyl chains, a phospholipid bilayer is a two-dimensional liquid without surface shear rigidity. Unlike the phospholipid component of a biological membrane, the composite red cell membrane exhibits solid elastic properties; that is, the ability to support membrane shear force resultants in proportion to shear deformation or uniaxial extension in the membrane plane. Consequently, the red cell membrane surface must be characterized by an additional equation of state as discussed in Section 4.11,

$$dT_s = \left(\frac{\partial T_s}{\partial \tilde{\lambda}}\right)_T d\tilde{\lambda} + \left(\frac{\partial T_s}{\partial T}\right)_{\tilde{\lambda}} dT \tag{5.5.7}$$

In the differential form of the deviatoric equation of state, Equation 5.5.7, we assume that the surface area remains constant. This is reasonable for the red cell membrane since the fractional changes in area are less than 10^{-4} for the small force resultants necessary to produce shear deformation. We can also neglect the work and thermoelastic change that are associated with the small area change.

In Section 5.3, we showed that the red cell membrane exhibits first order hyperelasticity in its solid regime of material behavior. The first order deviatoric equation of state at constant temperature is the elastic constitutive relation,

$$T_s = \frac{\mu}{2} (\tilde{\lambda}^2 - \tilde{\lambda}^{-2}) \tag{5.5.8}$$

where the shear modulus is assumed to be independent of extension. Using this relation, we developed the coefficients for the differential expansion of the deviatoric equation of state in Section 4.11 as

$$\left(\frac{\partial T_s}{\partial \tilde{\lambda}}\right)_T = \mu \, (\tilde{\lambda} + \tilde{\lambda}^{-3})$$

$$\left(\frac{\partial T_s}{\partial T}\right)_{\tilde{\lambda}} = \left(\frac{d\mu}{dT}\right) \frac{(\tilde{\lambda}^2 - \tilde{\lambda}^{-2})}{2} \tag{5.5.9}$$

or

$$\left(\frac{\partial \, \ell n \, T_s}{\partial T}\right)_{\tilde{\lambda}} = \frac{d \, \ell n \, \mu}{dT} \tag{5.5.10}$$

Equation 5.5.10 is the fractional change in shear resultant with temperature for a specific or constant material extension, i.e., the thermoelastic change for the tension deviator. The thermoelastic effect is related to the reversible heat of extension by Equation 4.11.36,

$$T \, \left(\frac{\partial \tilde{S}}{\partial \tilde{\lambda}}\right)_T = -T \, \frac{d\mu}{dT} \, (\tilde{\lambda} - \tilde{\lambda}^{-3}) \tag{5.5.11}$$

As we showed in Section 4.7, a simple two-dimensional, entropic elastomer would have a positive thermoelastic change, inversely proportional to absolute temperature,

$$\left(\frac{\partial \, \ell n \, T_s}{\partial T}\right)_{\tilde{\lambda}} = \frac{1}{T}$$

and, likewise, the membrane shear modulus would increase with temperature in such an entropic material model. This is expected since an elastomer derives its free energy change from the reduction of configuration entropy and thus would exhibit a heat of extension that is equal and opposite in sign to its shear modulus times $(\tilde{\lambda} - \tilde{\lambda}^{-3})$.

Because of the hyperelastic behavior exhibited by the red cell membrane and the magnitude of its shear modulus, it was anticipated that the membrane shear elasticity was due to the negative changes in configurational entropy when the material was deformed, i.e., due to ordering of the structural component. However, recent micropipet experiments by Waugh (1977)[91] have shown that the hyperelastic red cell membrane is not this type of simple entropic elastomer. The membrane elastic shear modulus decreases with temperature (see Figure 5.31). The configurational entropy of the membrane increases with extension. Extension of the membrane is endothermic, i.e., heat is taken up by the membrane. The heat of extension is calculated from Figure 5.31 and Equation 5.5.11 to be

$$T \, \left(\frac{\partial \tilde{S}}{\partial \tilde{\lambda}}\right)_{T,\alpha} = 2 \times 10^{-2} \cdot (\tilde{\lambda} - \tilde{\lambda}^{-3}) \; erg/cm^2$$

with the temperature gradient of the shear modulus of -6×10^{-5} dyn/cm°C (Waugh, 1977).[91] Also, the reversible change in internal energy density with extension at constant area is given by Equation 4.11.36,

$$\left(\frac{\partial \tilde{E}}{\partial \tilde{\lambda}}\right)_{T,\alpha} = 2.6 \times 10^{-2} \cdot (\tilde{\lambda} - \tilde{\lambda}^{-3}) \; erg/cm^2$$

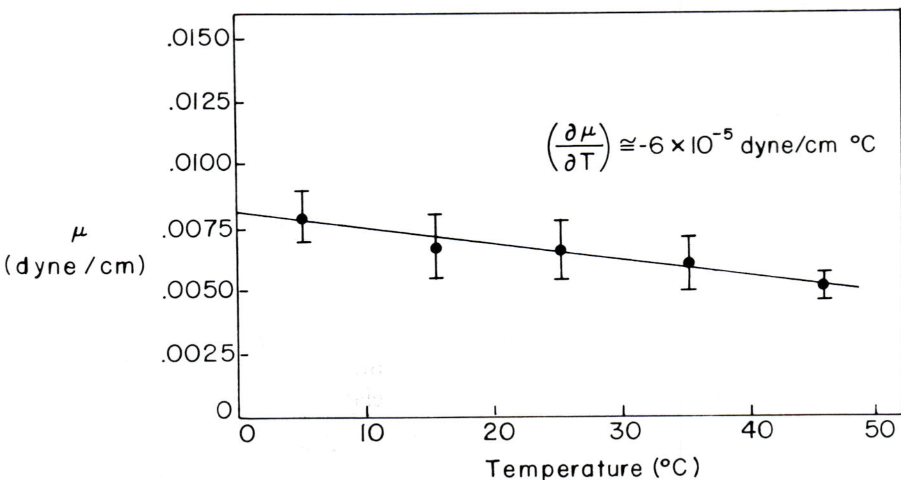

FIGURE 5.31.   The membrane shear modulus measured by micropipet aspiration of red blood cells as a function of temperature. Points are the average of about 25 cell moduli at each of the five experimental temperatures. Brackets are plus or minus one standard deviation. The solid line is the linear regression to the data. (From Waugh, R. E., Temperature Dependence of the Elastic Properties of Red Blood Cell Membrane, Ph.D. dissertation, Duke University, Durham, N.C., 1977. With permission.)

The difference between these two relations is the free energy density change at constant temperature and density which is produced by extension,

$$\left(\frac{\partial \tilde{F}}{\partial \tilde{\lambda}}\right)_{T,\alpha} = \left(\frac{\partial \tilde{E}}{\partial \tilde{\lambda}}\right)_{T,\alpha} - T \left(\frac{\partial \tilde{S}}{\partial \tilde{\lambda}}\right)_{T,\alpha} = \mu\,(\tilde{\lambda} - \tilde{\lambda}^{-3})$$

(5.5.12)

where the coefficient, μ, is the surface elastic shear modulus. It is apparent that the small shear modulus is the difference between two contributions of nearly equal magnitude. Consequently, the resolution of the temperature dependence of the shear modulus is critical in the determination of the free energy decomposition.

Biochemical and ultrastructural evidence strongly support the thesis that the spectrin material (adsorbed on or bound to the cytoplasmic face of the red cell membrane) is responsible for the shear elastic or solid behavior of the red cell membrane. As yet, little is known about the direct relation of spectrin to integral membrane proteins and the amphiphilic lipid components themselves. It is clear, however, that spectrin prefers the interfacial phase where it is adsorbed onto the membrane surface as opposed to dissolving in an aqueous phase of moderate ionic strength ($>0.01\ M$). The natural state of spectrin in a normal red cell membrane is disrupted by temperatures in the range of 45 to 50°C, low pH (≈ 5), and very low ionic strength aqueous media (e.g., distilled water).* At high temperature or low pH, spectrin still resists solution in the cytoplasmic phase. Instead it self-aggregates, forming patches or clumps on the membrane surface. At low ionic strengths, spectrin will go into solution but some self-aggregation persists. The implication is that the conformation of the spectrin molecule is very sensitive to the balance of hydrophobic and electrostatic interactions. Furthermore, it

* ATP depletion also appears to affect the spectrin configuration, but the evidence is deduced from red cell crenation (echinocyte formation).

appears that for the normal environment present in the red cell cytoplasm (pH and ionic composition), spectrin favors an adsorbed interfacial phase or layer configuration. Such a configuration is a moderately ordered state because the molecule is restricted to essentially two dimensions. The thermoelastic measurements for membrane shear deformation show that the configurational entropy is increased by extension relative to the natural state. This indicates that extension forces the molecular complexes to disorder or utilize more degrees of freedom. If the concept of a spectrin layer is valid, we expect that portions of the molecule would have to rotate out of the plane (e.g., by trans-gauche rotation) when extended with the constraints imposed by constant surface area and molecular flexibility. Another aspect of extension at constant surface area is that the material is compressed in the direction normal to the axis of extension (and in the plane of the membrane). The compression or condensation of dense surface material such as spectrin (while it is extended in the other direction) could easily disrupt the organization or order of the molecular arrangement. In any case, such actions essentially force spectrin to go partially into solution in the cytoplasmic interior.

Recent evidence (Bennett and Branton, 1977)[2] indicates that spectrin may be bound to an integral membrane protein or fragment of protein. In addition, spectrin exists in at least dimeric form as well as monomeric units when eluted from the red cell membrane and may be associated with other membrane proteins (Steck, 1975).[86] Hence, extensional deformation of the red cell membrane (at constant surface area) could produce reversible (since it is elastic) changes in the association of these various membrane proteins by disrupting the normally ordered subsurface structure. This, too, is consistent with the observed entropy increase.

The likelihood that spectrin is the primary structural element for surface shear rigidity and the plausibility of the previous arguments indicate that the heat of extension and internal energy density change should be considered on a per mole basis of spectrin material. The surface density of spectrin for the red cell membrane is on the order of 10^{-7} g/cm^2, and its molecular weight is on the order of 10^5 daltons. Therefore, the calculated molar density is about 10^{-12} mol/cm^2. With this value plus the measurements for the heat of extension of the red cell membrane and internal energy density change produced by extension, the values distributed per mole of spectrin are obtained. First, the reversible heat of extension is calculated to be

$$T \left(\frac{\partial \tilde{S}}{\partial \tilde{\lambda}} \right)_T \sim 2 \times 10^{10} \cdot (\tilde{\lambda} - \tilde{\lambda}^{-3}) \; \text{erg/mol}$$

at 25°C. Next, the internal energy of extension is obtained as

$$\left(\frac{\partial \tilde{E}}{\partial \tilde{\lambda}} \right)_T \sim 2.6 \times 10^{10} \cdot (\tilde{\lambda} - \tilde{\lambda}^{-3}) \; \text{erg/mol}$$

at 25°C. In terms of calories per mole, these functions are

$$T \left(\frac{\partial \tilde{S}}{\partial \tilde{\lambda}} \right)_T \sim 500 \cdot (\tilde{\lambda} - \tilde{\lambda}^{-3}) \; \text{cal/mol}$$

$$\left(\frac{\partial \tilde{E}}{\partial \tilde{\lambda}} \right)_T \sim 650 \cdot (\tilde{\lambda} - \tilde{\lambda}^{-3}) \; \text{cal/mol}$$

When the interfacial phase is mixed with the adjacent aqueous phase, the increase in heat content due to the utilization of more degrees of freedom by the large molecule would be on the order of the gas constant, 8.3×10^7 erg/K/mol, times the absolute temperature, i.e.,

$$N_A \cdot k \cdot T \sim 2.5 \times 10^{10} \text{ erg/mol}$$

$$= 600 \text{ cal/mol}$$

This is the same order as the measured value for the heat of extension per mole of spectrin material at constant surface density. As indicated originally, a natural tendency to increase entropy is represented by a positive heat of extension. In equilibrium, this is opposed by the internal energy or enthalpy required to mix two components such as spectrin and the aqueous phase. The energy changes are expected to be the result of exposure of normally sheltered hydrophobic regions of spectrin to the aqueous phase. Thus, we are prompted to ask the following questions: how much of the spectrin molecule is hydrophobic; what is the free energy for transfer of a hydrophobic group to an aqueous phase; and, therefore, what proportion of the hydrophobic region would necessarily be exposed to account for the measured internal energy change that is produced by membrane extension?

Estimates of the number of hydrophobic groups are available, e.g., Marchesi et al. (1969),[57] but the free energy of transfer from a lipid layer interface to an aqueous phase is not known. The free energy of transfer has been measured (Nozaki and Tanford, 1971)[65] for various amino acid residues from organic solvents such as ethanol and dioxane to water. These results, combined with the composition data, give an estimate (Waugh, 1977)[91] of the total energy potentially available in the transfer of spectrin residues to the aqueous phase. It is about 300 kcal/mol! This is a tremendously large value compared to that measured for red cell extensional deformation, 600 to 700 cal/mol. Consequently, it appears that a half percent increase in exposure of the hydrophobic portion of spectrin could account for the energy stored in shear deformation. This subtlety demonstrates the valuable insight that can be derived from thermomechanical experiments in conjunction with biochemical analysis. However, it is emphasized that extensive data must be obtained for membranes of well-defined compositions in order to make reliable deductions. The previous discussion on spectrin and the red cell is clearly speculative, but we hope that it will stimulate further studies.

5.6 Thermoelasticity and Area Compressibility of Multilamellar Lipid Phases and Water

Throughout our discussions, we have alluded to and made use of the amphiphilic nature of phospholipid molecules. Again, we recall that these molecules are characterized by a polar head group which has a high affinity for aqueous solutions and two acyl chains (hydrocarbon polymers) that are attached to the glycerol backbone of the polar end. The hydrocarbon chains are essentially insoluble in the aqueous media. This amphiphilic feature of lipids and other membrane molecules (e.g., proteins, cholesterol, etc.) is responsible for the natural association that forms bilamellar structures, e.g., single-walled vesicles and plasma membranes. As yet, we know little about the mechanical behavior of the two-dimensional liquid material that makes up the continuous structure. In previous examples, we have abstracted lipid bilayer properties from the behavior of insoluble lipid monolayers spread at oil-water interfaces. However, there have been recent experiments that provide data for calculation of the elastic area compressibility modulus and thermal area expansivity for a lipid bilayer directly. In addition to the calculation of elastic and thermoelastic properties, we will be able to examine the separation of the free energy density due to surface expansion into an area dependent part and a part which depends only on the solution properties of the aqueous environment. We postulated this partition in Section 4.6. The result is that the isotropic tension, $\bar{T}$, is equal to the difference, $\gamma - \pi$, between the interfacial interaction and the internal surface pressure. The net force resultant is the isotropic tension, which is zero in the absence of externally applied forces.

The particular experiments that we will describe take advantage of the regular lamellar repeat distance of the multibilayer and water strata. In these experiments, the thickness of a single bilayer and the water gap distance between bilayers are derived from X-ray diffraction measurements as a function of water content. When water is removed, an osmotic stress is created because of the natural affinity of the polar head groups for water, i.e., the activity of water in the system is reduced. The osmotic stress acts to condense the lipid surface as well as to bring the bilayers closer together. This is analogous to squeezing water out of an elastic sponge. In this case, water is drawn out of the lipid multibilayer sponge by reducing the activity of water in the environment external to the multibilayer-water mixture.

The affinity of lipids for water has been known for many years (Elworthy, 1961; Luzzati, 1968).[14,56] The dependence of the weight fraction of water was measured as a function of the vapor pressure of water in equilibrium with the mixed phase of lipid and water (Elworthy, 1961).[14] With gravimetrically determined lipid/water weight fractions, the bilayer repeat spacing was measured by X-ray diffraction of the regular multilamellar structure. Then, the X-ray and weight fraction data were combined to provide the bilayer thickness and water gap distance between bilayers as a function of lipid weight fraction (Reiss-Husson, 1967 and LeNeveu et al., 1977)[53,72] as shown in Figure 5.32 for egg lecithin. Experimental data for the work required to remove water from the mixed phase of lecithin lamelli and water has been reported recently (LeNeveu et al., 1977).[53] The technique was to use osmotic pressure created by solutions of high molecular weight dextran to draw water out of the interbilayer space (since dextran cannot penetrate this space). With these results, the work per molar volume of water required to dehydrate the lecithin multilayers may be determined as a function of the water gap distance between bilayers (LeNeveu et al., 1977).[53]

The change in chemical potential of water specifies the work done per mole of water exchanged with the mixture at equilibrium. Consequently, when divided by the partial molar volume of water, the work per molar volume of water has units of pressure. LeNeveu et al. have presented their data accordingly as shown in Figure 5.33. The vapor pressure measurements of Elworthy may also be used to calculate the work required to remove 1 mol of water from the multibilayer and water mixture. The vapor pressure data give values that are plotted as open circles in Figure 5.33 and the stars are for the osmotic pressure measurements with dextran solutions reported by LeNeveu et al. (1977).[53] There is an apparent discontinuity between these two sets of results at the water spacing of 17 Å. The reason for this difference is not clear. It is difficult to evaluate because the weight fractions of lipid/water in the multilamellar phase are not known accurately when suspended in the dextran environment.

From the combined X-ray diffraction and gravimetric data, it has long been recognized that the bilayer thickness increases with progressive dehydration (Reiss-Husson, 1967; Luzzati, 1968; LeNeveu et al., 1977).[53,56,72] Since the lipids are essentially volumetrically incompressible (Liu and Kay, 1977),[55] the increase in bilayer thickness is accompanied by a commensurate decrease in the area of the surface occupied by a lipid molecule. Consequently, it is expected that this surface compression will produce an increased surface pressure which is associated with a surface elastic free energy increase. We will now consider the magnitude of the surface pressure change for the bilayer and the origin for the lateral compression of the lipid component. From this development, we will show that the elastic area compressibility modulus is close to the

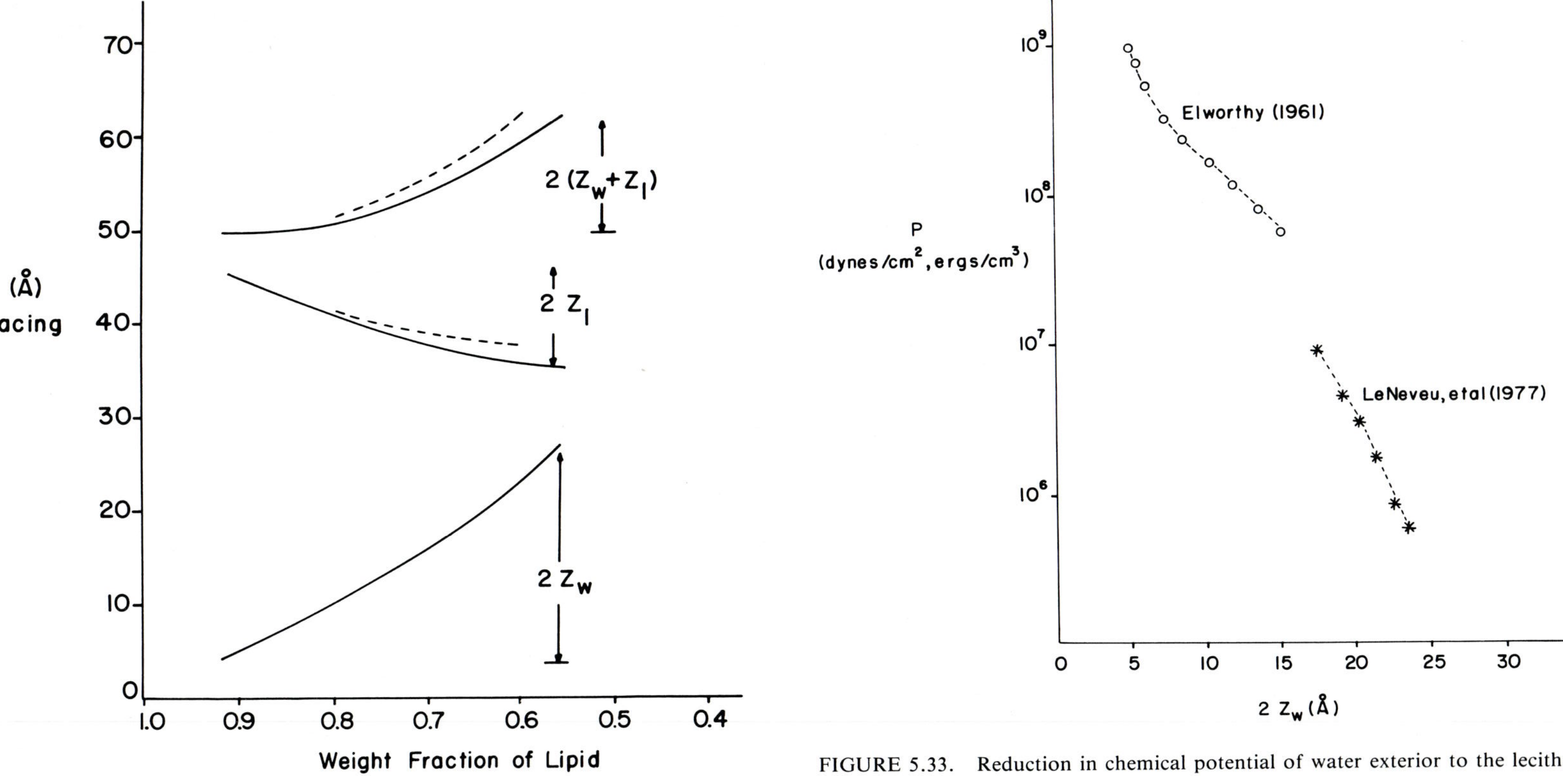

FIGURE 5.32.　Combined results from X-ray diffraction measurements of lamellar repeat spacing and gravimetric measurements of lipid weight fraction. The lamellar repeat spacing, $2 \cdot (Z_w + Z_\ell)$, is plotted as a function of weight fraction of lipid for egg lecithin. The bilayer thickness, $2 \cdot Z_\ell$, and the water spacing, $2 \cdot Z_w$, between bilayers are deduced from the relative volume fractions of lecithin and water, respectively. The solid curves are for data by LeNeveu et al. (1977).[53] The dashed curves are for data by Reiss-Husson (1967).[72]

FIGURE 5.33.　Reduction in chemical potential of water exterior to the lecithin/water mixture per molar volume of water, P, plotted against the water spacing determined from the lamellar repeat distance and gravimetric measurements, Figure 5.32. The open circles are from experiments on vapor hydration of lecithin, published by Elworthy (1961). The stars are osmotic pressures of dextran solutions used to dehydrate lecithin/water mixtures, LeNeveu et al. (1977).[53] P has pressure units of dyn/cm² or erg/cm³.

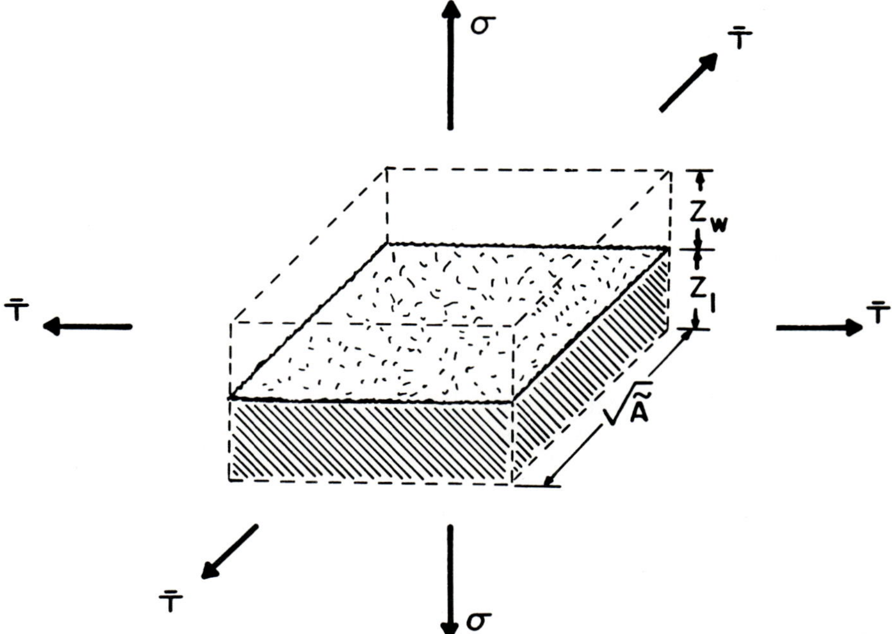

FIGURE 5.34.   Schematic illustration of the unit cell per lipid molecule with surface area, A; water half-space, Z_w; and single layer (half-bilayer) thickness, Z_ℓ. Also shown are the principal force resultants that characterize the composite system: $\bar{T}$ is the isotropic tension (force per unit edge length) that acts parallel to the plane of the lipid layer; σ is the stress resultant (force per unit area) that acts along the outward normal to the unit cell, defined by the lamellar plane.

value abstracted from monolayer data. Similar methods of analysis are being used by Parsegian and coworkers (1978)[67] to investigate both surface elasticity and repulsion between bilayer surfaces across the water gap. The first step in our analysis is to examine the work required to deform the lipid and water structure. The isothermal work is equal to the differential change in free energy defined by the combined first and second laws of thermodynamics.

The work of water exchange to or from the lamellar lipid-water configuration can be expressed in terms of the work required to deform a unit cell of the multibilayer system. The unit cell is defined by a single lipid layer thickness, Z_ℓ, plus half the water space, Z_w, between layers, and the surface area, $\tilde{A}$, of a lipid molecule in the layer. The unit cell with appropriate dimensions is schematically illustrated in Figure 5.34. At constant temperature, the work is equal to the change in free energy produced by deformation of the lipid layer plus the free energy required to change the space between bilayers.* The total work per molecule of lipid is given by

$$\tilde{W} = \int \left(\frac{\partial \tilde{F}}{\partial \tilde{A}}\right)_{T,Z_w} d\tilde{A} + \int \left(\frac{\partial \tilde{F}}{\partial Z_w}\right)_{T,\tilde{A}} dZ_w \tag{5.6.1}$$

where the partial derivatives of the free energy per molecule, $\tilde{F}$, are taken along independent paths for the equilibrium process; T is the temperature. The partial derivative of the free energy with respect to area defines the isotropic tension (force per unit

* Since the lipid component is essentially volumetrically incompressible, only the change in area per molecule need be considered (Evans and Waugh, 1977; Evans and Hochmuth, 1978).[22,24] However, if large hydrostatic pressures are employed, this treatment must be modified to include the volumetric compression of each phase.

length) that acts in the plane of the lipid layer as developed in Section 4.4 (Evans and Waugh, 1977; Evans and Hochmuth, 1978),[22,24]

$$\overline{T} \equiv \left(\frac{\partial \widetilde{F}}{\partial \widetilde{A}} \right)_{T,Z_w} \tag{5.6.2}$$

The partial derivative of the free energy with respect to the half thickness of the water space defines a surface traction or stress resultant (force per unit area) that acts along the outward normal to the water space, perpendicular to the plane of the lipid layer,

$$\sigma \cdot A \equiv \left(\frac{\partial \widetilde{F}}{\partial Z_w} \right)_{T,\widetilde{A}} \tag{5.6.3}$$

Figure 5.34 shows the unit cell with the force resultants defined by Equations 5.6.2 and 5.6.3. Equation 5.6.1 is now written as

$$\widetilde{W} = \int \overline{T} \cdot d\widetilde{A} + \int \sigma \cdot \widetilde{A} \cdot dZ_w \tag{5.6.4}$$

There are two volume conservation equations that must be satisfied. The first specifies the change in volume of the unit cell,

$$d\widetilde{V} = (Z_\varrho + Z_w) \cdot d\widetilde{A} + \widetilde{A} \cdot d(Z_\varrho + Z_w) \tag{5.6.5}$$

and the second represents the volumetric incompressibility of the lipid component,

$$0 = Z_\varrho \cdot d\widetilde{A} + \widetilde{A} \cdot dZ_\varrho \tag{5.6.6}$$

The work that is required to remove water may be called the work of lipid hydration. It is given by the change in chemical potential, $\Delta\mu$, of the water exterior to the lipid/water mixture times the number of moles of water, dn_w, that are exchanged between the mixture and the environment in the process. Thus,

$$W = - \int \Delta\mu \cdot dn_w$$

This can be expressed on the basis of one molecule of lipid as

$$\widetilde{W} = - \int \Delta\mu \cdot d\widetilde{n}_w \tag{5.6.7}$$

where $d\widetilde{n}_w$ is the number of moles of water exchanged per lipid molecule. Therefore,

$$\nu_w \cdot d\widetilde{n}_w = - d\widetilde{V}$$

and

$$\widetilde{W} = \int \frac{\Delta\mu}{\nu_w} \cdot d\widetilde{V}$$

where ν_w is the partial molar volume of water. The ratio, $\Delta\mu/\nu_w$, has units of pressure (dyn/cm² or erg/cm³) and will be identified by the variable, $-P$, such that P is a positive quantity. We now use the volume conservation equations 5.6.5 and 5.6.6 to express the work per molecule of lipid as

$$\widetilde{W} = \int (-P) \cdot Z_w \cdot d\widetilde{A} + \int (-P) \cdot \widetilde{A} \cdot dZ_w \tag{5.6.8}$$

The work of hydration is equal to the work of deformation of the unit cell. Therefore, we equate the two expressions for total work per molecule of lipid, Equations 5.6.4 and 5.6.8, to obtain the following relation:

$$0 = \int (\overline{T} + P \cdot Z_w) \cdot d\widetilde{A} + \int (\sigma + P) \cdot \widetilde{A} \circ dZ_w$$

If there are no constraining forces that act at the boundaries of the lipid layers, then the change in area per molecule and change in water thickness may be assumed to be independent functions. Therefore, the coefficients of $d\widetilde{A}$ and dZ_w must be zero. This gives the result that

$$P = \frac{-\overline{T}}{Z_w} = -\sigma \tag{5.6.9}$$

Consequently, the isotropic tension in the lipid layer can be obtained from the change in chemical potential of the exterior environment by

$$\overline{T} = \frac{\Delta\mu \cdot Z_w}{\nu_w} \tag{5.6.10}$$

We see from Equation 5.6.9 that the work of removal of water creates a lower hydrostatic pressure in the water gap that is in equilibrium with the repulsion between layers, σ, and in equilibrium with the force resultant in the plane of a layer, $\overline{T}$, distributed across the water gap. We examine next the compressibility of the bilayer surface and the effects of alterations in the chemical environmnt. Therefore, we focus on the isotropic tension or force resultant in the plane of the lipid layer.

The isotropic tension of the lipid layer is obtained from the partial differentiation of the free energy per lipid molecule with respect to area change at constant distance between layers. For small changes in area per molecule, the isotropic tension can be represented by the difference between the interfacial free energy density, γ, of water that exchanges with the interstitial hydrocarbon interface and the lateral surface pressure, π, which is the time averaged exchange of lateral momentum between amphiphiles alone as we discussed in Section 4.6 (Evans and Waugh, 1977).[24]

$$\overline{T} = \gamma - \pi \tag{5.6.11}$$

The superposition of free energy contributions of lipid-lipid interaction plus the free exchange of water with the interstitial hydrocarbon interface between polar head groups has been utilized by Tanford (1974)[89] in the successful treatment of micelle formation and by Defay and Prigogine (1966)[13] in the thermodynamics of insoluble monolayers. The primary basis for this decomposition is that the concentration of amphiphiles in the exterior environment can be neglected and the surface layer can be considered as a closed system, but water is free to exchange with the interstitial hydrocarbon interface as the area per molecule changes. We will present evidence here to support the free energy partition.

Surface pressure is assumed to depend on temperature and surface area per lipid molecule in a surface equation of state, $\pi = f(T, \widetilde{A})$. Effectively, γ is an interfacial tension, independent of area change, which represents the hydrophobic interaction. Although by definition it does not depend on area per molecule, γ will depend on the state of the aqueous region with which water is ultimately exchanged. This is the result of the assumption that water is free to exchange with the interfacial hydrocarbon region. Consequently, a change in interfacial free energy density, $\Delta\gamma$, is determined as if the two regions were in direct contact (this is the principle of Gibbs, 1961).[33] For instance, it is anticipated that temperature, pressure, and exterior solution composition will affect the interfacial interaction of water with the hydrocarbon region between polar head groups. Such changes are observed for interfacial tension of free water interfaces.

The hypothesis represented by Equation 5.6.11, the chemical character that is assumed for the interfacial free energy density, γ, and the surface pressure, π, require verification. Surface pressure is an intensive force resultant specific to an anisotropic, lamellar liquid. In monolayers, the surface pressure is defined by the change in surface tension of the substrate interface with the presence of the insoluble monolayer. The general definition that we use is the time averaged exchange of lateral momentum between amphiphiles in the surface. From the perspective of chemical equilibrium, the surface pressure may be viewed as associated with the reduced activity of water in an interfacial phase as developed in Section 4.6 (Defay and Prigogine, 1966; Evans and Waugh, 1977).[13,24] The monolayer definition and measurement will not necessarily coincide with surface pressure changes in bilayer structures without substrates. However, the change in surface pressure produced by a change in area per molecule at constant temperature is not yet known for bilayer membranes. Consequently, monolayer experiments provide the only data for *a priori* estimation of the elastic energy necessary to deform lipid bilayers. There is support for the use of monolayer data. For example, estimates of elastic area compressibility modulus, $-\tilde{A}(\partial\pi/\partial\tilde{A})_T$, have been made from electrocompression, micropipet aspiration of large vesicles, and bending fluctuations of cylindrical bilayer tubes (see Section 4.9 and Evans and Hochmuth, 1978, for discussion).[22] These estimates give magnitudes on the order of 10^2 dyn/cm. Such estimates are consistent with modulus determinations from monolayer data of surface pressure vs. area per molecule. Even though surface pressure data is different for air-water and oil-water interfaces, little difference is seen if a limited range is taken relative to an initial area per molecule of 70 $\overset{\circ}{A}{}^2$ for lecithin. Figure 5.35 shows the change in surface pressure vs. reduction in area per molecule relative to 70 $\overset{\circ}{A}{}^2$ for egg lecithin spread at both air-water and oil-water interfaces. These data are for room temperature ($\sim$25°C), replotted from Van Deenan et al. (1962),[90] and Yue et al. (1975),[98] respectively. We estimate the initial area per molecule using 1267 $\overset{\circ}{A}{}^3$ for the lecithin molecular volume and the bilayer thickness of 36 $\overset{\circ}{A}$ determined from LeNeveu et al. (1977).[53] Knowing the reduction in surface area per lipid molecule relative to the initial value of 70 $\overset{\circ}{A}{}^2$, we can determine the change in surface pressure. The elastic area compressibility modulus of a single lipid layer is about 80 dyn/cm for a surface area per molecule of 70 $\overset{\circ}{A}{}^2$ as calculated from Figure 5.35.

Changes in the interaction of water with hydrocarbon in bilayer membranes also remain to be measured for alterations in aqueous environment. However, it is possible to use data obtained for surface tension of free water interfaces to test the hypothesis that we posed earlier. If the state of water in the exterior environment is altered, then it is expected that the interfacial free energy density for water interaction with the interstitial hydrocarbon will also change. Thus, the differential form for the isotropic tension is given by

$$d\overline{T} = d\gamma - d\pi$$

and

$$d\overline{T} = \frac{\partial\gamma}{\partial\phi}\cdot d\phi - K_T\cdot\frac{d\tilde{A}}{\tilde{A}} \tag{5.6.12}$$

where ϕ is a variable that characterizes the state of water in the external medium, e.g., temperature, pressure, medium composition, etc. The isothermal modulus of compressibility is defined by

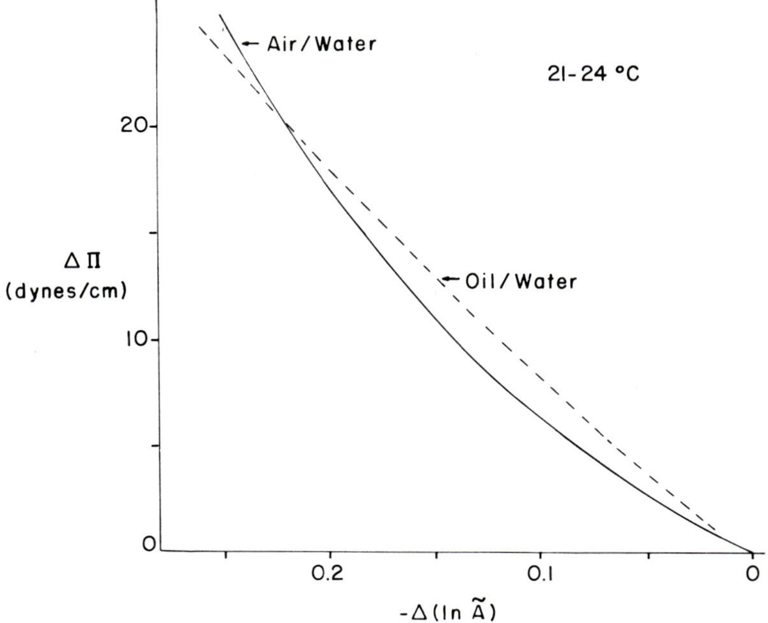

FIGURE 5.35.   Surface pressure increases, $\Delta\pi$, produced by reduction in area per molecule relative to 70 $Å^2$ for lecithin. The abscissa is the logarithm of the area per molecule divided by 70 $Å^2$, $\ln Ã - \ln 70$. The solid curve is data taken from the Van Deenan et al. (1962),[90] which also appears in Gaines (1966),[32] for a lecithin monolyer at an air-water interface (the particular lecithin was γ-stearoyl-β-oleyl-L-α-lecithin, an analog of egg lecithin). The dashed curve is data taken from Yue et al. (1975),[98] for a lecithin monolayer at an oil-water interface (the particular lecithin was 1,2 dioleoyl-lecithin with two unsaturated chains). The surface pressure at 70 $Å^2$ was about 19 dyn/cm for the air-water interface and 33 dyn/cm for the oil-water interface.

$$K_T \equiv -\tilde{A} \left(\frac{\partial \pi}{\partial \tilde{A}} \right)_T \tag{5.6.13}$$

which is also given by

$$K_T = \tilde{A} \left(\frac{\partial \overline{T}}{\partial \tilde{A}} \right)_{T,\gamma}$$

for the case where the state of water in the exterior environment is held constant. An example of a change of interfacial free energy density produced by solution changes is shown in Figure 5.36 for the increase in surface tension of sucrose-water solutions at 20°C (taken from *CRC Handbook of Chemistry and Physics*). A change in interfacial tension of about 2.1 dyn/cm occurs for a 40% by weight sucrose solution.

LeNeveu et al. (1977)[53] mixed lecithin with sucrose solutions. For constant lipid weight fraction, the water spacing between lecithin bilayers appeared to stay constant, but there was a slight increase in bilayer thickness, as show in Figure 5.37 (replotted from LeNeveu et al.). The bilayer thickness changed from 38.9 to 40 $Å$ as measured from the linear fit presented in the paper by LeNeveu et al. This change is equivalent to a 2.6% change in area per molecule according to the conservation equation, (5.6.6). If we assume (as did the previous authors) that water exchanges freely with the sucrose solution and that the sucrose is uniformly distributed in the mixture, there should be

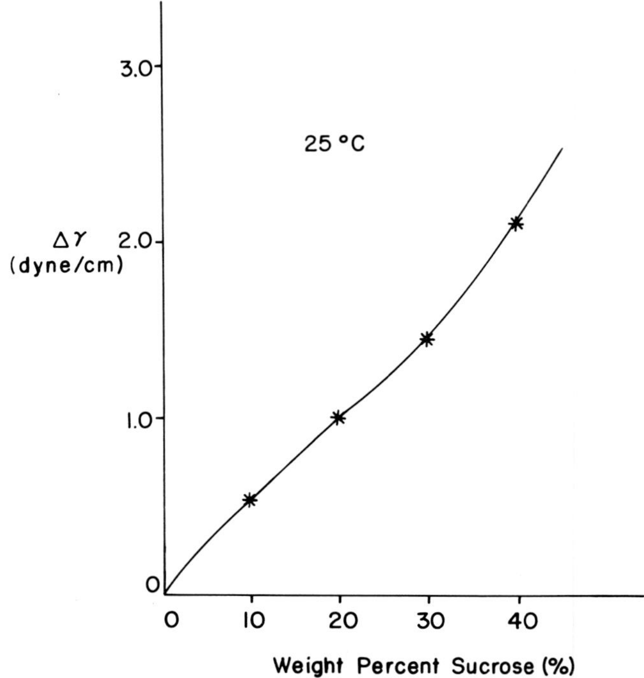

FIGURE 5.36.   Increase in interfacial tension or free energy density, $\Delta\gamma$, estimated by the increase in surface tension of sucrose solutions over pure water at an air interface. Data are for 20°C, taken from *CRC Handbook of Chemistry and Physics* (1972).[11]

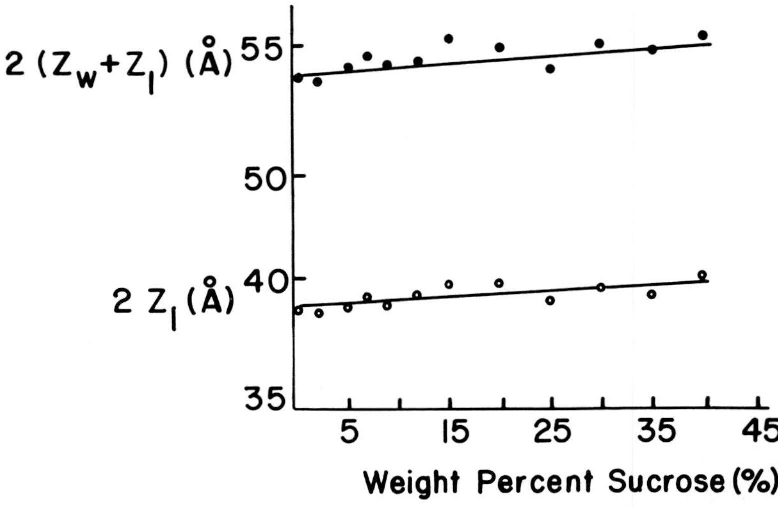

FIGURE 5.37.   Lamellar repeat spacing, $2 \cdot (Z_w + Z_\ell)$, measured by X-ray diffraction for a fixed volume fraction of lipid (0.72) as a function of sucrose concentration. The bilayer thickness, $2 \cdot Z_\ell$, is obtained from the volume fraction of lipid and the repeat spacing. This data is taken directly from the article by LeNeveu et al. (1977).[53]

no change in force resultant in the bilayer surface, i.e., zero tension change. Therefore, from Equation 5.6.12, the change in interfacial free energy density divided by the fractional change in surface area gives an estimate of the surface compressibility modulus, K_T, for a lipid layer,

$$0 = \frac{\partial \gamma}{\partial \phi} \, d\phi - K_T \, d \, (\ell n \, \tilde{A})$$

If we use the data for interfacial tension change of a sucrose-water solution, we would estimate that the compressibility modulus is

$$K_T \cong 80 \ \text{dyn/cm}$$

in order to account for the fractional change in surface area. This value agrees with the value derived from the lecithin monolayer data, Figure 5.35. A lecithin bilayer is represented by twice the single layer modulus of surface compressibility, $2K_T \simeq 160$ dyn/cm.

Another example of the effect of the adjacent aqueous environment is the effect of temperature change. A model for the change in interfacial free energy density, γ, is the decrease in interfacial tension at oil-water interfaces with increase in temperature. The temperature coefficient depends on the hydrocarbon chain length for paraffin oils (Adam, 1968).[1] The nominal value for the temperature gradient of the interfacial tension is -0.1 dyn/cm°C. In addition, Costello and Gulik-Krzywicki[10] have measured the bilayer repeat distance with X-ray diffraction as a function of temperature. The ratio of lipid to water was fixed for each sample, then the X-ray patterns were observed for each sample as a function of both increase and decrease in temperature. The heating and cooling phases were essentially the same, well within the limits of experimental resolution. Thus, their method provides the relative changes in bilayer thickness and water gap as a function of temperature. The fractional change in bilayer thickness with increase in temperature was determined to be in the range of -2 to $2.7 \times 10^{-3}/°C$ for egg lecithin multibilayers and water. Since the volumetric thermal expansivity is much smaller than this value (Liu and Kay, 1977),[55] the fractional change in bilayer thickness is a good measure of the opposite fractional change in area, i.e., the thermal area expansivity. As we discussed in Section 4.11, the thermal area expansivity that is observed in an experiment implicitly represents both the heat of expansion of membrane amphiphiles and the heat of expansion of the interfaces, e.g.,

$$\frac{1}{\tilde{A}} \left(\frac{\partial \tilde{A}}{\partial T} \right) = \frac{\pi}{K_T \cdot T} - \frac{1}{K_T} \frac{d\gamma}{dT} \tag{5.6.14}$$

where K_T is the isothermal modulus of compressibility. Using the data described above and monolayer data, we can evaluate Equation 5.6.14. For instance, in a lecithin monolayer, the surface pressure at 70 Å^2 per molecule is about 19 dyn/cm for air-water interface and about 33 dyn/cm for oil-water. The modulus of compressibility for a single layer (half of a bilayer value) is about 80 to 90 dyn/cm for either interface type (Figure 5.35). Therefore, the first term in Equation 5.6.14 is between 0.8 to $1.3 \times 10^{-3}/°C$, which is considerably less than the measured value from X-ray diffraction. The temperature gradient of oil-water or air-water interfacial tension is on the order of -0.1 dyn/cm°C. Using this value, we estimate the second term in Equation 5.6.14 to be 1.2 to $1.3 \times 10^{-3}/°C$. Summing the two contributions, we see that the predicted thermal area expansivity for lecithin is 2.0 to $2.6 \times 10^{-3}/°C$, which agrees well with the X-ray data. We now synthesize the effects of water removal and changes in interfacial free energy density, γ, to evaluate the vapor hydration or dehydration of multibilayer and water mixtures. This will provide a correlation of bilayer

TABLE 5.3

Vapor Pressure and Gravimetric Data for Egg Lecithin from Elworthy[1961]

Relative vapor pressure X	Water/lipid weight ratio	Water spacing $2 \cdot Z_w$ (Å)
0.50	0.10	5.0
0.60	0.113	5.3
0.70	0.135	6.0
0.80	0.163	7.1
0.85	0.20	8.5
0.90	0.25	10.3
0.92	0.29	11.8
0.945	0.34	13.5
0.96	0.39	15.0

compressibility in the multilamellar configuration with monolayer compressibility data.

Elworthy (1961)[14] used sulfuric acid solutions to establish the vapor pressure of water to which a fixed quantity of lecithin was exposed in a separate weighing bottle. The lecithin absorbed water and was allowed to reach equilibrium. The weight fractions of lipid and water were determined simultaneously for these equilibrium mixtures. The work per molar volume of water exchanged with the lecithin and water mixture at equilibrium is determined by the chemical potential change of water vapor relative to the reference state. The work per molar volume of water exchanged at equilibrium is calculated with the following relation plus the relative vapor pressure data of Elworthy (1961):[14]

$$P = - \frac{k \cdot T \cdot N_A}{v_w} \cdot \ell n\ X$$

where k is Boltzmann's constant, N_A is Avogadro's number, and X is the relative vapor pressure of the sulfuric acid solutions. At 25°C, this relation is

$$P = -1.38 \times 10^9 \cdot \ell n\ X \quad (\text{erg/cm}^3)$$

Table 5.3 contains a list of relative vapor pressures as a function of water: lipid weight ratios taken from the article of Elworthy (1961)[14] for egg lecithin at 25°C. Also listed in Table 5.3 are the corresponding values of water spacing determined from the data of LeNeveu et al. (1977).[53] The water gap space between bilayers is calculated from X-ray diffraction and gravimetric measurements. The open circles in Figure 5.33 are Elworthy's data plotted vs. water spacing from LeNeveu et al. (1977).[53] The work per molar volume of water exchanged at equilibrium is given by the combination of Equations 5.6.9 and 5.6.11 as

$$P = \Delta\pi/Z_w \tag{5.6.15}$$

Table 5.4 gives the corresponding values for the work per molar volume of water, P, taken from Elworthy's vapor pressure data as a function of the calculated water gap, $2 \cdot Z_w$. Thus, the change in surface pressure may be calculated with the values in Table 5.4 and Equation 5.6.15,

$$\Delta\pi = Z_w \cdot P$$

TABLE 5.4

Surface Pressure Changes Calculated from Work per Molar Volume of Water Removed from Lecithin Multilayers

Water space $2 \cdot Z_w(\text{Å})$	Work per molar volume of water $P(\text{dyn/cm}^2)$	Surface pressure increase $\Delta\pi(\text{dyn/cm})$
5.0	9.5×10^8	23.8
5.3	7.0×10^8	18.7
6.0	5.0×10^8	14.9
7.1	3.0×10^8	10.9
8.5	2.3×10^8	9.7
10.3	1.5×10^8	7.8
11.8	1.1×10^8	6.6
13.5	0.77×10^8	5.2
15.0	0.54×10^8	4.1

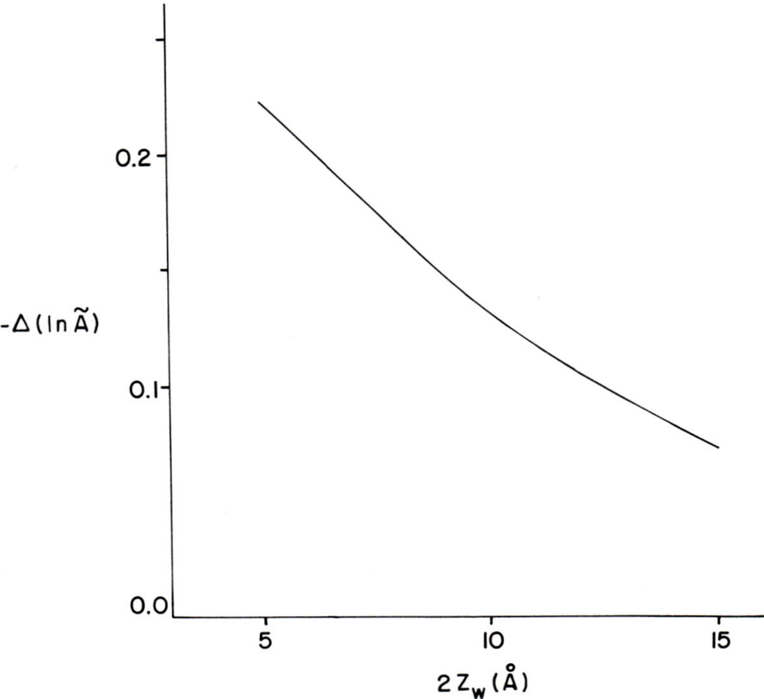

FIGURE 5.38.   The differential change in logarithm of the area per lecithin molecule, $-\Delta(\ell n\ \tilde{A})$, is plotted vs. the water space, $2 \cdot Z_w$ between bilayers. The differential change in logarithm of the area per molecule is obtained using the bilayer thickness and conservation equation: $\Delta(\ell n\ Z_\ell) = -\Delta(\ell n\ \tilde{A})$. The values are calculated with the data of LeNeveu et al. (1977),[53] as presented in Figure 5.32.

The results are also listed in Table 5.4. For an initial bilayer thickness of 36 Å, the change in logarithm of the surface area is determined by the bilayer thickness data in Figure 5.32 and the conservation equation 5.6.6. Figure 5.38 is a plot of the differential change in the logarithm of the area per molecule as a function of the water space, $2 \cdot Z_w$, between bilayers. Now, the surface pressure increase can be plotted as a function

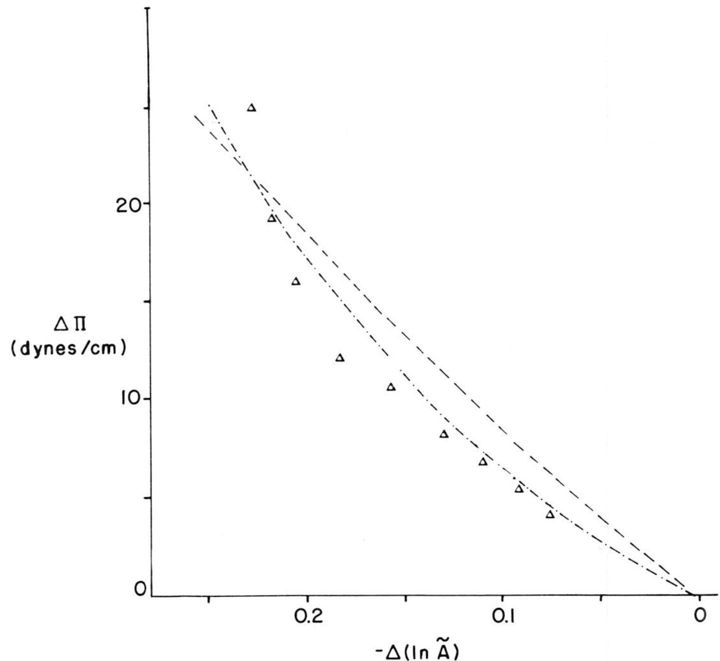

FIGURE 5.39.   The increase in surface pressure vs. the differential change in logarithm of area per molecule is presented for the analytical results (triangles) determined from the vapor hydration of lecithin (Elworthy, 1961).[14] These results are values plotted from Table 5.4 and Figure 5.38. The analytical results correlate well with the surface pressure data obtained from monolayer experiments (Figure 5.35), shown here as broken curves.

of the change in logarithm of the surface area. Figure 5.39 presents the change in surface pressure calculated with Equation 5.6.15 and the values calculated for the vapor hydration experiments (shown as triangles). Also plotted are the lecithin monolayer surface pressure relations from Figure 5.35. The correlation is good and the modulus of surface compressibility, K_T, for a lecithin layer is the order of 80 dyn/cm, or 160 dyn/cm for a bilayer, which agrees with the value deduced from the change in bilayer thickness produced by sucrose solutions.

It appears that the surface compressibility modulus calculated from vapor hydration experiments on lecithin-water multibilayer systems correlates well with lecithin monolayer data. By decomposing the free energy of the amphiphilic surface into the contribution of lipid-lipid interaction plus the contribution of the interfacial interaction of water with the exposed hydrocarbon interstices, we have shown that it is also possible to use the change in bilayer thickness produced by sucrose solutions to calculate the surface compressibility modulus. This method gives a value for a single lecithin layer which closely agrees with the monolayer data and the hydration experiments. In addition, the fractional change in lecithin bilayer thickness with change in temperature has been shown to correlate with the free energy decomposition hypothesis. The surface compressibility modulus for a lecithin bilayer will be twice the value for a single layer, giving a result of about 160 dyn/cm. Coupled with the thermal area expansivity, $(1/\tilde{A})(\partial \tilde{A}/\partial T)$, of 2 to 2.7 × 10⁻³/°C, the reversible heat of expansion at constant temperature for a lecithin bilayer can be calculated by

$$T \left(\frac{\partial \tilde{S}}{\partial \tilde{A}} \right)_T = T \cdot \frac{K_T}{\tilde{A}} \cdot \left(\frac{\partial \tilde{A}}{\partial T} \right)$$

where $\tilde{S}$ is the entropy per molecule of lecithin. The heat of expansion of the bilayer is calculated to be 100 to 135 erg/cm² or 5.2 to 7.0 kcal/mol of lecithin. The heat of expansion is similar in magnitude to the value obtained with thermoelastic experiments on the red cell membrane, Section 5.5.

5.7 Viscoelastic Recovery of and Response to Membrane Extension

Even though a membrane material resumes its initial shape after external forces are removed, the deformation process or cycle may have been irreversible in a thermodynamic sense. The application of force and the associated material deformation take place over a finite time period which implies that the rate of deformation is always nonzero. Thus, some internal dissipation of work by heat generation always occurs; this heat is lost to the environment. The loss of mechanical work is apparent as hysteresis for a cyclic process at constant temperature. In a mechanical experiment, it is even possible that the material response may be dominated by frictional or dissipative force resultants if sufficiently large rates of deformation are involved. The material may still be considered as an elastic solid if its original shape can be recovered, but the constitutive behavior must reflect the time-dependent or viscoelastic nature of the real material.

In Section 4.12, we developed first-order constitutive relations for viscoelastic behavior of membranes. These equations represent simple superposition of reversible (elastic) energy storage and dissipative (viscous) processes. The dissipation and the frictional force components that result are determined by the inelastic kinetic processes within the membrane molecular structure. The coefficients of viscosity characterize the dissipation that is produced in response to continuous rates of deformation, i.e., long range compared to the microscopic structure. In other words, spatial averages of the kinetic processes are inherent in the continuum properties. By contrast, diffusion constants for kinetic motions within the membrane material also represent dissipative or irreversible behavior, but on a microscopic scale. Diffusion constants and material viscosities can be related (e.g., Stokes-Einstein equation) in some cases for homogeneous, single component fluids. However, for heterogeneous materials such as membranes, this connection cannot be directly established. Even for a single-component bilayer membrane, the diffusion constant is not easily related to the surface viscosity (see Evans and Hochmuth, 1978, for discussion).[22] Here we will only consider the dissipative effects measured by mechanical experiment. Mechanical experiments yield coefficients of viscosity which characterize the membrane as a surface continuum. In order to obtain a value for a membrane coefficient of viscosity, we must first test the postulated viscoelastic constitutive relation, Equation 4.12.18. As is true with elastic constitutive relations, the behavior of a material may be significantly more complicated than the first-order equation, (4.12.18), e.g., the viscosity may depend on deformation and rate of deformation, or the superposition of elastic and viscous processes may not be valid. However, the resolution of micromechanical experiments on cell membranes limits the potential for evaluation of more complicated relations. As we will see for red blood cell membrane, the first order viscoelastic relation models the observed behavior quite well.

Measuring small (less than 10^{-6} dyn) time-dependent forces and displacements ($\sim$ 10^{-4} cm) is extremely difficult. This is common when we study the viscoelasticity of cell membrane surfaces. One very useful approach is to observe the time-dependent recovery of an initially stretched membrane. The problem is greatly simplified because the net level of force resultants in the membrane can be taken as zero, i.e., the elastic restoring forces are balanced by the internal frictional (viscous) forces. With Equation 4.12.18, this is written as

$$T_{ij} \equiv 0 = \overline{T}\delta_{ij} + \frac{2\mu}{(1+\alpha)^2}\,\hat{\epsilon}_{ij} + 2\eta_e\,\widetilde{V}_{ij}$$

Both the isotropic tension and the shear resultant component are identically zero. The result is the time-dependent recovery equations in Section 4.12,

$$\left[\alpha\,(1+\alpha)\right]^{-1}\frac{\partial\alpha}{\partial t} = -\frac{K}{\kappa}$$

$$4\,(1+\alpha)\left[\widetilde{\lambda}\,(\widetilde{\lambda}^2 - \widetilde{\lambda}^{-2})\right]^{-1}\frac{\partial\widetilde{\lambda}}{\partial t} = \frac{-\mu}{\eta_e} \qquad (5.7.1)$$

As an example of membrane viscoelastic recovery, we will consider the extensional recovery of red cell membrane. The area dilation recovery has not been observed because it is faster than the experimental capability to change the applied forces (e.g., suction pressure in a micropipet aspiration experiment), and it is further complicated by the extremely small displacements involved. Indeed, we estimated that the rate of area recovery for a pure lipid bilayer (Example 1, Section 4.12) to be very fast, $\sim 10^{-8}$ sec. On the other hand, recovery of extensional deformations of red cell membrane has been studied and provides an interesting illustration of the constitutive relation for shear viscoelasticity. Neglecting area changes, we expect the extension ratio at any location on the membrane surface to exhibit the following time-dependent behavior from the partial integration of Equation 5.7.1,

$$\widetilde{\lambda}(\xi_i,t) = \left[\frac{\Lambda(\xi_i) + e^{-\widetilde{t}}}{\Lambda(\xi_i) - e^{-\widetilde{t}}}\right]^{1/2}$$

$$\Lambda(\xi_i) \equiv \frac{\widetilde{\lambda}_0^2\,(\xi_i) + 1}{\widetilde{\lambda}_0^2\,(\xi_i) - 1} \qquad (5.7.2)$$

where the spatial coordinates of a material location are presented by ξ_i, and $\widetilde{\lambda}_0$ is the local extension ratio at the initial instant in time where the forces are set to zero (i.e., the material is released). The dimensionless time, $\widetilde{t}$, is the quotient of the real time divided by the material recovery time constant, t_c,

$$t_c \equiv \eta_e/\mu$$

and

$$\widetilde{t} \equiv t/t_c$$

Consequently, each extended element of membrane surface will recover its initial shape in accordance with the nonlinear recovery equation (5.7.2). Figure 5.40 shows the time-dependent recovery of an element for a specific initial value of extension ratio. The time axis is in units of dimensionless time, $\widetilde{t}$. If the membrane recovers in this manner, then we can measure the intrinsic material time constant for the recovery process and with the previously measured value for membrane elastic shear modulus, we can deduce the value for the coefficient of surface viscosity in shear for the membrane,

$$\eta_e = \mu \cdot t_c$$

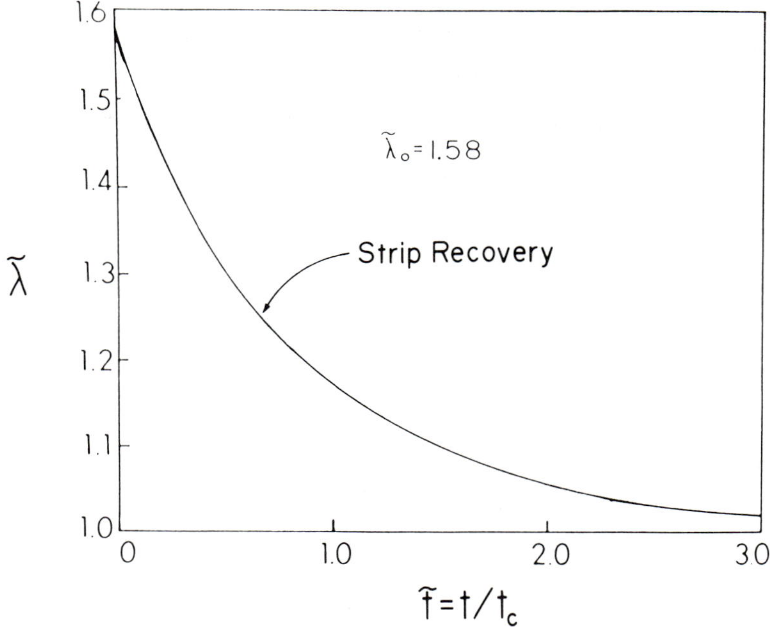

FIGURE 5.40.   The time dependent recovery of a membrane element for a specific initial value of extension ratio. The time axis is in dimensionless units, $\tilde{t} \equiv t/t_c$, where the recovery time constant, t_c, is equal to η_e/μ. This curve describes the extensional recovery of membrane material after it is released.

This coefficient of viscosity represents dissipation in the material as a solid structure.

Recently, Hochmuth has developed a novel experiment for extensional recovery studies on red cell membrane (Hochmuth et al., 1978).[45] This experiment involves pulling a flaccid red cell disk at diametrically opposed locations on the rim of the cell. The cell is attached to glass on one side and pulled with a micropipet from the opposite side. The cell is then released and the recovery is observed with high speed photography. Figure 5.41 is a series of photographs taken of a cell during the recovery phase (kindly provided by Professor R. M. Hochmuth, Duke University, Durham, N.C.). Figure 5.42 shows the data for cell end-to-end length as a function of time for a single cell experiment (taken from Hochmuth et al., 1978).[45] Two major problems must be considered in the analysis of this experiment: the membrane extension (extension ratio, $\tilde{\lambda}$) is nonuniform over the cell surface; dissipation also occurs in the interior and exterior aqueous phases, which may artifactually increase the observed recovery time constant. In principle, the solution given by Equation 5.7.2 provides for the nonuniform membrane extension. However, it is difficult to solve the nonlinear equations of mechanical equilibrium for the curved surface coupled with the elastic constitutive equation, which determines the cell geometry and extension ratio distribution. It is possible, on the other hand, to investigate the effects of nonuniform extension with a flat surface or disk model of the membrane surfaces. A similar approach was used by Evans (1973)[15] to model the elastic behavior of fluid shear deformed red cells that are point attached to a glass substrate. The approach is to treat the disk as a set of small sections (slices or elements) that support uniaxial tension along the direction of extension. By conserving area (two-dimensional incompressibility) and balancing the forces on the element, we obtain the initial, static equilibrium shape with the hyperelastic constitutive relation (Figure 5.43 outlines the procedure). Now, we know the initial extension ratio distribution. Next, the recovery of the elements in the deformed disk is simply

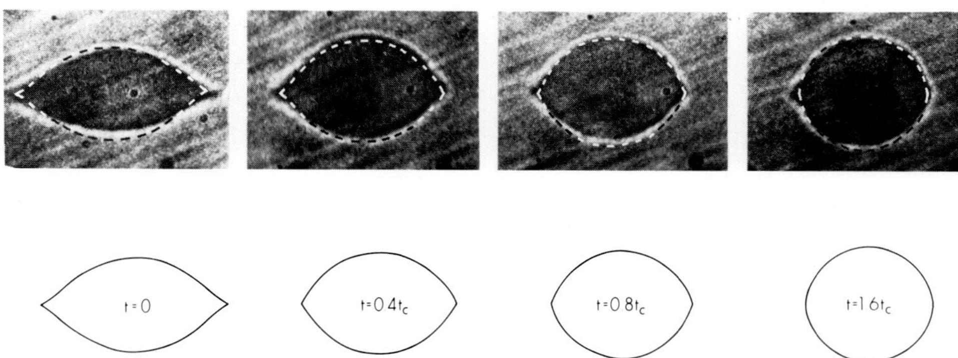

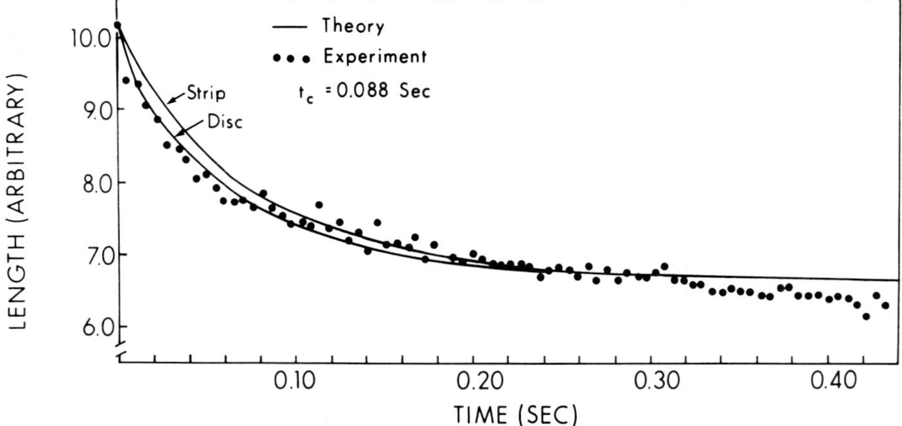

FIGURE 5.41.   Comparison of the extensional recovery predicted for a viscoelastic disk model to that of a series of photographs taken of an actual red blood cell during recovery. The correlation provides a time constant, t_c, for the membrane recovery of 0.10 sec in this case. The dashed lines are an overlay of the model behavior onto the outline of the cell.

FIGURE 5.42.   Extensional recovery of a red cell compared to that of a viscoelastic membrane strip and disk where the resting lengths of the strip and disk are chosen as the cell length after three to four time constants. The viscoelastic constitutive equation, Equation 4.12.17, is a good first-order approximation to red cell membrane behavior in the time dependent elastic domain.

prescribed by Equation 5.7.2. The results are overlaid on the photographs of a red cell experiment in Figure 5.41 and are also shown as the line drawings directly below. Since the extension ratio is large near the attachment locations, the regions proximal to these points will recover faster than the central portion. Therefore, the total end-to-end length of the cell is expected to recover faster than the simple uniform extension of a membrane element or strip (given in Figure 5.40). The effect of the nonuniform extension is not appreciable. Figure 5.42 shows the comparison between the recovery of both a uniformly extended strip and a nonuniformly extended disk with the experimental results. It appears that the disk recovery models the red cell experiment somewhat better. In either case, the viscoelastic constitutive relation, (4.12.17), is a good first order approximation to red cell membrane behavior in the time-dependent, elastic domain.

The data provided by Hochmuth et al. (1978),[45] give an average value of 0.1 sec for the material time constant, t_c, at 25°C. Using this value and the membrane elastic shear

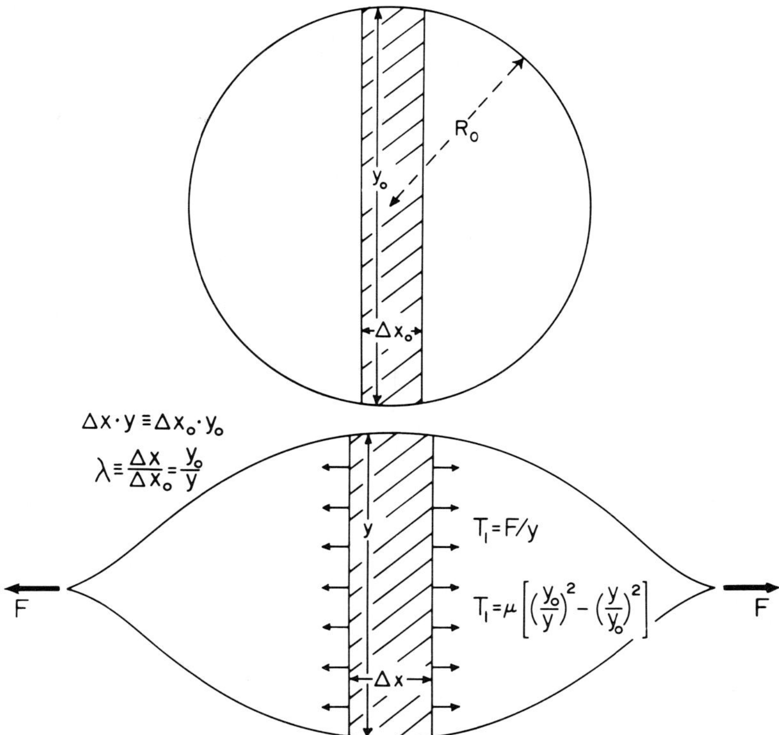

FIGURE 5.43. A diagrammatic outline of the procedure used to model the non-uniform extension of a membrane disk. The approach is to treat the disk as a set of small sections (slices or elements) that support uniaxial tension along the direction of extension. Local areas of elements are conserved during extension and recovery. The initial static force distribution is given by the principal tension, T_1 and the elastic constitutive relation as shown. After release (i.e., $T_1 \equiv 0$), the local extension ratio ($\lambda = y_o/y$) is prescribed by the time dependent function, Equation 5.6.2, which is illustrated in Figure 5.40.

modulus measured by Waugh (1977),[91] $\mu = 6 \times 10^{-3}$ dyn/cm, we determine the coefficient of membrane surface shear viscosity to be

$$\eta_e = t_c \cdot \mu = 6 \times 10^{-4} \text{ dyn-sec/cm}$$

which is in units of surface poise (dyn-sec/cm).

Clearly, there is dissipation in the interior hemoglobin solution, as well as the extracellular medium, that acts to impede the whole cell recovery process. It is possible to estimate the magnitude of such effects by assuming that the rate of deformation (shear rate) is the same in the cytoplasm as in the membrane. This model is a reasonable approximation, since the projected cell surface area remains essentially constant, which implies that the mean thickness is constant during extension and recovery (due to constant cell volume). Therefore, the maximum shear in the liquid interior primarily occurs on planes normal to the disk surface at $\pm 45°$ to the direction of extension and with the same rate of deformation as the membrane surface. In this case, the quotient,

$$\frac{\tilde{\eta} \cdot h}{\eta_e}$$

is the ratio of dissipation in the cytoplasmic interior to the membrane dissipation, $\tilde{\eta}$ is the viscosity of the cytoplasm (poise), and h is the cell thickness from disk surface to surface (cm). Using a conservative estimate for the hemoglobin solution viscosity, 10^{-1} poise (dyn-sec/cm^2), and a thickness of the order 10^{-4} cm, we can estimate the ratio of cytoplasmic dissipation to membrane material dissipation with the value for membrane viscosity just derived from recovery, 6×10^{-4} surface poise,

$$\frac{\tilde{\eta} \cdot h}{\eta_e} \sim 2 \times 10^{-2}$$

The extracellular aqueous phase is neglected because its viscosity is nearly an order of magnitude less than the interior hemoglobin solution. The ratio shows that we need only to consider membrane dissipation. Hochmuth has obtained direct evidence that hemoglobin dissipation can be neglected by observing recovery of red cell ghosts (cells which have been lysed to release the hemoglobin and are subsequently resealed).

Since the lipid component of biological membranes appears to exist in a liquid state, we naturally desire to compare the surface viscosity values of red cell membrane to those of a single component or pure lipid bilayer membrane. Unfortunately, no direct measurements of surface viscosity have been made on lipid bilayer membranes. Lateral diffusion measurements in lipid bilayer membranes have been made (Wu et al., 1977).[95] With these measurements and a surface mobility relation developed by Saffman (1976),[77a] it is possible to deduce a value of surface viscosity for lipid bilayers (see Evans and Hochmuth for discussion, 1978).[22] The value obtained is on the order of 4 $\times 10^{-6}$ surface poise: two orders of magnitude lower than the red cell membrane surface viscosity. This evidence implies that the heterogeneous composite of a biological membrane has significantly larger dissipative mechanisms than the pure lipid membrane system.

The large dissipation in the solid membrane material limits the rate of response of the membrane to applied forces. This effect has important implications for the dynamic response of red cells as they enter small capillaries in the microcirculation. For this reason, Chien and co-workers (1978)[7] have been studying the response of the red cell membrane to a sudden change in suction pressure which is applied to the surface by a micropipet. The technique is similar to the static method outlined in Section 5.3 for measurement of surface elasticity. Here, we will outline an approach that can be used to analyze the time-dependent response of the membrane projection to the sudden pressure increase in the pipet. We will present results from a computer solution based on the analysis. Chien et al. (1978)[7] present an approximate analysis, using a similar approach.

In order to analyze the time-dependent response to a step change in suction pressure, we follow the procedure that was used in Section 5.3 for the micropipet aspiration of a flaccid red cell. The cell surface is approximated by a flat sheet, since the membrane shear force drops off as $1/r^2$ from the pipet tip. The viscoelastic constitutive relation is used with the boundary condition at the pipet tip and the geometric description of deformation in the surface to provide a dynamic equation for the time-dependent length of the cell projection. We summarize these relations in the following equations: first the boundary condition,

$$\frac{\Delta P \cdot R_p}{4} = \int_{R_p}^{\infty} T_s \circ \frac{dr}{r} \tag{5.7.3}$$

next the constitutive behavior,

$$T_s = \frac{\mu}{2} (\tilde{\lambda}^2 - \tilde{\lambda}^{-2}) + 2\eta_e \frac{\partial \ln \tilde{\lambda}}{\partial t}$$

and the time-dependent state of deformation in the membrane surface exterior to the pipet,

$$\tilde{\lambda}^2 = 1 + \left(\frac{R_p}{r}\right)^2 \left(\frac{2L}{R_p} - 1\right) \qquad L \geq R_p$$

when the projection length exceeds the pipet radius. If we combine the last two relations and substitute them into Equation 5.7.3, we obtain the sum of two integral terms. The first integral term is the result that we derived for the elastic behavior in micropipet aspiration of the flaccid cell. The term is essentially linear in the projection length, L. The second term is the integral of the rate of shear deformation over the membrane surface,

$$2\eta_e \int_{R_p}^{\infty} \frac{\partial(\ell n\,\tilde{\lambda})}{\partial t} \; \frac{dr}{r} = 2\eta_e \int_{R_p}^{\infty} V_s \; \frac{dr}{r}$$

Here, it is convenient to introduce the Eulerian expression for rate of deformation. From the incompressibility of the membrane surface, this relation may be written in terms of the velocity at the base of the pipet, v_0 and the instantaneous radius of the material point (see Example 2 of Section 2.6),

$$2\eta_e \int_{R_p}^{\infty} V_s \; \frac{dr}{r} = 2\eta_e \int_{R_p}^{\infty} \frac{v_0 \cdot R_p}{r^3} \; dr = \frac{\eta_e\, v_0}{R_p}$$

$$(5.7.4)$$

For $L \geq R_p$, the velocity of the membrane at the pipet entrance is simply the time rate of change of the projection length,

$$v_0 = \frac{dL}{dt} \qquad L \geq R_p$$

Therefore, the dynamic relation between projection length and time is given by

$$\frac{\Delta P \cdot R_p}{\mu} = \left[\left(\frac{2L}{R_p} - 1\right) + \ell n\left(\frac{2L}{R_p}\right)\right] + \frac{4}{R_p} \frac{dL}{d\tilde{t}}$$

$$(5.7.5)$$

for a step change in suction pressure and for lengths greater than the pipet radius. The time, $\tilde{t} = t/t_c$, is in units of the characteristic response time, $t_c = \eta_e/\mu$, which is also the recovery time constant.

We see that Equation 5.7.5 gives esentially a first order exponential response of the length to the application of the suction pressure. This observation is interesting because the deformation is nonuniform over the membrane surface, and the differential equation for the dynamic response is nonlinear in the extension ratio. The explicit solution of the equations for response of the membrane surface to the step change in pipet suction requires a definition of the projection height of the membrane cap in the range, $0 \leq L < R_p$. Using spherical segments to model this range, we obtain the solution by computer techniques. Figure 5.44 is the computer solution for the viscoelastic response of the membrane surface to a sudden application of pipet suction pressure. We see that the internal dissipation of mechanical energy as heat limits the rate at which the membrane material can deform as a solid structure.

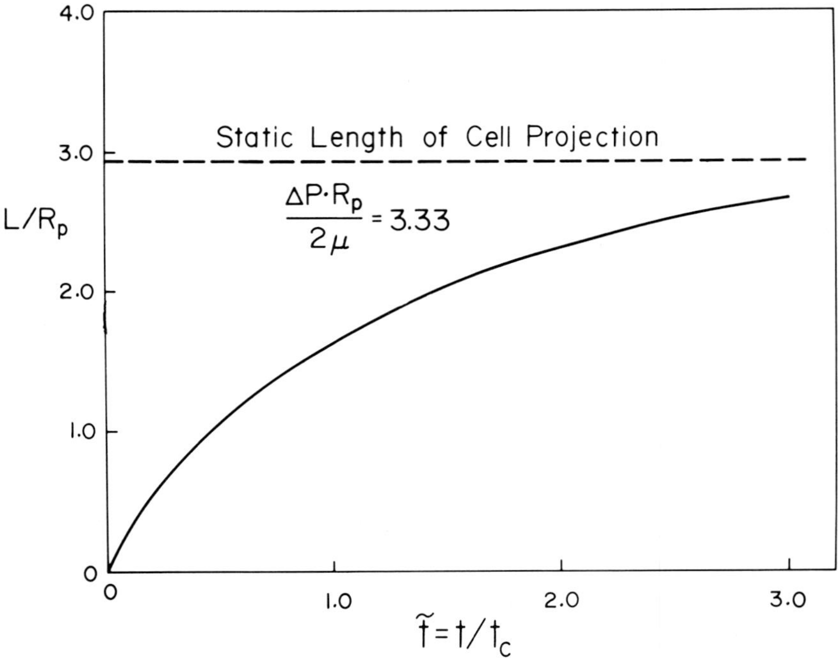

FIGURE 5.44.   The computer solution for the viscoelastic response of a flat membrane surface to a sudden application of pipet suction pressure. The internal dissipation of mechanical energy as heat limits the rate at which the membrane material can deform as a solid structure. The time is normalized by the material response time, $t_c = \eta_c/\mu$.

5.8 Relaxation and Viscoplastic Flow of Membrane

In the previous section, we discussed dissipation that occurs in recoverable deformation. We assumed that after release, the deformed material returns to its original undeformed state. Such behavior is characteristic of an elastic solid and implies that no irreversible changes in structure have taken place. On the other hand, if the material force resultants are of sufficient magnitude or duration, the material structure can permanently deform by structural rearrangement. The relaxation of the force resultants is accomplished by plastic flow of the material. If the force resultant level is maintained by external forces, then the material will continue to flow in a liquid manner. In Section 4.12, we discussed two representative constitutive relations for relaxation and viscoplastic flow processes. In this section, we will describe observations of relaxation and plastic flow of red cell membranes. We will apply the appropriate constitutive relation to the analysis of each experiment, which will provide estimates of the characteristic time limit for solid material behavior and the surface viscosity of the liquid flow of red cell membrane.

Relaxation and creep have been observed for red cell membranes (Evans and LaCelle, 1975).[22a] These experiments were cursory investigations, but they demonstrated the semisolid transition that represents the relaxation and reorganization of molecular structure (specifically spectrin and its associated protein components). The experiments involved the micropipet aspiration of flaccid red cells. The cells were held in the pipet for fixed periods of time and with fixed suction pressures. Then the cells were expelled from the pipet, and the height of the residual membrane bump was measured. The membrane was permanently deformed with residual bump heights that increased in proportion to the length of time the cell was held in the pipet.

To analyze this process, we again use the procedure that was developed in Section

5.3 for micropipet deformation of membrane surfaces of flaccid cells. Here, we employ the constitutive relation for relaxation and creep of the membrane surface, Equation 4.12.29,

$$\frac{\partial \ln \tilde{\lambda}}{\partial t} = f(T_s) \frac{\partial T_s}{\partial t} + \frac{T_s}{2\eta_c} \tag{5.8.1}$$

where the function, $f(T_s)$, is the inverse elastic relation between extension and maximum shear resultant. For a first-order hyperelastic membrane, this function is

$$f(T_s) = \frac{1}{2T_s} \left[\frac{1}{\sqrt{1 + u^2}} \right]$$

where $u \equiv \mu/T_s$. Equation 5.8.1 is the Lagrangian representation of the relaxation and creep behavior. It specifies the time rate of change of the extension ratio at a specific location in the material surface. For the viewpoint of a reference frame fixed in space, the Eulerian form of this equation may be expressed by

$$\frac{-v}{r} \frac{dr}{ds} = f(T_s) \left[\frac{dT_s}{dt} + v \frac{\partial T_s}{\partial s} \right] + \frac{T_s}{2\eta_c} \tag{5.8.2}$$

where s is the curvilinear distance along the membrane meridian as illustrated in Figure 5.16, v is the velocity along the meridian, and the time derivative is the time rate of change as viewed from a stationary or spatial reference frame. We recognize that Equation 5.8.2 is reduced to two different expressions: one for membrane material in the pipet,

$$0 = f(T_s) \left[\frac{dT_s}{dt} + v_0 \frac{dT_s}{dz} \right] + \frac{T_s}{2\eta_c} \tag{5.8.3}$$

where v_0 is the uniform velocity of the projection as it moves up the pipet and z is the axial coordinate of a membrane location in the pipet. The other expression is for the membrane surface exterior to the pipet,

$$\frac{v_0 \cdot R_p}{r^2} = f(T_s) \left[\frac{dT_s}{dt} - \frac{v_0 \cdot R_p}{r} \frac{dT_s}{dr} \right] + \frac{T_s}{2\eta_c} \tag{5.8.4}$$

where the surface incompressibility condition ($-v_0 \cdot R_p = v \cdot r$) has been introduced. Now, we recall the integral relation for the boundary condition,

$$\frac{\Delta P \cdot R_p}{4} = \int_{R_p}^{\infty} T_s \frac{dr}{r} \tag{5.8.5}$$

Along with Equations 5.8.3 and 5.8.4, the integral relation, (5.8.5), specifies the time-dependent length of the membrane projection, since $v_0 = dL/dt$. These equations are functions of the fixed spatial coordinates and time. The equations may be solved iteratively on a computer. For any time, the distribution of the maximum shear resultant is known as a function of the spatial coordinates, r and z. Thus, with the inverse elastic relation, it is possible to factor out the recoverable deformation when the surface is released from the pipet. This leaves the residual or permanent deformation. Figure 5.45 shows the results that are calculated for the residual membrane profiles as a function of the length of time the surface is held in the pipet. The time is in units of the

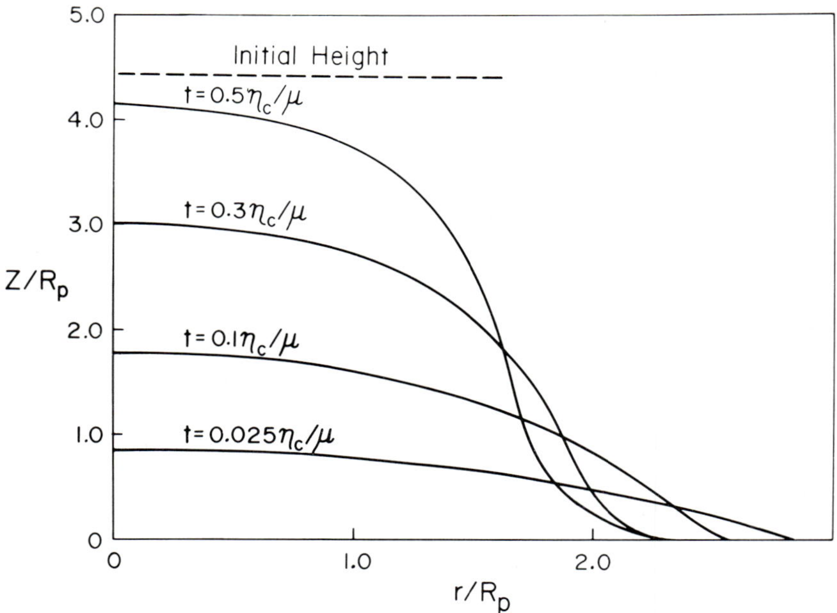

FIGURE 5.45.   Calculated results for the residual membrane profiles as a function of the length of time the surface is held in the pipet. The time periods are in units of the characteristic time constant, η_c/μ, for the relaxation process.

characteristic time constant, η_c/μ, which represents the relaxation process. The same results are obtained if we use the Lagrangian expressions for constitutive behavior, force distribution, and time-dependent extension of material. We will outline the procedure in Example 1 at the end of this section. With the preliminary data of Evans and LaCelle (1975),[22a] we estimate the characteristic time constant, η_c/μ, to be on the order of 10^3 sec and the coefficient of viscosity for the creep and relaxation process to be on the order of 6 dyn-sec/cm (surface poise). This viscosity coefficient is five orders of magnitude larger than the surface viscosity determined for heat dissipation in the solid regime, which indicates that different kinetic processes are involved in each case. From the results in Figure 5.45, we see that the time limit for solid material behavior (i.e., recoverable deformations) is on the order of 0.025 η_c/μ for maximum extension ratios of 3:1. The value is estimated to be about 100 sec for red cell membrane. If the extension ratios are less than 3:1, then the time limit for solid behavior is progressively longer than 100 sec.

Another type of material alteration that is commonly observed with biological membranes is continuous plastic flow (or failure). The irrecoverable flow of material commences when the membrane shear resultant exceeds a threshold or yield shear. As a specific example, we will consider the plastic flow and growth of a microfilament or tether that is pulled from a red blood cell membrane. The tether is produced by fluid shear deformation of the cell which is attached to a glass substrate. Figure 5.46 is taken from Evans and Hochmuth (1977)[21] and shows a scanning electron micrograph of the deformed cell and the membrane tether (original micrograph was furnished by Dr. J. R. Williamson, Washington University). The experimental observations are summarized in Figure 5.47 (taken from Hochmuth et al., 1976).[44a] The evidence is that the cell can elastically support the total force of the fluid shear stress for forces up to 10^{-6} dyn. Two material characteristics are apparent in these observations: (1) there is a yield shear, $\hat{T}_s$, above which the membrane begins to flow immediately in an irrecoverable

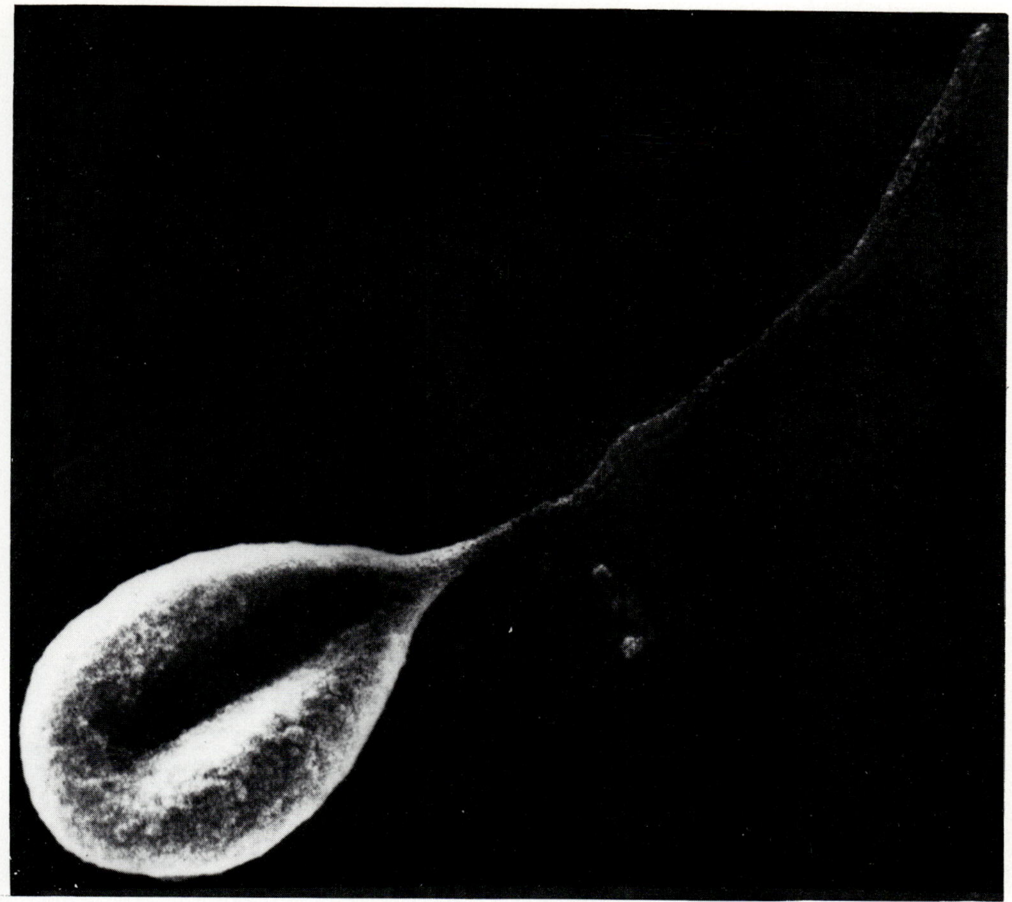

FIGURE 5.46.   A scanning electron micrograph of a fluid-shear deformed red blood cell and a membrane microfilament or tether. (From Evans, E. A. and Hochmuth, R. M., *J. Membr. Biol.*, 30, 351, 1977. With permission.)

manner; (2) there is a coefficient of surface viscosity, η_p, that represents the dissipation in the material flow process. The surface viscosity in plastic flow characterizes the dissipative processes that are produced when the material behaves in a liquid-like manner. However, plastic flow differs from ideal liquid behavior in that there is a nonzero yield shear. We will outline the methods used by Evans and Hochmuth (1976)[20] to analyze the plastic growth of red cell membrane microfilaments.

First, a constitutive relation which represents the observed behavior is chosen for viscoplastic flow of the membrane surface. The following discontinuous behavior was postulated for the membrane shear resultant in Section 4.12: (1) $T_s < \hat{T}_s$: viscoelastic and (2) $T_s > \hat{T}_s$: plastic flow with the rate of deformation given by

$$2\eta_p \tilde{V}_{ij} = F \cdot \tilde{T}_{ij}$$

where the function, F, is a yield function that depends on the ratio of the maximum shear resultant to the yield shear, $T_s/\hat{T}_s$. The simplest form of this relation is the analog of a Bingham plastic,

$$F = 1 - \frac{\hat{T}_s}{T_s} \qquad \text{for } T_s > \hat{T}_s$$

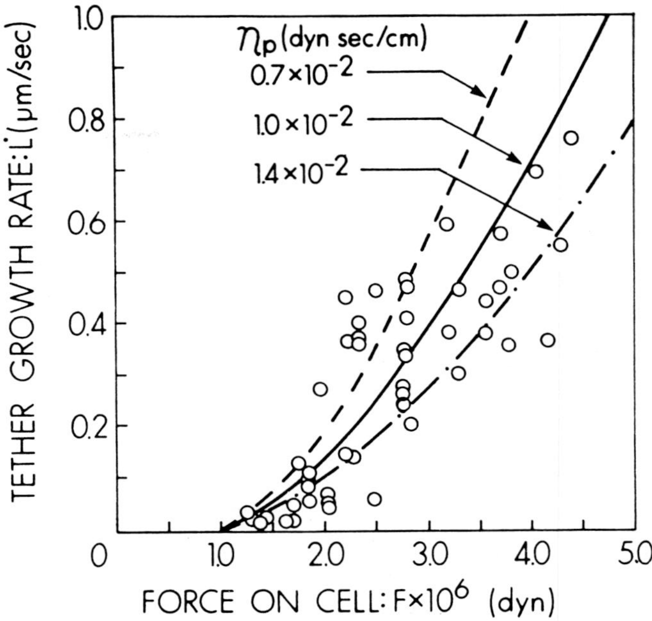

FIGURE 5.47. Experimental observations of continuous, plastic flow (or failure) of red cell membranes. The data show that the cells elastically support the total force of the fluid shear stress for forces up to 10^{-6} dyn. Above this level, the membrane yields and plastically flows with the microfilament growth rates given in this figure. (From Hochmuth, R. M., Evans, E. A., and Colvard, D. F., *Microvasc. Res.*, 11, 155, 1976. With permission.)

The result is Equation 4.12.32, which gives the rate of permanent shear deformation in proportion to the excess shear resultant. The proportionality is determined by the coefficient of surface viscosity for plastic flow,

$$V_s = \frac{1}{2\eta_p}\,(T_s - \hat{T}_s)$$

(5.8.6)

for $T_s > \hat{T}_s$. In Equation 5.8.6, we see the characteristic yield and steady growth that is observed for a membrane tether pulled from a red cell membrane.

Figure 5.48 diagrams the concept of a dissipation region where the plastic yield occurs and membrane material flows to the tether (taken from Evans and Hochmuth, 1976).[20] The cell body is assumed (as observed) to behave in an elastic manner. The total force, F_σ, that acts on the cell surface is given by the extracellular fluid shear stress, σ_s, times the exposed surface area, A_c,

$$F_\sigma = \sigma_s \cdot A_c$$

(5.8.7)

Now, we can describe the mechanical equilibrium of the cell membrane that supports the force, F_σ. If we assume that the hydrostatic pressure in the cell is equal to the external pressure, this axial force must be balanced by the meridional tension times the circumference of any membrane ring in the dissipation region,

$$2\pi r\,\cos\theta\,T_m = F_\sigma$$

(5.8.8)

where the coordinates are shown in Figure 5.48. We assume cylindrical symmetry with the principal axes of the force resultants preserved along the curvilinear coordinates.

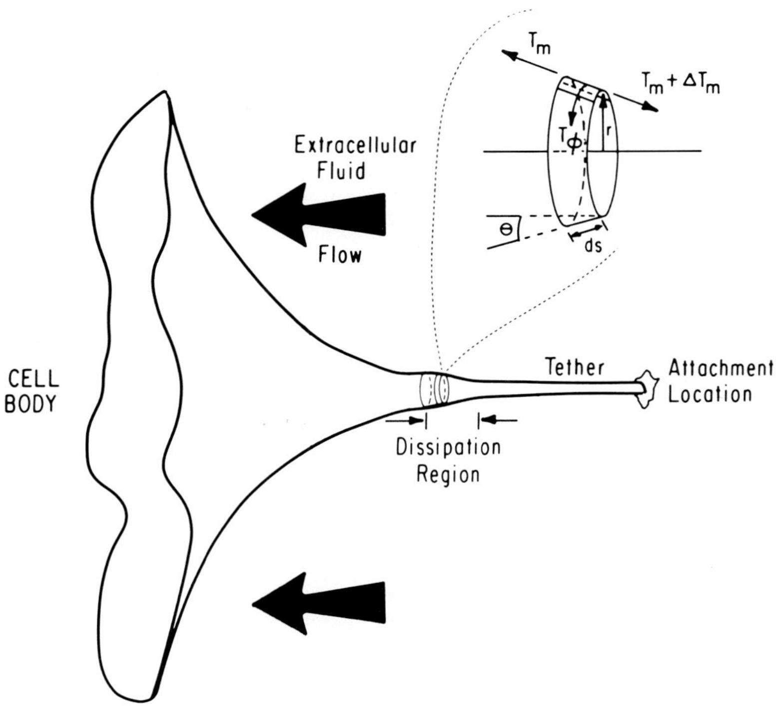

FIGURE 5.48 .   Schematic illustration of the microtether and the region of viscous dissipation or "necking" region where the plastic flow occurs. The geometry of a membrane flow element is shown in the enlarged view. (From Evans, E. A. and Hochmuth, R. M., *Biophys. J.,* 16, 13, 1976. With permission.)

The second equation of mechanical equilibrium may be chosen as the balance of forces tangent to the surface to be satisfied with Equation 5.8.8,

$$r \, \frac{dT_m}{ds} + (T_m - T_\phi) \, \frac{dr}{ds} = 0 \tag{5.8.9}$$

Implicit in the force equilibrium equations, (5.8.8) and (5.8.9), is the assumption that the membrane surface is thin in comparison to the cross-sectional dimension, r. This assumption is not valid in the vicinity of the permanent cylindrical tether. However, the nature of the molecular structure is unknown in this region, and we rely on the first order approximation to model the flow behavior within the yield and dissipation region. The observed feature of a fixed cylindrical radius for the tether is used in the model for the plastic flow process. Specifically, the model simply considers the yield of a cylindrical membrane surface with radius, r_o, at the cell body followed by extrusional flow to a minimum radius given by the permanent tether dimension, r_t. In the region between the cell body and the permanent tether, the membrane surface equilibrium is modeled by the equations of mechanical equilibrium Equations 5.8.8 and 5.8.9. The membrane force resultants are produced by the material dissipation and are assumed to behave as the constitutive relation (Equation 5.8.6).

Equations 5.8.8 and 5.8.9 may be expressed in terms of isotropic and deviatoric force resultants as

$$\overline{T} + T_s = \frac{F_\sigma}{2\pi r \, \cos\theta}$$

$$ r \frac{d\overline{T}}{ds} + r \frac{dT_s}{ds} + 2T_s \frac{dr}{ds} = 0 $$

We combine these to obtain

$$ -2T_s \frac{dr}{ds} = \frac{F_\sigma}{2\pi} \frac{d}{ds} \left(\frac{1}{r \cos \theta} \right) $$

or

$$ T_s = - \frac{F_\sigma}{4\pi} \frac{d}{dr} \left(\frac{1}{r \cos \theta} \right) \tag{5.8.10} $$

Equation 5.8.10 gives the distribution of the deviatoric force resultant at any location on the dissipation surface. For the steady flow of material that has exceeded the yield shear, Equations 5.8.6 and 5.8.10 prescribe the surface kinematics. Consequently, the local rate of shear deformation is given by

$$ V_s = \left(\frac{\hat{T}_s}{2\eta_p} \right) \left[- \frac{F_\sigma}{4\eta \hat{T}_s} \frac{d}{dr} \left(\frac{1}{r \cos \theta} \right) - 1 \right] \tag{5.8.11} $$

At this point, we introduce the Eulerian rate of deformation into Equation 5.8.11. Since the red cell membrane is essentially an incompressible surface, we may use the expression for the rate of shear deformation of a conical surface element, which was derived in Example 2 of Section 2.5,

$$ V_s = \frac{v_s}{r} \frac{dr}{ds} \tag{5.8.12} $$

v_s is the flow velocity for material along the meridional coordinate, s. The surface incompressibility also gives the continuity relation for conservation of material flow,

$$ r \cdot v_s = r_t \cdot v_{s_t} = \text{constant} $$

i.e., the velocity increases with decrease in radius. The final velocity, v_{s_t}, at $r = r_t$, is used to establish the magnitude of surface flow. With the minimum tether radius, r_t, we combine Equations 5.8.11, 5.8.12, and the continuity relation to give

$$ \left(\frac{2\eta_p}{\hat{T}_s} \right) \left(\frac{v_{s_t}}{r_t} \right) \frac{\sin \theta}{\tilde{r}^3} = - \frac{d}{d\tilde{r}} \left[\frac{\tilde{F}_\sigma}{\tilde{r} \cos \theta} + \ell n \, \tilde{r} \right] \tag{5.8.13} $$

where the geometric relation, $dr/ds = \sin \theta$, has been used. The radius, $\tilde{r}$, is in units of tether radii,

$$ \tilde{r} \equiv r/r_t $$

The total force is also in dimensionless form,

$$ \tilde{F}_\sigma \equiv \frac{F_\sigma}{4\pi r_t \cdot \hat{T}_s} \tag{5.8.14} $$

This is the ratio of the force on the cell to the critical force threshold below which no tether growth will occur (Evans and Hochmuth, 1976).[20] From the force balance on the ends of the cylindrical dissipation region, the ration, $\tilde{F}_\sigma$, may be shown to equal the ratio of the radius, r_0, for the yield location to the minimum tether radius, r_t,

$$\frac{r_0}{r_t} = \tilde{F}_\sigma$$

The rate of tether growth, $\Delta L/\Delta t$, is measured between the attachment point and the cell body. Therefore, it is equal to the difference between the initial material velocity, v_{s_0} at $r = r_0$, and the final velocity, v_{s_t} at $r = r_t$, for the tether. With the continuity relation, we determine the tether growth rate from the final velocity and the ratio of radii,

$$\frac{\Delta L}{\Delta t} = v_{s_t}\left(1 - \frac{r_t}{r_0}\right)$$

This equation depends on the force ratio, $\tilde{F}_\sigma$,

$$\frac{\Delta L}{\Delta t} = v_{s_t}\left(1 - \tilde{F}_\sigma^{-1}\right) \tag{5.8.15}$$

Equation 5.8.15 expresses the nature of the yield process: the tether growth rate increases with force levels greater than the threshold value, $\tilde{F}_\sigma > 1$, as is shown in Figure 5.47. A dimensionless growth parameter, G_t, can be defined by

$$G_t \equiv \frac{8\,\pi\eta_p}{F_\sigma^{crit}} \cdot \frac{\Delta L}{\Delta t} \tag{5.8.16}$$

The critical or threshold force, F_σ^{crit}, is given by

$$F_\sigma^{crit} = 4\pi r_t \hat{T}_s$$

and is measured by the intercept in Figure 5.47 at 10^{-6} dyn for the red cell membrane. Using the dimensionless tether growth parameter, Equation 5.8.16, and Equation 5.8.15, we can express the kinematic relation for flow of surface material through the dissipation region as the following:

$$- G_t\,\frac{\sin\theta}{\tilde{r}^3} = (1 - \tilde{F}_\sigma^{-1})\,\frac{d}{d\tilde{r}}\left[\frac{\tilde{F}_\sigma}{\tilde{r}\,\cos\theta} + \ell n\,\tilde{r}\right] \tag{5.8.17}$$

Given a specific force ratio, $\tilde{F}_\sigma$, there is a unique value for the dimensionless growth rate, G_t, which will give a solution for the geometry of the dissipation region with the two conditions on the angles at the ends of the cylinder model, i.e.,

$$\cos\theta \equiv 1$$

for $r = r_t$ and for $r = r_0$. Details of the numerical solution of Equation 5.8.17 and additional discussion are contained in the article by Evans and Hochmuth (1976).[20] The results for the dimensionless tether growth rate as a function of the force ratio, $\tilde{F}_\sigma$, are shown in Figure 5.49, and the geometry of dissipation region is shown in Figure 5.50 as predicted by the computer solution to Equation 5.8.17.

The prediction of tether growth rate is given by

$$\frac{\Delta L}{\Delta t} = \frac{F_\sigma^{crit}}{8\pi\eta_p} \cdot G_t\left(\frac{F_\sigma}{F_\sigma^{crit}}\right)$$

Consequently, knowing the critical force threshold for tether growth, F_σ^{crit}, and the data for growth rate vs. applied force, F_σ, we can use the predicted relation to calculate the coefficient of surface viscosity in plastic flow. Figure 5.47 shows the predicted

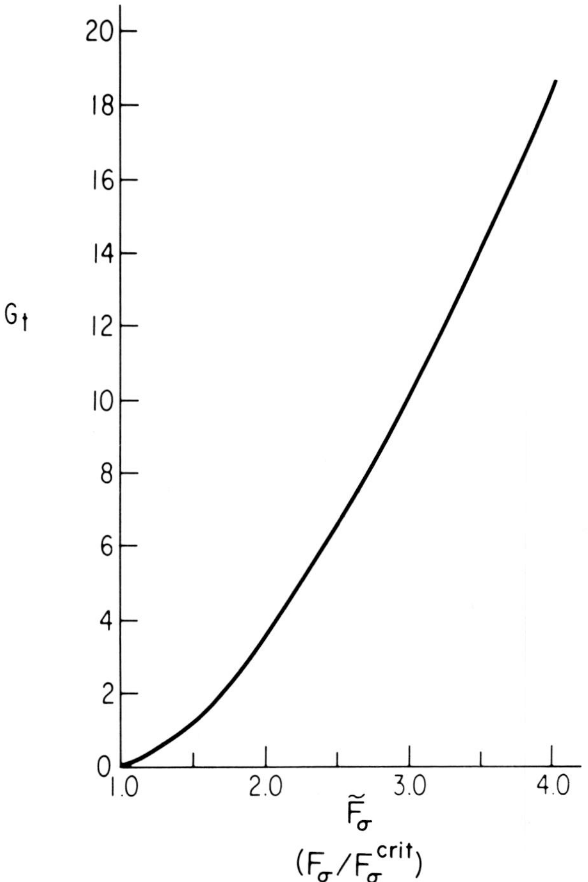

FIGURE 5.49.   Results for the dimensionless tether growth rate, G_t, as a function of the force ratio, $\tilde{F}_\sigma$; $\tilde{F}_\sigma = 1$ is the yield condition. (From Evans, E. A. and Hochmuth, R. M., *Biophys. J.*, 16, 13, 1976. With permission.)

curves for three values of surface viscosity in comparison to the actual data taken on red cell tether growth experiments (Hochmuth et al., 1976).[44a] The coefficient for surface viscosity is calculated to be on the order of 10^{-2} surface poise (dyn-sec/cm). This value is 20 times greater than the value deduced for red cell membrane viscosity from viscoelastic recovery in the solid domain.

In principle, it is possible to calculate the yield shear, $\hat{T}_s$, that characterizes the plastic failure of the membrane material from the value of the critical force threshold, 10^{-6} dyn,

$$\hat{T}_s = \frac{F_\sigma^{crit}}{4\pi r_t}$$

but it is impossible to accurately specify the radius of the structural material layer in the tether region. The scanning electron micrographs indicate that the tether diameters are less than 10^{-5} cm. For a range of tether diameters from 2 to 8×10^{-6} cm, the yield shear values range from 8×10^{-2} to 2×10^{-2} dyn/cm. If it is assumed that the membrane is perfectly elastic up to the yield level (as given by the elastic constitutive relation [Equation 4.4.12]), then it is possible to estimate the maximum elastic extension ratio at which yield occurs from

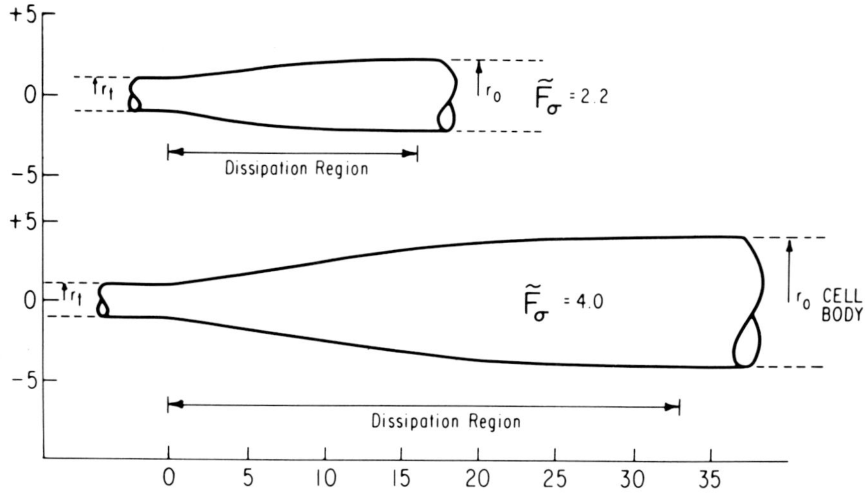

FIGURE 5.50. The geometry of the dissipation region for plastic flow as predicted by the computer solution to Equation 5.8.17. (From Evans, E. A. and Hochmuth, R. M., *Biophys. J.*, 16, 13, 1976. With permission.)

$$\hat{T}_s \sim \frac{\mu}{2} \, (\hat{\lambda}^2 - \hat{\lambda}^{-2})$$

which gives the range of

$$3 < \hat{\lambda} < 5$$

This is consistent with the observations of Evans and LaCelle (1975)[22a] that departure from elastic behavior occurred rapidly for extension ratios greater than 3:1. In contrast to relaxation of membrane force resultants, the plastic flow process that we have just described occurs immediately if the magnitude of the shear resultant exceeds the yield shear. However, these mechanisms of permanent deformation are not distinct except in the conceptual limits of either subyield shear resultants with long duration or immediate shear resultants in excess of yield.

Example 1

In the analysis of relaxation and creep of membrane surface that is aspirated with a micropipet, we used Eulerian representations for variables in the constitutive relation which are based on fixed spatial coordinates. Here we will outline the equivalent approach for Lagrangian variables which are defined local to the material surface. For membrane surface in the pipet (i.e., the projection into the pipet), the constitutive behavior is simply given by the relaxation equation,

$$f(T_s) \, \frac{\partial T_s}{\partial t} = - \, \frac{T_s}{2\eta_c}$$

since the rate of deformation is identically zero. (Note: we are only considering projection lengths greater than or equal to one pipet radius, $L \geq R_p$.) For the membrane surface outside the pipet, the constitutive equation (5.8.1) is used,

$$\frac{\partial \ln \tilde{\lambda}}{\partial t} = f(T_s) \, \frac{\partial T_s}{\partial t} + \frac{T_s}{2\eta_c} \tag{5.8.18}$$

The state of deformation in the exterior surface is defined with reference to the initial coordinate, r_0, for the undeformed surface,

$$\tilde{\lambda}^2 = \frac{\tilde{r}_0^2}{\tilde{r}_0^2 - 2\tilde{L} + 1} \tag{5.8.19}$$

where $\tilde{r}_0$ and $\tilde{L}$ are the radial coordinate and projection length scaled by the pipet radius, i.e., $\tilde{r}_0 \equiv r_0/R_p$ and $\tilde{L} \equiv L/R_p$. The rate of deformation is obtained by taking the time derivative of the extension ratio,

$$\frac{\partial \ln \tilde{\lambda}}{\partial t} = \frac{\partial \tilde{L}}{\partial t} / [\tilde{r}_0^2 - (2\tilde{L} - 1)] \tag{5.8.20}$$

The last step is to transform the integral relation for the boundary condition into a Lagrangian form. This is given by

$$\frac{\Delta P \cdot R_p}{4} = \int_{R_p}^{\infty} T_s \, \frac{dr}{r} = \int_{\tilde{R}_0(t)}^{\infty} \tilde{\lambda}^2 T_s \, \frac{d\tilde{r}_0}{\tilde{r}_0} \tag{5.8.21}$$

where $\tilde{R}_0(t)$ is the location in the undeformed state that corresponds to the instantaneous position at the entrance to the pipet, $r = R_p$. Hence,

$$\tilde{R}_0 \equiv \frac{R_0}{R_p} = 2\tilde{L} \tag{5.8.22}$$

The method of solution is simple. First, a value for the time rate of change of the projection length is assumed. Then, the trial distribution of the extension ratio and the membrane shear resultant are calculated for a small time increment with Equations 5.8.18 through 5.8.20. These trial functions are used in Equation 5.8.21 to test the results. The process is performed iteratively until the proper choice of $\partial \tilde{L}/\partial t$ is found such that the integral relation for the boundary condition is satisfied. As expected, we obtain the same results using this Lagrangian approach as we do with the Eulerian method.

EPILOGUE

In our presentation, we have emphasized the rational methods of physical mechanics and empiricism of thermodynamics as they relate to the study of membrane biophysics. To facilitate this study, we developed analytical procedures that are designed to separate the intrinsic properties of the membrane as a continuous material from the extrinsic morphology of the cellular envelope or capsule. Through specific applications of these methods, we have attempted to provide a contemporary view of the material structure and properties that is derived from mechanical experiments. We focused primarily on red cell membranes and lipid bilayer systems with a brief discussion of the sea urchin egg membrane-cortex as an example of a more complex membrane system. The study of red cell membranes and lipid bilayer systems is parochial by nature and does not adequately represent the diversity of biomembrane materials. However, it does provide scientific insight into important aspects which are common to cell membranes. Membranes, in general, are thin anisotropic composites of amphiphilic molecules (e.g., lipids and proteins) which often possess peripheral structural components and which may interact chemically with the aqueous environment with negligible exchange of amphiphilic components.

The ultimate goal in science is reductionism; that is, to take complex phenomena and establish the simplest set of determinants for these natural occurrences. Thus, our goal as biophysicists is to determine the physical character of the protein and lipid structure which forms a membrane material. We recognize that physical data are not enough. We must utilize evidence provided by the study of molecular chemistry and the study of the ultrastructural arrangement of molecular components. No discipline in natural science can stand alone. For instance, we investigate the whole membrane structure as a continuous material which homogenizes molecular irregularity and character; biochemists disrupt the structure to analyze the chemistry of small concentrations of the molecular constituents; ultrastructuralists observe the detailed arrangements of molecules with static or frozen microscopic images; and spectroscopists analyze the behavior of select probes as influenced by the molecular structure. All of these approaches must be synthesized if we hope to gain a unified and informed view of natural membrane systems.

In order to transform this ideal into a reality, the language barriers between scientists must be adequately reduced; that is, biophysicists must learn about biochemistry and methods of high resolution microscopy. Likewise, biochemists and ultrastructuralists must develop a sufficient level of mathematical skill to benefit from the methods of physical analysis. We hope that our tutorial presentation will contribute to this process.

GENERAL SYMBOLS

Geometry and Deformation

A_{ij}	General matrix
A_0, A	Surface area in initial and deformed state
$\tilde{A}$	Surface area per molecule
$\tilde{A}_e$	Excluded surface area per molecule
$\dot{A}$	Flux of surface area
a_i	Initial coordinate along ith axis in a plane tangent to the surface
B_{ij}	Finger's strain matrix
C_0	Equilibrium curvature in natural state
C_1, C_2	Changes in principal curvatures of the surface
D_{ij}	Velocity gradient matrix
e_{ij}	Almansi strain matrix (Eulerian)
e_s	Maximum Eulerian shear strain
h	Distance between two monolayers
I	Invariant of a matrix
$\tilde{I}$	Invariant of a deviator matrix
L	Length
n	Surface normal
n_j	Unit vector along jth axis
$R_{\ell j}$	Rotation matrix
R_1, R_2	Principal radii of curvature of the surface
R_c	Cylinder radius
R_p	Pipet radius

R_s	Sphere radius
r, r_0	Radial coordinate for a point in the deformed and undeformed state
s	Curvilinear distance along a meridian of an axisymmetric surface
ds^2	Metric for the deformed surface
ds_0^2	Metric for the undeformed surface
V_{ij}	Rate of deformation matrix
$\overline{V}_{ij}$	Isotropic rate of dilatation matrix
$\tilde{V}_{ij}$	Deviatoric rate of shear deformation matrix
V_s	Maximum rate of shear deformation
v_i	Velocity
v_r, v_s	Radial and meridional velocity
x_i	Coordinate of the deformed state along ith axis in a plane tangent to the surface
z	Axis of symmetry and axial coordinate
α', α	Fractional change in surface area with respect to deformed and initial area
β', β	Rotationally invariant deformation parameter that is a quadratic function of the extension ratios with respect to deformed and initial dimensions
$\langle \beta \rangle$	Average deformation parameter which characterizes extensional deformation at constant area
ε_{ij}	Green's strain matrix (Lagrangian)
$\varepsilon_1, \varepsilon_2$	Principal or characteristic values of Lagrangian strain matrix
$\overline{\varepsilon}$	Isotropic or mean Lagrangian strain
$\tilde{\varepsilon}_{ij}$	Deviatoric Lagrangian strain matrix
ε_s	Maximum Lagrangian shear strain
ξ_i	Cartesian spatial coordinates
θ	Polar angle relative to the axis of symmetry
λ_i	Surface extension ratio along ith principal axis
λ_i^ℓ	Surface extension ratio for ℓth constituent layer along ith axis
$\tilde{\lambda}$	Surface extension ratio at constant area
ϕ	Azimuthal angle about the axis of symmetry; shear deformation angle between initial and instantaneous material elements; coordinate rotation angle about the normal to the surface

Forces, Moments, Resultants, and Stresses

F_i	Force component in ith spatial direction
f_i	Body forces in ith spatial direction
M_{ij}	Moment resultant that acts on the edge of the material element parallel to the ith axis and which acts to twist about the jth axis through the centroid ($i \neq j$)
M_i	Principal moment resultant that acts on the edge of the material element normal to the ith axis and which acts to twist about the edge of the element
M_i^B	Principal moment resultants due to bending
M_i^C	Principal moment resultants induced by change in chemical state
P, P'	Hydrostatic pressure on upper and lower surface
Q_i	Transverse shear resultant that acts on edge normal to ith axis
Q_T	Transverse shear resultant
T_{ij}	Force resultant matrix

$\tilde{T}_{ij}$	Deviatoric force resultant matrix
T_1, T_2	Principal membrane tensions
T_m	Principal tension tangent to the meridian in curvilinear coordinates with cylindrical symmetry
T_ϕ	Principal tension tangent to the latitude circle (circumferential tension)
$\overline{T}$	Isotropic tension
$\overline{T}_0$	Initial isotropic tension
$\overline{T}^i$	Contribution to isotropic tension by the ith interface
T_s	Maximum shear resultant
$\hat{T}_s$	Yield shear
σ_i	Tractions that act on a surface, e.g., fluid shear and normal stresses
σ	Normal stress that acts on a surface

Thermodynamic Variables

C_α	Specific heat at constant surface area
$C_{\overline{T}}$	Specific heat at constant isotropic tension
C_n	Specific heat at constant surface pressure
$E, \tilde{E}$	Energy, energy density
$F, \tilde{F}$	Helmholtz free energy, Helmholtz free energy density
$\tilde{F}_B$	Bending free energy density
$\tilde{F}_c$	Curvature free energy density
$\tilde{F}_H$	Free energy density of hydrophobic interaction
$\tilde{F}_p$	Free energy density due to interactions between membrane amphiphiles
$\tilde{F}_s$	Free energy density of shear deformation
$\tilde{F}_w$	Free energy density due to interactions between the aqueous media and the interfacial groups
$\tilde{H}, \tilde{H}'$	Enthalpic densities
Q	Heat
$S, \tilde{S}$	Entropy, entropy density
$\dot{s}$	Rate of entropy production in an irreversible process; dissipation function
T	Temperature
$W, \tilde{W}$	Mechanical work, work density
γ	Interfacial free energy density (effective interfacial tension)
Ω	Number of system configurations
Ω_N	Density of molecular configurations for Nth molecule
$\tilde{n}_w^s$	Partial molar density of water in interfacial phase (per unit area)
$\phi_w^0, \phi_w^s, \phi_w^{0,s}$	Chemical potential for water in bulk phase, interfacial phase, standard chemical potential in the interfacial phase

Membrane Material Properties

B	Curvature or bending elastic modulus; coefficient of bending resistance or rigidity
B_{ij}	Matrix of bending moduli
F	Inverse elastic function; yield function
K	Isothermal area compressibility modulus
L_{pq}	Coupling coefficients that relate flows to forces in the Onsager theory of irreversible processes
$\overline{L}_{11}$	Coupling coefficient related to area dilation viscosity, κ

$\bar{L}_{22}$	Coupling coefficient related to surface shear viscosity, η
t_c	Characteristic time constant for extensional recovery or response
Γ	Induced moment coefficient
η	Viscous coefficient of surface shear deformation
η_e, η_c, η_p	Viscous coefficient of shear deformation for solid, semisolid, and liquid regimes of material behavior
$\tilde{\eta}_{HC}$	Viscocity of hydrocarbon interior of lipid bilayer
$\varkappa$	Viscous coefficient for energy dissipation by finite rates of area dilation or condensation
μ	Surface elastic shear modulus of elasticity
$\Delta\Phi_p, J_p$	Conjugate force and flow in Onsager equations

Miscellaneous Variables and Notation

k	Boltzmann's constant
M_w	Molecular weight
N_A	Avogadro's number
$\tilde{R}$	Gas constant
$(\ )_T$	Temperature as a subscript, denotes isothermal process
t	Time
δ_{km}	Identity matrix
ϱ	Surface density (mass/unit area)
σ^2	Gaussian variance
$\Delta(\ )$	Difference
$<\ >$	Ensemble average
$(\cdot)$	Time rate of change
$\partial(\ )/\partial t$	The time rate of change measured at a fixed material location
$d(\ )/dt$	The time rate of change measured at a location fixed in space
$(\partial/\partial X_i)_{X_{j\neq i}}$	Partial differentiation with respect to the independent variable, X_i, and with the variables, $X_{j\neq i}$, held constant
$\delta(\ )$	Variation
$d(\ )$	Differential change
$(\hat{\ })$	Per unit initial (undeformed) area; per molecule
$(\ )^e$	Elastic part
$(\ )^v$	Viscous part
$(\ )^p$	Plastic part
$(\ )^i$	For the ith interface
$(\ )^S$	Symmetric part
$(\ )^A$	Antisymmetric part
$(\ )^R$	Rigid rotation
$(\ )^D$	Deformation, independent of rigid rotation
$(\ )^{IRR}$	Irreversible process
$(\ )^{REV}$	Reversible process
$(\ )^\ell$	ℓth layer in the composite strata
$(\ )_m$	Component in meridional direction
$(\ )_\phi$	Component in azimuthal direction

APPENDIX

A.1 Minimum Energy Deformation of Red Cell Membrane

In Sections 5.2 through 5.4, we presented examples of mechanical experiments which demonstrated that the red blood cell membrane can store energy in a conservative manner. Subsequently, we showed that this elastic solid behavior is limited by the magnitude and duration of applied forces. The significant mechanical feature of the red cell membrane is that its resistance to area change is four to five orders of magnitude greater than its resistance to in-plane extensional (shear) deformation; bending rigidity or resistance to curvature change appears to be an even lower magnitude contribution. Thus, nonspherical membrane shapes are very flaccid and easily deformed by surface extension and curvature changes while the membrane area remains constant. Elastic deformations of the red cell are related to the forces applied to the cell through a convolution of the shear and bending elastic moduli with the cell shape. Therefore, the deformation of the red cell provides a direct measure of the forces that act on the cell in static equilibrium. The equilibrium shape of the unsupported cell membrane contour is established by the first law of thermodynamics for isothermal, reversible processes, i.e., the variation in the free energy of the membrane minus the variation in work of surface tractions which act on the membrane is zero. Such an approach was described in Section 5.4 as used by Zarda et al. (1977)[100] to study osmotic swelling of red cells. Zarda et al., use a finite element technique where the membrane is partitioned into sequential elements with several nodal positions inside each element. Within each element, the tension and moment resultants plus the surface geometry are approximated by polynomials of the degree appropriate to the number of nodal positions; continuity at element interfaces is assured by adding constraint equations for element curvature at the interface. The requirement of surface incompressibility (i.e., constant local area) involves additional constraint equations. The result is a system of equations of very large dimensionality. In addition, these authors have only considered free energy density functions that are local in nature. In other words, the variation of the total free energy integral becomes the integral of the local variation of the free energy density. We discussed local and nonlocal features of curvature elastic energy in Section 4.10. Based on physical structure, for example, the bending elastic energy of a phospholipid bilayer is best represented by a nonlocal energy functional because its component layers can slip relative to each other. On the other hand, a membrane with strongly associated material (like the peripheral protein structure of a red cell membrane) will be characterized by local and nonlocal bending energy functionals in addition to the shear elastic energy contribution.

In this section, we will outline another variational method (Evans, 1979)[103] which can be used to determine minimum energy (i.e., equilibrium) contours for deformed red cell membrane surfaces. Here, the axisymmetric surface geometry is represented by intrinsic or curvilinear coordinates which are defined such that the first and second derivatives are continuous over the entire contour. The angular coordinate values at only a few points on the contour and derivatives at the ends are the free variables in the variational procedure. The first and second variations of the free energy functional are derived analytically; therefore, the integrands are continuous functions over the entire contour. Nonlocal energy functionals are also considered, specifically nonlocal bending or curvature energy densities. The method has been applied to the analysis of micropipet aspiration of a flaccid red cell and red cell membrane adhesion to an adjacent surface (Evans, 1979).[103] We will present some of the results and compare them to the approximation developed in Section 5.3 for the micropipet experiment.

A.2 Virtual Work of Surface Tractions and Membrane Equilibrium

Near equilibrium, the virtual work done on the cell membrane (as the result of a small virtual deformation of the shape) is the integral over the surface of the surface tractions times the local virtual displacement of the surface

$$\delta W = \int (\sigma_i \, \delta \zeta_i) \, dA \tag{A.2.1}$$

where σ_i are the components of the traction which act on the surface; $\delta\zeta_i$ are the components of the local virtual displacement. In the micropipet aspiration experiment, only uniform pressure exists in the fluid phases; thus, the integral can be expressed in terms of the external fluid pressure, P_o, and the pressure in the pipet, P_p,

$$\delta W = -P_o \int_o (\delta\zeta_n) \, dA - P_p \int_p (\delta\zeta_n) \, dA$$

where the integrals represent the cell segments outside and inside the pipet, respectively; $\delta\zeta_n$ is the displacement normal to the surface. (Friction at the pipet wall is assumed to be negligible.) It is recognized that the integrals are simply the virtual changes in the volumes of the cell segments outside and inside the pipet, respectively,

$$\delta W = -P_o \cdot \delta V_o - P_p \cdot \delta V_p$$

Since the cellular contents are an incompressible liquid solution, these volume variations are equal but of opposite sign; thus,

$$\delta W = \Delta P \cdot \delta V_p$$

where $\Delta P \equiv P_o - P_p$. The variation of the volume in the pipet is given by the variation of the projection length, L, in the pipet,

$$\delta V_p = (\pi R_p{}^2) \, \delta L$$

with R_p as the pipet radius.* Consequently, the variation in work is produced by the virtual displacement of a pipet suction force,

$$\delta W = (\pi R_p{}^2) \, \Delta P \cdot \delta L \tag{A.2.2}$$

---

*   This assumes that the projection length, L, is greater than the pipet radius, R_p, and that the shape of the projection cap is constant (e.g., a hemisphere).

For an isothermal, equilibrium process, the variation in work is equal to the variation in Helmholtz free energy of the system (see Section 4.3). Since the cytoplasm is an incompressible liquid, the free energy variation is the membrane free energy variation, δF; hence,

$$\delta F = \delta W = (\pi R_p^2)\, \Delta P \cdot \delta L \tag{A.2.3}$$

The isothermal free energy is an elastic (conservative) potential function; thus, the pipet suction force can be obtained from equilibrium variations in the free energy with respect to projection length inside the pipet,

$$\left(\frac{\partial F}{\partial L}\right) = (\pi R_p^2)\, \Delta P \tag{A.2.4}$$

Also, for a fixed projection length, we see that the variation in free energy is stationary,

$$(\delta F)_L = 0 \tag{A.2.5}$$

Equations A.2.4 and A.2.5 provide the analytical recipes for determining the pipet suction pressure and equilibrium deformation of the membrane. Equation A.2.5 is the statement that the unsupported membrane contour will be a minimum energy shape for a specific value of the projection length.

A.3 Minimum Energy Deformation

In order to determine the contour that will minimize the membrane free energy, we must define a function of several variables, a_i, which will describe the axisymmetric surface geometry; then, the membrane free energy will be made stationary with respect to these variables (for an excellent development of the methods of the calculus of variations, see Courant and Hilbert, *Methods of Mathematical Physics,* Vol. 1, 1966). The procedure is to expand the free energy in a Taylor series about a location defined by a specific set of independent variables, $\bar{a}_i$; to second order, the expansion is

$$F = F_0 + \sum_{i=1}^{N} \left.\frac{\partial F}{\partial a_i}\right|_{\bar{a}_i} (a_i - \bar{a}_i) + \sum_{i=1}^{N}\sum_{j=1}^{N} \left.\frac{1}{2}\frac{\partial^2 F}{\partial a_i \partial a_j}\right|_{\bar{a}_i,\, \bar{a}_j} (a_i - \bar{a}_i)(a_j - \bar{a}_j) + \ldots \tag{A.3.1}$$

where the free energy, F_0, and its derivatives are evaluated at the location given by the set, $\bar{a}_i$, of N variables. If the location of a minimum is nearby, then Equation A.2.5 is satisfied by the simultaneous solution to the system of equations

$$\frac{\partial F}{\partial a_i} = 0 = \left.\frac{\partial F}{\partial a_i}\right|_{\bar{a}_i} + \sum_{j=1}^{N} \left.\frac{\partial^2 F}{\partial a_i \partial a_j}\right|_{\bar{a}_i,\, \bar{a}_j} (a_j - \bar{a}_j) + \ldots \tag{A.3.2}$$

where the location of the minimum is defined by the set of variables, a_i. Since the variable set, a_i, determines the shape of the membrane surface, the partial derivatives in Equation A.3.2 represent the first and second variations of the free energy with respect to variations of the surface geometry. Often a particular problem involves subsidiary or constraint conditions (e.g., the cell volume is constant because of incompressibility). In such cases, the system of equations is enlarged and the set of independent variables will include a Lagrange multiplier, g_i, for each subsidiary condition, $G_i(a_j)$. Thus, Equation A.3.2 becomes

$$G_i(a_j) \equiv 0$$

$$\frac{\partial F}{\partial a_i} + \sum_{k=1}^{N_c} g_k \cdot \frac{\partial G_k}{\partial a_i} = 0$$

$$(A.3.3)$$

which represents a system of $(N + N_c)$ equations in $(N + N_c)$ unknowns; N_c is the number of subsidiary conditions. With the definition of matrix and vector variables, we express the system of equations, (A.3.3), to first order in terms of appropriate summations as

$$G_i = 0$$

$$0 = \overline{F}_i + \sum_{j=1}^{N} \overline{F}_{ij}(a_j - \overline{a}_j) + \sum_{k=1}^{N_c} \overline{G}_{ki}\, g_k$$

$$(A.3.4)$$

where

$$\overline{F}_i \equiv \frac{\partial F}{\partial a_i}\bigg|_{\overline{a}_i}$$

$$\overline{F}_{ij} \equiv \frac{\partial^2 F}{\partial a_i \partial a_j}\bigg|_{\overline{a}_i, \overline{a}_j}$$

$$\overline{G}_{ki} \equiv \frac{\partial G_k}{\partial a_i}\bigg|_{\overline{a}_i}$$

The task now is to introduce the elastic free energy functional for the red cell membrane. As discussed in Sections 5.2 and 5.3, experiments have shown that the red cell membrane has a great resistance to area dilation and much lower shear and bending rigidities. Thus, the red cell membrane can be treated as a two dimensionally incompressible surface material, i.e., it deforms at constant area. As such, local deformations of membrane can be described by a single surface extension ratio, λ, and by changes in the principal curvatures, C_1 and C_2, that characterize the surface. Figure A.1 illustrates the deformation of an element of the axisymmetric membrane surface.

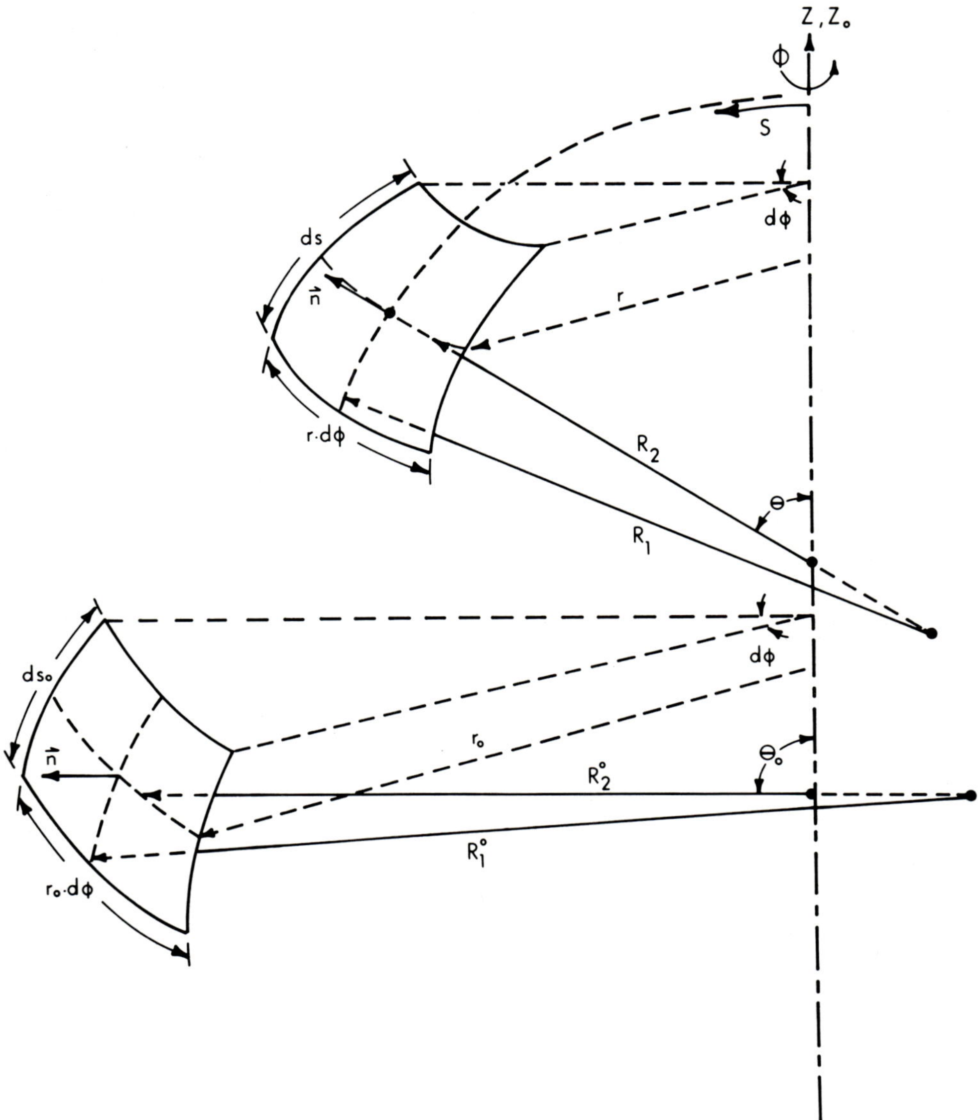

FIGURE A.1. Illustration of the initial (undeformed) and final (deformed) geometries of a specific element of the axisymmetric membrane surface. The spatial coordinates are (r_0, z_0, ϕ) and (r, z, ϕ) for the initial and final states, respectively; the corresponding curvilinear coordinates are (s_0, θ_0, ϕ) and (s, θ, ϕ). The principal radii of curvature for the surface element are labeled (R_1^0, R_2^0) and (R_1, R_2); R_1(or R_1^0) is the radius of curvature that describes a local arc of the surface meridian; R_2 (or R_2^0) describes a local arc of the surface in the plane which contains the surface normal and is orthogonal to the meridian contour.

From Sections 4.8, 4.9, and 4.10, we take the first order elastic free energy functional for the membrane as the superposition of shear and curvature elastic effects,

$$
F = \underbrace{\int \frac{\mu}{2} (\lambda^2 + \lambda^{-2} - 2)dA_0}_{\text{``Shear''}} + \underbrace{\int \frac{B}{2} (C_1 + C_2)^2 dA_0}_{\text{``Bending''}}
$$

$$+ \int \Gamma(C_1 + C_2)dA_0 + \underbrace{\frac{\overline{B}}{2} \left| \int (C_1 + C_2)dA_0 \right|^2}$$

$$\underbrace{}_{\text{``Chemical''}} \qquad \underbrace{\phantom{\frac{\overline{B}}{2} \left| \int (C_1 + C_2)dA_0 \right|^2}}_{\text{``Nonlocal Bending''}} \qquad\qquad \text{(A.3.5)}$$

where μ is the elastic shear modulus; B is the coefficient of local bending rigidity; Γ is a chemically induced moment that can be caused by changes in the surface equilibrium; and $\overline{B}$ is the nonlocal coefficient of bending rigidity. The free energy per unit area (the integrand) is defined relative to the undeformed or initial state (area); for the constant area case, the initial and final (deformed) differential areas are equal, i.e., $dA_0 \equiv dA$. Of these material properties, only the elastic shear modulus has been measured experimentally; we estimated the bending rigidity in Section 4.10 from theoretical models which appear to correlate with meager experimental evidence (Evans and Hochmuth, 1978).[22] Furthermore, no discriminatory data are available for separation of local and nonlocal bending effects. We recall that local bending corresponds to a membrane material with rigidly connected layers of molecules which cannot slip relative to one another whereas nonlocal bending represents associated layers which can slip locally (e.g., a phospholipid bilayer where the terminal ends of acyl chains are unconnected). The difference between local and nonlocal bending energies is apparent when we take the variation of the free energy,

$$\delta F = \int \left[\mu(\lambda - \lambda^{-3})\delta\lambda + (B \cdot C_1 + B \cdot C_2 + \Gamma + \overline{M})(\delta C_1 + \delta C_2) \right] dA_0 \qquad\qquad \text{(A.3.6)}$$

where

$$\overline{M} \equiv \overline{B} \int (C_1 + C_2)dA_0 \qquad\qquad \text{(A.3.7)}$$

Note, the integrand includes a parameter, $\overline{M}$, that is evaluated by the global integration of the mean curvature (a type of ``global'' bending moment), whereas the local bending moment, $B(C_1 + C_2)$, is involved in the integration over the contour. The physical significance is illustrated by bending a tablet of paper: if the sheets of the paper are glued together, the extrinsic bending rigidity of the tablet is very large; on the other hand, the tablet is very flexible if the sheets are unconnected.

The first and second variations are related to the first and second partial derivatives through the following summations:

$$\delta F = \sum_{i=1}^{N} \frac{\partial F}{\partial a_i} \cdot \delta a_i$$

$$\delta^2 F = \sum_{i=1}^{N} \sum_{j=1}^{N} \frac{\partial^2 F}{\partial a_i \partial a_j} \cdot \delta a_i \cdot \delta a_j$$

With the elastic free energy functional, these are expressed as a set of integrals of partial derivatives,

$$\frac{\partial F}{\partial a_i} = \int \left[\mu(\lambda - \lambda^{-3}) \frac{\partial \lambda}{\partial a_i} + (B \cdot C_1 + B \cdot C_2 + \Gamma + \overline{M}) \left(\frac{\partial C_1}{\partial a_i} + \frac{\partial C_2}{\partial a_i} \right) \right] dA_0 \qquad (A.3.8)$$

and

$$\frac{\partial^2 F}{\partial a_i \partial a_j} = \int \left[\mu \left((\lambda - \lambda^{-3}) \frac{\partial^2 \lambda}{\partial a_i \partial a_j} + (1 + 3\lambda^{-4}) \frac{\partial \lambda}{\partial a_i} \frac{\partial \lambda}{\partial a_j} \right) \right.$$

$$+ (B \cdot C_1 + B \cdot C_2 + \Gamma + \overline{M}) \left(\frac{\partial^2 C_1}{\partial a_i \partial a_j} + \frac{\partial^2 C_2}{\partial a_i \partial a_j} \right)$$

$$+ B \left(\frac{\partial C_1}{\partial a_i} + \frac{\partial C_2}{\partial a_i} \right) \left(\frac{\partial C_1}{\partial a_j} + \frac{\partial C_2}{\partial a_j} \right) + \left. \frac{\partial \overline{M}}{\partial a_j} \left(\frac{\partial C_1}{\partial a_i} + \frac{\partial C_2}{\partial a_i} \right) \right] dA_0 \qquad (A.3.9)$$

where

$$\frac{\partial \overline{M}}{\partial a_j} = \overline{B} \int \left(\frac{\partial C_1}{\partial a_j} + \frac{\partial C_2}{\partial a_j} \right) dA_0 \qquad (A.3.10)$$

In order to evaluate the necessary partial derivatives in Equations A.2.4 and A.3.8 and to solve the system of Equations A.3.4 for the minimum, we define the axisymmetric geometry in curvilinear coordinates (s, θ, ϕ) and (s_0, θ_0, ϕ_0) for the final (deformed) and initial contours, respectively. The distance along the meridian (contour generator) is defined as "s"; the coordinate "θ" is defined as the angle between the outward normal to the surface and the axis of symmetry; the coordinate "ϕ" is the azimuthal angle. Figure A.1 illustrates these coordinates for initial and deformed surface elements. The cylindrical coordinates (r,z) can be expressed as parametric functions of the coordinate, s. Likewise, for an axisymmetric surface, the angle θ can be expressed as a function of position and along the meridian,

$$r = f(s)$$

$$z = f(s)$$

$$\theta = f(s)$$

where

$$\frac{dr}{ds} = \cos \theta$$

$$\frac{dz}{ds} = -\sin \theta \qquad (A.3.11)$$

The principal curvature changes of the surface between the initial and deformed geometries are given by

$$C_1 = \frac{1}{R_1} - \frac{1}{R_1^\circ} = \frac{d\theta}{ds} - \frac{d\theta_0}{ds_0}$$

$$C_2 = \frac{1}{R_2} - \frac{1}{R_2^\circ} = \frac{\sin\theta}{r} - \frac{\sin\theta_0}{r_0} \qquad (A.3.12)$$

The constant area restriction is explicitly satisfied by the relation

$$r_0 \cdot ds_0 \equiv r \cdot ds \qquad (A.3.13)$$

where the subscript, 0, refers to the initial, undeformed surface geometry. We recall that the membrane extension ratio is defined by the ratio of differential lengths along meridians of the deformed and undeformed surfaces

$$\lambda \equiv \frac{ds}{ds_0}$$

This specifies the planar deformation of the axisymmetric surface. Because of the constant area requirement, the membrane extension ratio is equivalently given by the ratio of radial coordinates for corresponding surface locations

$$\lambda = \frac{r_0}{r} \qquad (A.3.14)$$

The surface deformation and spatial coordinates of the final shape can be expressed in terms of the initial surface geometry (presumed to be known); the differential relations between the initial and final geometries are given by

$$\frac{dr}{ds} = \frac{1}{\lambda}\frac{d}{ds_0}\left(\frac{r_0}{\lambda}\right) = \cos\theta$$

$$\frac{d\lambda}{ds} = (\lambda \cdot \cos\theta_0 - \lambda^3 \cdot \cos\theta)/r_0$$

$$\frac{dz}{ds} = -\lambda \cdot \sin\theta$$

$$\theta = f(s_0) \qquad (A.3.15)$$

The curvilinear coordinate, θ, as a function of surface position specifies the shape of the axisymmetric contour. Thus, we need a functional form for the angle, θ, in terms of independent variables, (a_i, s_0), which satisfies the necessary continuity conditions. In order for the membrane force resultants (tensions and transverse shear) to be continuous, the angle with its first and second derivatives must be continuous. It is possible to define a piecewise continuous polynomial in s_o that satisfies these criteria (Conte and deBoor, 1972);[101] this is a sequence of fifth order polynomials, each defined for a specific region (say the pth) by

$$\theta_p(s_o) = \sum_{n=1}^{5} C_{pn}(s_o - s_{op})^n \qquad (A.3.16)$$

where

$$\theta_p(s_{op}) = \theta_{p-1}(s_{op})$$

$$\left. \frac{d\theta_p}{ds_o} \right|_{s_{op}} = \left. \frac{d\theta_{p-1}}{ds_o} \right|_{s_{op}}$$

$$\left. \frac{d^2\theta_p}{ds_o^2} \right|_{s_{op}} = \left. \frac{d^2\theta_{p-1}}{ds_o^2} \right|_{s_{op}}$$

and s_{op} are the region boundaries for each polynomial of the sequence. The coefficients C_{pn} are linear combinations of a set of variables, a_i, given by,

$$a_i = \begin{cases} \theta(s_{op}) \\[1em] \left. \dfrac{d\theta}{ds_o} \right|_{s_o = 0} \,, \quad \left. \dfrac{d\theta}{ds_o} \right|_{s_o = s_{max}} \\[1em] \left. \dfrac{d^2\theta}{ds_o^2} \right|_{s_o = 0} \,, \quad \left. \dfrac{d^2\theta}{ds_o^2} \right|_{s_o = s_{max}} \end{cases} \qquad (A.3.17)$$

In terms of the shape variables, a_i, the geometric variations involved in the free energy minimization are given by a set of nonlinear differential equations

$$\frac{d}{ds_o}\left(\frac{\partial\lambda}{\partial a_i} \right) = \left(\frac{\partial\lambda}{\partial a_i} \cdot \cos\theta_0 - 3\lambda^2 \cdot \frac{\partial\lambda}{\partial a_i} \cdot \cos\theta + \lambda^3 \cdot \sin\theta \cdot \frac{\partial\theta}{\partial a_i} \right) / r_o$$

$$\frac{d}{ds_0}\left(\frac{\partial z}{\partial a_i}\right) = -\frac{\partial \lambda}{\partial a_i}\cdot\sin\theta - \lambda\cos\theta\cdot\frac{\partial\theta}{\partial a_i}$$

$$\frac{\partial C_1}{\partial a_i} = -\frac{1}{\lambda^2}\cdot\frac{\partial\lambda}{\partial a_i}\cdot\frac{d\theta}{ds_0} + \frac{1}{\lambda}\cdot\frac{d}{ds_0}\left(\frac{\partial\theta}{\partial a_i}\right)$$

$$\frac{\partial C_2}{\partial a_i} = \frac{1}{r_0}\cdot\frac{\partial\lambda}{\partial a_i}\cdot\sin\theta + \frac{\lambda}{r_0}\cdot\cos\theta\cdot\frac{\partial\theta}{\partial a_i} \tag{A.3.18}$$

The second partial derivatives are similar differential equations. Geometric constraints and partial derivatives are illustrated by the particular relation for constant volume

$$G = V_0 + \int z\cdot\cos\theta\cdot dA_0$$

$$\frac{\partial G}{\partial a_i} = \int\left(\frac{\partial z}{\partial a_i}\cdot\cos\theta - z\sin\theta\cdot\frac{\partial\theta}{\partial a_i}\right)dA_0 \tag{A.3.19}$$

where V_0 is the cell volume. The partial derivatives of the coordinate, θ, with respect to the independent variables, a_i, are simply fixed matrices, which are determined by the derivatives of the linear combinations, C_{pn}, i.e.,

$$\frac{\partial C_{pn}}{\partial a_i}$$

The computational method for obtaining a minimum energy contour involves an algorithm based on the Newton-Raphson technique of successive approximation to the variables, a_i. The matrices, $\overline{F}_{ij}$ and $\overline{G}_{ki}$, and vectors, $\overline{F}_i$ and $\overline{G}_i$, are evaluated with an initial "guess", $\overline{a}_i$; then successive "guesses", a_i, are obtained from the system of Equations A.3.4, until the system convergence criteria are satisfied, e.g.,

$$\sum_{i=1}^{N}\left|\overline{F}_i\overline{F}_i\right| + \sum_{i=1}^{N_c}\left|\overline{G}_i\overline{G}_i\right| \text{ or } \sum_{i=1}^{N}\left|(a_i - \overline{a}_i)(a_i - \overline{a}_i)\right| \sim 0$$

Equations A.3.8, A.3.9, A.3.15, and A.3.16 are integrated simultaneously to provide the required matrices and vectors used in Equation A.3.4. The system of equations usually involves no more than three free angles (with the angles at the ends fixed), a Lagrange multiplier (for constant volume), plus the appropriate derivatives at the ends as free variables to give a dimensionality of (7×7).

A.4 Micropipet Aspiration of a Flaccid Red Blood Cell

Micropipet aspiration of a flaccid red cell is shown in Figure 5.18C. The aspirated

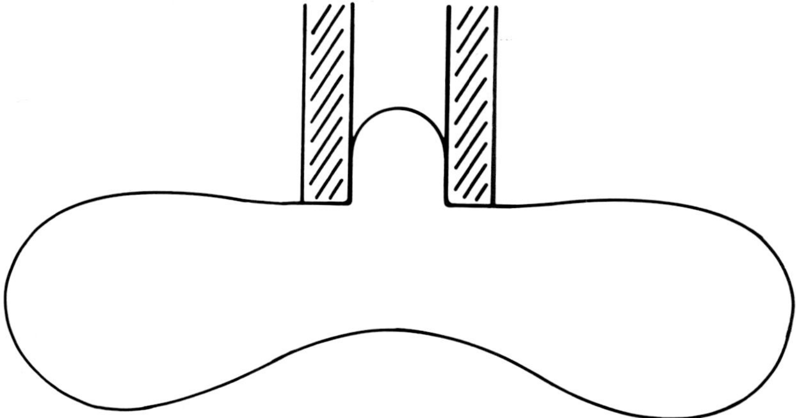

FIGURE A.2.   Minimum energy contour for an aspiration length to pipet radius ratio of 2 to 1; the contour was required to lie along the pipet front face and inner wall. The bending to shear rigidity ratio, $\tilde{B}$, was 10^{-3}.

cell projection length exhibits a linear dependence on suction pressure like the sample data in Figure 5.18A. In Section 5.3, we analyzed the experiment using an approximation of the surface as a flat sheet (since the membrane force resultants drop off inversely with the square of the distance from the pipet entrance); neglecting bending rigidity, this analysis yielded a linear relationship with a first order elastic constitutive relation. Correlation of red cell aspiration data with the approximate analysis provided a value for the elastic shear modulus of 6.6×10^{-3} dyne/cm at $25°$ (Waugh, 1977).[91] Even though observations indicate that bending rigidity is small, no direct evaluation of bending effects has been previously undertaken for the aspiration experiment.

With the minimum energy method described here, the micropipet aspiration experiment has been modeled for a two order of magnitude range of bending to shear rigidity ratio, i.e., $\tilde{B} \equiv B/\mu \cdot R_0^2 \sim 10^{-4}$ to 10^{-2} (Evans, 1979).[103] For an elastic shear modulus of 6.6×10^{-3} dyne/cm and an outer radius of the initial cell cross section equal to 3.91×10^{-4} cm, the elastic bending modulus would vary between $10^{-13} - 10^{-11}$ ergs (dyne-cm) for the above range. The outer radius, R_0, is the value appropriate to the initial cell cross section shown in Figure 5.22A (taken from Evans and Fung, 1972).[18] With this initial cross section and an intermediate value of 10^{-3} for $\tilde{B}$, Figure A.2 is a minimum energy contour for an aspiration length to pipet radius ratio, L/R_p, of 2 to 1 where the contour was required to lie along the pipet surface on the front face and inner wall as shown; this constraint was accomplished by setting the curvilinear coordinate, θ, equal to zero along the front face of the pipet and equal to $\pi/2$ along the inner wall. Since the shape of the projection cap inside the pipet has little influence on the total membrane free energy, the cap was defined to be a spherical segment for convenience. In order to assess the importance of the corner at the entrance to the inner cylinder of the pipet, an alternate pipet entrance condition was investigated for the contour which only required that the contour be constrained to the inner wall but was totally unsupported outside; Figure A.3 is the minimum energy contour. The total membrane free energy was different for the two different entrance conditions (illustrated in Figures A.2 ad A.3); however, the derivative of the free energy with respect to the aspiration length, which is the pipet suction pressure, was essentially the same for both conditions. Also, for the two orders of magnitude range of bending to shear rigidities investigated, the radius of curvature of the membrane at the corner of the pipet mouth was

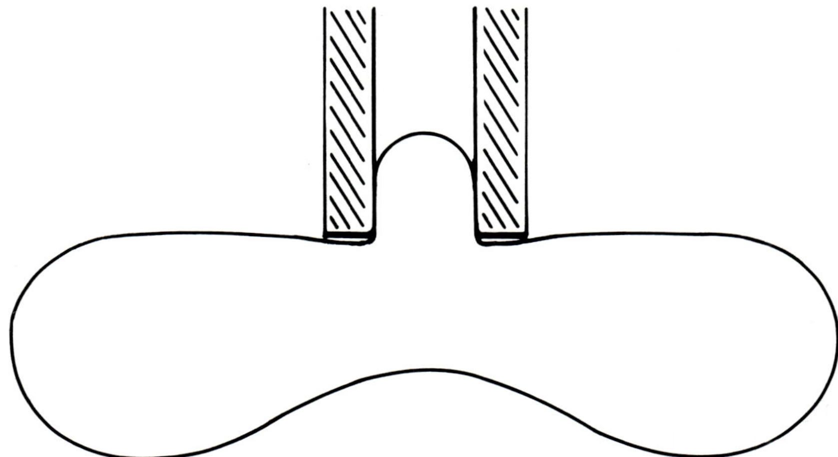

FIGURE A.3.   Minimum energy contour for the same aspiration length to pipet radius ratio as A.2, but the contour was only required to lie along the inner wall of the pipet and was totally unsupported outside. The bending to shear rigidity ratio was the same value as that for A.2.

very small such that the curvilinear coordinate, θ, went to zero within a very short distance from the corner. The radius of curvature for the contour meridian at the corner varied between 0.01 — 0.03 times the outer radius of the initial cell cross section; this range is equivalent to $0.04 - 0.12 \times 10^{-4}$cm.

The pipet suction pressure is the derivative of the equilibrium free energy of the membrane with respect to aspirated length; in a dimensionless form, this relationship is

$$\frac{\Delta P \cdot R_p}{2\mu} = \frac{R_0{}^2}{2\pi R_p{}^2} \left(\frac{\partial \tilde{F}}{\partial \tilde{L}} \right) \tag{A.4.1}$$

where the free energy has been normalized by $\mu R_0{}^2$ and the cell projection length by R_p,

$$\tilde{F} \equiv \frac{F}{\mu R_0{}^2}$$

$$\tilde{L} \equiv \frac{L}{R_p}$$

Equation A.4.1 is a dimensionless form of the approximate membrane tension local to the pipet entrance derived in Section 5.3. For a pipet diameter to cell diameter ratio of 0.13 to 1 (which corresponds to a pipet diameter of 10^{-4}cm), Figure A.4 shows the results of the dimensionless membrane tension vs. the dimensionless cell projection length; the effect of bending rigidity is demonstrated by the three curves which correspond to values of $\tilde{B}$ equal to 10^{-4}, 10^{-3}, 10^{-2}.

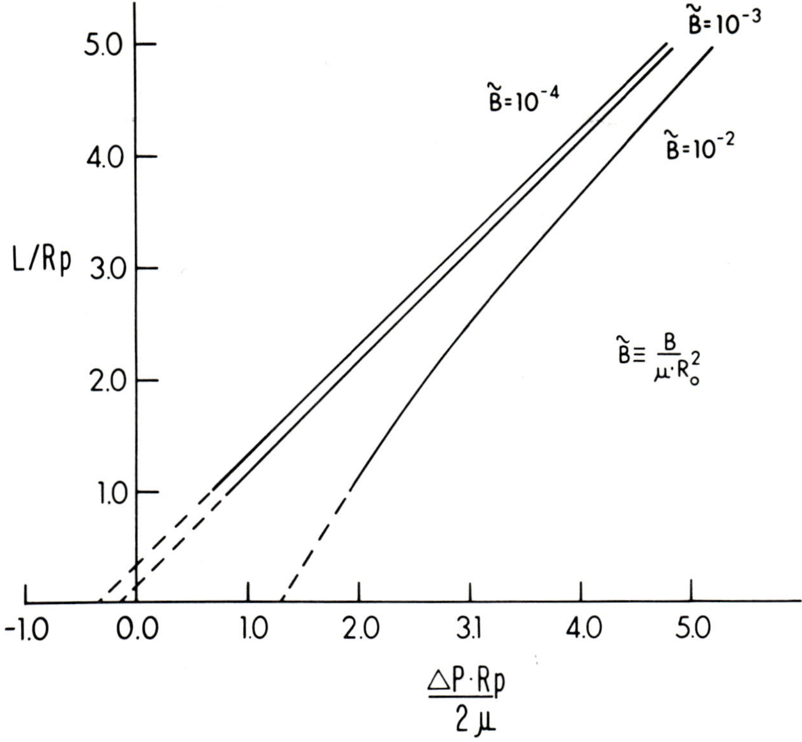

FIGURE A.4.   The results of aspiration length to pipet radius ratio vs. the suction pressure times pipet radius, normalized by twice the elastic shear modulus. The dimensionless tension, $\Delta P \cdot R_p/2\mu$, is essentially a linear function of the aspiration length to pipet radius ratio. A two order of magnitude range of bending to shear rigidity ratio, $\tilde{B} = B/\mu \cdot R_o^2$, is represented by the three curves. For an elastic shear modulus of 6.6 $\times$ 10^{-3} dyne/cm, the elastic bending modulus would range between 10^{-13} to 10^{-11} dyne-cm (ergs). The dashed lines are linear extrapolations of the measurable data range. As shown in Figure 5.18A, the extrapolated pressure intercept for aspiration of a flaccid red cell is negative which demonstrates that the upper bound for the bending modulus is 10^{-12} dyne-cm (ergs).

Each of these curves begins at the origin; the dashed lines represent the linear extrapolation of the measurable data range. Because of the limitation of optical resolution, data cannot be obtained accurately for values of the projection length less than 0.5 $\times$ 10^{-4} cm ($\tilde{L} = 1$). It is apparent, however, that the extrapolated intercept will correspond to a positive pressure if the value of $\tilde{B}$ exceeds 10^{-3}. As shown in Figure 5.18A, the extrapolated pressure intercept is negative which demonstrates that the upper bound for $\tilde{B}$ would be 10^{-3} for the red cell membrane. Again, this value corresponds to a bending modulus of 10^{-12} ergs for a shear modulus of 6.6 $\times$ 10^{-3} dyne/cm. The slope of the dimensionless tension vs. the dimensionless length provides the elastic shear modulus since the curves are nearly linear; the slopes of the two curves for $\tilde{B} = 10^{-4}$, 10^{-3} are about 10% less than the slope obtained from the approximate solution given by Equation 5.3.10. In addition, a twofold increase in pipet diameter (1 $\times$ 10^{-4} to 2 $\times$ 10^{-4} cm) has only a very slight effect on the slope (less than a few percent) and shifts the curves slightly to the left.

REFERENCES

1. **Adam, N. K.,** *The Physics and Chemistry of Surfaces,* Dover Publications, New York, 1968.
2. **Bennett, V. and Branton, D.,** Selective association of spectrin with the cytoplasmic surface of human erythrocyte plasma membranes. Quantitative determination with purified (32P) spectrin, *J. Biol. Chem.,* 252, 2753, 1977.
3. **Bessis, M.,** *Living Blood Cells and Their Ultrastructure,* Springer-Verlag, Berlin, 1973.
4. **Bull, B. S. and Brailsford, J. D.,** The relative importance of bending and shear in stabilizing the shape of the red blood cell, *Blood Cells,* 1, 323, 1975.
5. **Canham, P. B.,** The minimum energy of bending as a possible explanation of the biconcave shape of the human red blood cell, *J. Theor. Biol.,* 36, 61, 1970.
5a. **Canham, P. B. and Parkinson, D. R.,** The area and volume of single human erythrocytes during gradual osmotic swelling to hemolysis, *Can. J. Physiol. Pharmacol.,* 48, 369, 1970.
6. **Chien, S.,** Biophysical behavior of red cells in suspensions, in *The Red Cell,* Vol. 2, 2nd ed., Surgenor, D. M., Ed., Academic Press, New York, 1975, chap. 26, 1031.
7. **Chien, S., Jan, K-M., Skalak, R., and Tozeren, A.,** Theoretical nd experimental studies on viscoelastic properties of red cell membrane, *Biophys. J.,* submitted.
8. **Chien, S., Usami, S., Jan, K-M., and Skalak, R.,** Macrorheological and microrheological correlation of blood flow in the macrocirculation and microcirculation, in *Rheology of Biological Systems,* Gabelnick, H. L. and Litt, M., Eds., Charles C Thomas, Springfield, Ill., 1973, chap. 2, 12.
9. **Cole, K. S.,** Surface forces of the *Arbacia* egg, *J. Cell. Comp. Physiol.,* 1, 1, 1932.
10. **Costello, M. J. and Gulik-Krzywicki, T.,** unpublished X-ray diffraction data from personal communication.
11. **CRC Handbook of Chemistry and Physics,** Weast, R. C., Ed., The Chemical Rubber Co., Cleveland,
12. **Davies, J. T. and Rideal, E. K.,** *Interfacial Phenomena,* Academic Press, New York, 1961.
13. **Defay, R. and Prigogine, I.,** *Surface Tension and Adsorption,* John Wiley & Sons, New York, 1966.
13a. **Dueling, H. J. and Helfrich, W.,** Red blood cell shapes as explained on the basis of curvature elasticity, *Biophys. J.,* 16, 861, 1976.
13b. **Elgsaeter, A. and Branton, D.,** Intramembrane particle aggregation in erythrocyte ghosts, *J. Cell Biol.,* 63, 1018, 1974.
14. **Elworthy, P. H.,** The adsorption of water vapor by lecithin and lysolecithin, and the hydration of lysolecithin micelles, *J. Chem. Soc.,* p. 5385, 1961.
15. **Evans, E. A.,** A new material concept for the red cell membrane, *Biophys. J.,* 13, 926, 1973.
16. **Evans, E. A.,** New membrane concept applied to the analysis of fluid shear- and micropipette-deformed red blood cells, *Biophys. J.,* 13, 941, 1973.
17. **Evans, E. A.,** Bending resistance and chemically induced moments in membrane bilayers, *Biophys. J.,* 14, 923, 1974.
18. **Evans, E. A. and Fung, Y. C.,** Improved measurements of the erythrocyte geometry, *Microvasc. Res.,* 4, 335, 1972.
19. **Evans, E. A. and Hochmuth, R. M.,** Membrane viscoelasticity, *Biophys. J.,* 16, 1, 1976.
20. **Evans, E. A. and Hochmuth, R. M.,** Membrane viscoplastic flow, *Biophys. J.,* 16, 13, 1976.
21. **Evans, E. A. and Hochmuth, R. M.,** A solid-liquid composite model of the red cell membrane, *J. Membr. Biol.,* 30, 351, 1977.
22. **Evans, E. A. and Hochmuth, R. M.,** Mechano-chemical properties of membranes, in *Current Topics in Membranes and Transport,* Vol. 10, Kleinzeller, A. and Bronner, F., Eds., Academic Press, New York, 1978, 1.
22a. **Evans, E. A. and LaCelle, P. L.,** Intrinsic material properties of the erythrocyte membrane indicated by mechanical analysis of deformation, *Blood,* 45, 29, 1975.
22b. **Evans, E. A. and Leblond, P. F.,** Geometric properties of individual red blood cell discocyte spherocyte transformations, *Biorheology,* 10, 393, 1973.
23. **Evans, E. A. and Simon, S.,** Mechanics of electrocompression of lipid bilayer membranes, *Biophys. J.,* 15, 850, 1975.
24. **Evans, E. A. and Waugh, R.,** Mechano-chemistry of closed, vesicular membrane systems, *J. Colloid Interface Sci.,* 60, 286, 1977.
25. **Evans, E. A. and Waugh, R.,** Osmotic correction to elastic area compressibility measurements on red cell membrane, *Biophys. J.,* 20, 307, 1977.
26. **Evans, E. A. and Waugh, R.,** Mechano-calorimetry of red cell membrane, *Biophys. J.,* submitted.
27. **Evans, E. A., Waugh, R., and Melnik, L.,** Elastic area compressibility modulus of red cell membrane, *Biophys. J.,* 16, 585, 1976.
28. **Flügge, W.,** *Stresses in Shells,* Springer-Verlag, Berlin, 1973.

29. **Fung, Y. C.,** Theoretical considerations of the elasticity of red cells and small blood vessels, *Fed. Proc. Fed. Am. Soc. Exp. Biol.,* 25, 1761, 1966.

30. **Fung, Y. C. and Sechler, E. E.,** in *Structural Mechanics,* Goodier, J. N. and Hoff, N. J., Eds., Pergamon Press, New York, 1960, 115.

31. **Fung, Y. C. and Tong, P.,** Theory of sphering of red blood cells, *Biophys. J.,* 8, 175, 1968.

32. **Gaines, G. L.,** *Insoluble Monolayers at Liquid-Gas Interfaces,* John Wiley & Sons, Ne York, 1966, 257.

33. **Gibbs, J. W.,** *The Scientific Papers of J. Willard Gibbs,* Vol. 1, Dover, New York, 1961, 258.

34. **Goldsmith, H. L. and Skalak, R.,** Hemodynamics, *Annu. Rev. Fluid Mech.,* 7, 213, 1975.

35. **Green, A. E., and Adkins, J. E.,** *Large Elastic Deformations,* Oxford University Press, London, 1970.

36. **Hamburger, H. J.,** *Pfluegers Arch. Gesante Physiol. Menschen Tiere,* 141, 230, 1895.

37. **Helfrich, W.,** Elastic properties of lipid bilayers: theory and possible experiments, *Z. Naturforsch.,* C28, 693, 1973.

38. **Helfrich, W.,** Blocked lipid exchange in bilayers and its possible influence on the shape of vesicles, *Z. Naturforsch.,* C29, 510, 1974.

39. **Helfrich, W. and Dueling, H. J.,** Some theoretical shapes of red blood cells, *J. Phys. (Paris) Colloq.,* 36, 1, 1975.

40. **Hiramoto, Y.,** Mechanical properties of sea urchin eggs. I. Surface force and elastic modulus of the cell membrane, *Exp. Cell Res.,* 32, 59, 1963.

41. **Hiramoto, Y.,** Mechanical properties of sea urchin eggs. II. Changes in mechanical properties from fertilization to cleavage, *Exp. Cell Res.,* 32, 76, 1963.

42. **Hiramoto, Y.,** Rheological properties of sea urchin eggs, *Biorheology,* 6, 201, 1970.

43. **Hochmuth, R. M. and Mohandas, N.,** Uniaxal loading of the red cell membrane, *J. Biomech.,* 5, 501, 1972.

44. **Hochmuth, R. M., Mohandas, N., and Blackshear, Jr., P. L.,** Measurement of the elastic modulus for red cell membrane using a fluid mechanical technique, *Biophys. J.,* 13, 747, 1973.

44a. **Hochmuth, R. M., Evans, E. A., and Colvard, D. F.,** Viscosity of human red cell membrane in plastic flow, *Microvasc. Res.,* 11, 155, 1976.

45. **Hochmuth, R. M., Worthy, P. R., and Evans, E. A.,** Red cell relaxation and the determination of membrane viscosity, *Biophys. J.,* submitted.

46. **Hoeber, T. W. and Hochmuth, R. M.,** Measurement of red cell modulus of elasticity by *in-vitro* and model cell experiments, *Trans. ASME J. Basic Eng.,* 92, 604, 1970.

47. **Katchalsky, A., Kedem, O., Klibansky, C., and DeVries, A.,** Rheological considerations of the haemolysing red blood cell, in *Flow Properties of Blood and Other Biological Systems,* Copley, A. L., and Stainsby, G., Eds., Pergamon Press, Oxford, 1960, 155.

48. **Katchalsky, A. and Curran, P. F.,** *Nonequilibrium Thermodynamics in Biophysics,* Harvard University Press, Cambridge, Mass., 1967.

49. **LaCelle, P. L.,** Alterations of deformability of the erythrocyte membrane in stored blood, *Transfusion (Philadelphia),* 9, 238, 1969.

50. **LaCelle, P. L.,** Alteration of membrane deformability in hemolytic anemias, *Semin. Hematol.,* 7, 355, 1970.

51. **Landau, L. D. and Lifshitz, E. M.,** *Theory of Elasticity,* translated by Sykes, J. B. and Reid, W. H., Pergamon Press, Oxford, 1970.

52. **Leblond, P.,** The discocyte-echinocyte transformation of the human red cell: deformability characteristics, in *Red Cell Shape,* Bessis, M., Weed, R. I., and Leblond, P., Eds., Springer-Verlag, Berlin, 1973, 95.

53. **LeNeveu, D. M., Rand, R. P., Parsegian, V. A., and Gingell, D.,** Measurement and modification of forces between lecithin bilayers, *Biophys. J.,* 18, 209, 1977.

54. **Lew, H. S.,** Electro-tension and torque in biological membranes modeled as a dipole sheet in a fluid conductor, *J. Biomech.,* 5, 399, 1972.

55. **Liu, N.-I. and Kay, R. L.,** Redetermination of the pressure dependence of the lipid bilayer phase transition, *Biochemistry,* 16, 3484, 1977.

56. **Luzzati, V.,** X-ray diffraction studies of lipid-water systems, in *Biological Membranes,* Chapman, D., Ed., Academic Press, New York, 1968, 71.

57. **Marchesi, V. T., Steers, E., Tillack, T. W., and Marchesi, S. L.,** Properties of spectrin: a fibrous protein isolated from red cell membranes in *Red Cell Membrane,* Jamieson, G. A. and Greenwalt, T. J., Eds., Lippincott, Philadelphia, 1969, 117.

58. **Marchesi, S. L., Steers, E., Marchesi, V. T., and Tillack, T. W.,** Physical and chemical properties of a protein isolated from red cell membranes, *Biochemistry,* 9, 50, 1970.

59. **Meares, P.,** *Polymers: Structure and Bulk Properties,* Van Nostrand Reinhold Company, London, 1965.

60. **Mela, M. J.,** Elastic mathematical theory of cells and mitochondria in swelling process. I. The membranous stresses and modulus of elasticity of sea urchin eggs, *Biophys. J.,* 7, 95, 1967.

61. **Mitchison, J. M. and Swann, M. M.,** The mechanical properties of the cell surface. I. The cell elastimeter, *J. Exp. Biol.,* 31, 443, 1954.

62. **Morse, P. M.,** *Thermal Physics,* W. A. Benjamin, New York, 1969.

63. **Naghdi, P. M.,** The theory of shells and plates, in *Handbuch der Physik,* Flugge, S., Ed., Springer-Verlag, Berlin, 1972.

64. **Norris, C. H.,** The tension at the surface, and other physical properties of the nucleated erythrocyte, *J. Cell. Comp. Physiol.,* 4, 117, 1939.

65. **Nozaki, Y. and Tanford, C.,** The solubility of amino acids and two glycine peptides in aqueous, ethanol and dioxane solutions, *J. Biol. Chem.,* 246, 2211, 1971.

66. **Onsager, L.,** Reciprocal relations in irreversible processes. I, II, *Phys. Rev.,* 37, 405, 1931; 38, 2265, 1931.

67. **Parsegian, V. A.,** personal communication, 1978; *Proc. Natl. Acad. Sci. U.S.A.,* in press; **McAlister, M., Fuller, N., Rand, R. P., and Parsegian, V. A.,** *Biophys. J.,* 21 (Abstr.), 213a, 1978.

68. **Ponder, E.,** *Hemolysis and Related Phenomena,* Grune & Stratton, New York, 1971.

69. **Prager, W.,** *Introduction to the Mechanics of Continua,* Dover, New York, 1961.

70. **Rand, R. P.,** Mechanical properties of the red cell membrane. II. Viscoelastic breakdown of the membrane, *Biophys. J.,* 4, 303, 1964.

71. **Rand, R. P. and Burton, A. C.,** Mechanical properties of the red cell membrane. I. Membrane stiffness and intracellular pressure, *Biophys. J.,* 4, 115, 1964.

72. **Reiss-Husson, F.,** Structure des phases liquide-cristallines de differents phospholipides, monoglycerides, sphingolipides, anhydres ou en presence d'eau, *J. Mol. Biol.,* 25, 363, 1967.

73. **Reissner, E.,** On the theory of thin elastic shells, H. Reissner Anniversary Volume, *Contributions to Applied Mechanics,* J. W. Edwards, Ann Arbor, Michigan, 1949, 231.

74. **Requena, J., Haydon, D. A., and Hladky, S. B.,** Lenses and the compression of black lipid membranes by an electric field, *Biophys. J.,* 15, 77, 1975.

75. **Reynolds, J. A., Gilbert, D. B., and Tanford, C.,** Empirical correlation between hydrophobic free energy and aqueous cavity surface area, *Proc. Natl. Acad. Sci. U.S.A.,* 71, 2925, 1974.

76. **Richards, T. W. and Carver, E. K.,** A critical study of the capillary rise method of determining surface tension with data for water, benzene, and dimethyl aniline, *J. Am. Chem.,* 78, 2469, 1921.

77. **Russell, B.,** *A History of Western Philosophy,* Simon and Schuster, New York, 1972.

77a. **Saffman, P. G.,** Brownian motion in thin sheets of viscous fluid, *J. Fluid Mech.,* 73, 593, 1976.

78. **Schmid-Schönbein, H. and Wells, R. E.,** Fluid drop-like transition of erythrocytes under shear, *Science,* 165, 288, 1969.

79. **Seifriz, W.,** The physical properties of erythrocytes, *Protoplasma,* 1, 345, 1926.

80. **Servuss, R. M., Harbich, W., and Helfrich, W.,** Measurement of the curvature elastic modulus of egg lecithin bilayers, *Biochem. Biophys. Acta,* 436, 900, 1976.

81. **Sheetz, M. P. and Singer, S. J.,** Biological membranes as bilayer couples, *Proc. Natl. Acad. Sci. U.S.A.,* 71, 4457, 1975.

82. **Singer, S. J.,** The molecular organization of membranes, *Annu. Rev. Biochem.,* 43, 805, 1974.

83. **Singer, S. J. and Nicolson, G. L.,** The fluid mosaic model of the structure of cell membranes, *Science,* 175, 720, 1972.

84. **Skalak, R., Tozeren, A., Zarda, R. P., and Chien, S.,** Strain energy function of red blood cell membranes, *Biophys. J.,* 13, 245, 1973.

85. **Skalak, R.,** Modelling the mechanical behavior of red blood cells, *Biorheology,* 10, 229, 1973.

86. **Steck, T. L.,** The organization of proteins in the human red cell membrane, *J. Cell Biol.,* 62, 1, 1974.

87. **Struik, D. J.,** *Differential Geometry,* Addison-Wesley, Reading, Mass., 1961.

88. **Tanford, C.,** *The Hydrophobic Effect,* John Wiley & Sons, New York, 1973.

89. **Tanford, C.,** Theory of micelle formation in aqueous solutions, *J. Phys. Chem.,* 78, 2469, 1974.

90. **Van Deenan, L. L. M., Houtsmuller, U. M. T., deHaas, G. H., and Mulder, E.,** Monomolecular layers of synthetic phosphatides, *J. Pharm. Pharmacol.,* 14, 429, 1962.

91. **Waugh, R. E.,** Temperature Dependence of the Elastic Properties of Red Blood Cell Membrane, Ph.D. dissertation, Duke University, Durham, N.C., 1977.

92. **Waugh, R. and Evans, E. A.,** Viscoelastic properties of erthrocyte membranes of different vertebrate animals, *Microvasc. Res.,* 12, 291, 1976.

93. **Waugh, R. and Evans, E. A.,** Temperature dependence of the elastic moduli of red blood cell membrane, *Biophys. J.,* submitted.

94. **White, S. H.,** A study of lipid bilayer membrane stability using precise measurements of specific capacitance, *Biophys. J.,* 10, 1127, 1970.

94a. **White, S. H.,** Comments on "electrical breakdown of bimolecular lipid membranes as an electrome-chanical instability," *Biophys. J.,* 14, 155, 1974.

95. **Wu, E-S., Jacobson, K., and Papahadjopoulos, D.,** Lateral diffusion in phospholipid multilayers, measured by fluorescence recovery after photobleaching, *Biochemistry,* 16, 3936, 1977.

96. **Yoneda, M.,** Tension at the surface of sea urchin egg: a critical examination of Cole's experiment, *J. Exp. Biol.,* 41, 893, 1964.

97. **Yoneda, M. and Dan, K.,** Tension at the surface of the dividing sea urchin egg, *J. Exp. Biol.,* 57, 575, 1972.

98. **Yue, B. Y., Jackson, C. M., Taylor, J. A. G., Mingins, J., and Pethica, B. A.,** Phospholipid mon-olayers at non-polar oil/water interfaces. I., *J. Chem. Soc. Faraday Trans.* 1, 72, 2685, 1975.

99. **Zarda, P. R.,** Large Deformations of an Elastic Shell in a Viscous Fluid, Ph.D. dissertation, Columbia University, New York, 1974.

100. **Zarda, P. R., Chien, S., and Skalak, R.,** Elastic deformations of red blood cells, *J. Biomech.,* 10, 211, 1977.

101. **Conte, S. D. and deBoor, C.,** *Elementary Numerical Analysis,* McGraw Hill, New York, 1972.

102. **Courant, R. and Hilbert, D.,** *Methods of Mathematical Physics,* Wiley Interscience, New York, 1966.

103. **Evans, E. A.,** Minimum energy analysis of membrane deformation applied to pipet aspiration and surface adhesion of red blood cells, *Biophys. J.,* submitted, 1979.

INDEX

A

Acceleration, zero, 50

Adiabatic case, 31, 74

Affinity of lipids for water, 191

Almansi strain tensor, 10

Amphiphilic layer system studies, 1—2, 5—6, 69, 85, 89—91, 97

Angle, rotation, see Rotation angle

Annular disc, flat, 60—61

Annular ring, 162—164

Annulus, conical, see Conical annulus

Area

    change in, see Change of area

    constant, see Constant area

    elastic, see Elastic area

    small, see Small area

Area compressibility, 190—203

    small, see Small area compressibility

Area compressibility modulus, 90—91, 110—111, 119—120, 184—185

    elastic, see Elastic area compressibility modulus

    isothermal, see Isothermal area compressibility modulus

    phospholipid bilayer, 90—91, 119

    red blood cell membrane, 119

Area expansivity, thermal, see Thermal area expansivity

Area specific functions, 73

Aspect Ratio, change of, 26—27

Aspriration length, see also Micropipet aspiration, 165—166, 235—237

Axes

    centroidal, 62—65

    coordinate, see Coordinate axes

    instantaneous, 46, 49

    orthoganol, 8

    principal, see Principal axes

    spatial, 53—54

    symmetry of, 40—41, 44

Axial force balance, 59, 61—62

Axially symmetric membrane envelope, 61—62

Axial symmetry, 65

Axisymmetric membrane surface

    equilibrium, 52—62

        with moment resultants, 65—67

    geometries of, 225—237

    radial expansion of, 11, 15—17

    surface density flow over, 40—41

    twist of, 11, 17—18

B

Balance of forces, see also Equilibrium, 66, 161

    axial, see Axial force balance

Bending, 3—5, 113—114, 169—180

    red cell membranes, 3—5

    shear rigidity vs., 169—180

Bending elastic energy, 169—180, 225

Bending elastic energy density, 106, 108

Bending energy, 68, 109—110

    local, 230

    nonlocal, 230

Bending energy density, 115

Bending free energy, changes in, 104—108, 110—112

Bending modulus, 109—111, 179—180

Bending moments, see also Couples, 68—69, 101—112, 114, 161

    isotropic, 109—110

Bending resistance, 68—69, 101—102, 112

    coefficient, 109—111

Biconcave disc shape, red blood cells, 4, 169—180

Bilayer

    lecithin, see Lecithin bilayer

    lipid, see Lipid bilayer

    phospholipid, see Phospholipid bilayer

Bingham material studies, 135, 138—140, 213—214

Biological membrane experiments, 141—142

Body forces, 63

Boltmann's equation, 92

Boundary conditions, 58—59, 89, 211, 220

Brownian motion, 154

Bulk aqueous phase, 87—89

Bulk isotropic malterial, 149—150

C

Cap, see Spheroidal cap

Carnot cycle, 71

Cell

    flaccid, see Flaccid cells

    nucleated, see Nucleated cells

    red blood, see Red blood cell membrane

Cell membrane

    deformation studies, 6—45

    elasticity studies, 141—220

    equilibrium studies, 46—67

    red blood, see Red blood cell membrane

    shape and properties, 4—5

    thermodynamic studies, 67—141

    thermoelasticity of, 180—190

    ultrastructure, 1—3

Centroidal axis, 62—65

Chain Rule, 10, 76, 122

Chains

    network, elastomers, 79, 94—97

    polymer, 129

Change of area, effects of, 3—4, 10, 41, 77, 97—98, 113, 115—116, 119—120

    fractional, see Fractional change of area

Change of aspect ratio, 26—27

Characteristic equations, 29

Chemical equilibrium, change of, 112, 117

Chemically induced curvature 112—117

Chemically, induced moments, 112—117
Chemical potential, water in interface, 87—88
Circular disc, 157—160
Closed membrane system studies, 68—69, 80, 85, 90
Compressibility, see Area compressibility; Small area compressibility
Compression
    along coordinate axis, 13—14
    between two flat surfaces, 143—145, 148—155
    force, vs. distance between plates, 152—153
Compression modulus
    isothermal, 119
    lecithin bilayer, 119
Conceptual isotherm, internal equation of state, 86—87
Condensation, isotropic, see Isotropic dilation and condensation
Cone
    conical annulus, see Conical annulus
    truncated, 59—60
Configurational entropy, 94—95
Conical annulus, 11, 17—18, 39—41
Conservation of energy, see also First law of thermodynamics, 67, 70—71
Conservative forces, 67—68
Constant area, effects of, 37—39, 44, 97—101, 118, 162—164, 232
Constant surface density, 41—45
Constitutive equations and relations, 67—70, 134, 138—139, 203—204, 219—220
    elastic, see Elastic constitutive relations
    hyperelastic, 99
    inelastic, see Inelastic constitutive equations
    isothermal, see Isothermal constitutive equations
Contact, area of, membrane and plates, 154—155
Continuity conditions and equation, 40—41, 43, 45, 232—233
Continuous plastic flow, 212—214
Continuum, cell structure, see Two-dimensional continuum
Coordinate axes
    curvilinear, see Curvilinear coordinates
    cylindrical, 52
    deformation, 34, 36, 38
    differential, see Differential coordinates
    equilibrium studies, 52—53
    extension and compression along, 13—15
    infinitesimal cartesian, 8—9
    initial, see Initial coordinates
    instantaneous, see Instantaneous coordinates
    isothermal constitutive equation studies, 79—81, 83—84
    magnitude of deformation, 18—19
    rotation of, 47—48
    spatial, 8—9
    tangent plane, 8—9
Coupled layers, see also Uncoupled layers, 48, 101—113
Couples, see also Bending moments; Force couples, 46—48, 50, 62, 68—69

Creep studies, 70, 135—139, 141, 210-212, 219—220
Cross derivatives, 122—123, 125
Curvature
    changes of, see also Bending, 8, 101—112
    differential, 106
    chemically induced, see Chemically induced curvature
    continuous, 46
    elastic, see Elastic curvature
    radii of, 52, 57, 68, 101—103, 106, 112
        principal, 52—53, 57
    spontaneous, see Spontaneous curvature
Curvature elastic modulus, see Elastic curvature modulus
Curvilinear coordinates, 52, 56—57, 65—66
Cyclic integral, 71—72
Cyclic process, 71
Cylinders, 58—59, 110
Cylindrical coordinates, 52
Cytoplasm, viscoelasticity studies, 207—208
Cytoskeleton (spectrin network),

D

Deflection of surface by rigid spherical particle, 143—145, 155
Deformation, 6—45, 67—85, 122—124
    bending, see Bending
    characteristics, 8—18
    chemically induced curvature and, 112—117
    elastomers, 92—93, 95, 97
Eularian strain components, see Eularian strain components
    extensional, see Extensional deformation
    finite, 78—79
    flat surfaces, 43—45
    heat of, 118, 123—124
    incremental, 74—75
    independent variables, 24—30
    intensive, see Intensive deformation
    invariants of strain, see Invariants of strain
    irrecoverable, see Irrecoverable flow and deformation of materials
    Lagrangian strain components, see Lagrangian strain components
    minimum area, 225, 227—234
    neutral surfaces, see Neutral surface deformation
    nonuniform, 41—45
    path-independent process, 74
    permanent, 69—71
    plastic, 69—70
    principal axes, see Principal axes
    rate of, see Rate of deformation
    recovery, 134
    red blood cell membrane, 3—6
    residual, 211—212
    reversible, 80
    shear, see Shear deformation
    small area, see Small area deformation

square elements, 23—25, 26
strain and, see Strain
symbols, 221—222
thermoelastic, see Thermoelasticity
time-dependent, 5—6
uniform, 15
variables, see Variables, deformation
viscoelastic, see Viscoelasticity
viscoplastic, see Viscoplasticity
work of, 193—195
Dehydration of lecithin, 191—192
Density, 50
   entropy, 73, 123, 129
   free energy, see Free energy density
   phospholipid bilayer, effect of pressure on,
      90—91
   surface, constant, see Constant surface density
Density flow, surface, 40—41
Density functions, 73
   probability, 92—94
Deviators and deviatoric parts, 22—24, 29—30,
      35, 49—50, 77, 84, 135, 143, 215—216
Dextran solution experiment, 191—192
Diagonal matrix, principal axis, 21—23
Differential
   elastic free energy, 112—113
   exact, 122, 131
Differential coordinates, 9—10, 17—24, 54, 56
   change in, 18—24
   initial state, 17
   instantaneous state, 17
Differential element, small, 53—55
Differential equations
   equilibrium, 47, 61—62
   isotropic tension, 196—197
   partial derivatives, 233—234
   state, thermoelastic studies, 118—123, 182,
      186—187
Differential increments, 56—57
Differential operator, 56
Differential work, 72—74
Diffusion constant, 203
Dilation, 123
   elastic area, see Elastic area, dilation
   isotropic, see Isotropic dilation and
     condensation
Dimensionless force vs. particle displacement,
     155—156
Dimensionless tether growth rate, 271—218
Disc, studies, 165—166, 205—207
   annular, 60—61
   biconcave, 4, 169—180
   circular, 157—160
   flat, 111—112
Discocyte, shape of, 4, 170—172
Discocyte-echinocyte-stomatocyte shape change,
     117
Discontinuity in membrane structure, 67—68
Displacement
   annular ring, 162—164
   cell projection into pipet, 147
   particle, 155—156

thermodynamics and, 67
Dissipation, 69—72, 129—142, 203, 207—208,
     213—217, 219
Distances
   between plates, 152—153
   end-to-end, 92—95
   incremental, 17
   layers separated by, 114—115
   mean square, 92—95
   point-to-point, 8—11
   water gap, between layers, 191—193, 199—202
Distribution functions
   Gaussian, 93—94
   network, 93
   probability, Gaussian, 93—94
Drops, studies, 152—153, 155—156
Dummy indices, 8
Dynamic euqations, deformation recovery, 134

E

Edges, 46—47, 49, 56, 64
Eigen value equations, 29
Elastic area, dilation, 142—147
   resistance to, 97—98
Elastic area compressibility modulus, 148,
     155—157, 190—193
   lecithin bilayer, 91
   phospholipid bilayer, 97
Elastic coefficients, 69
   red blood cell membrane, 5
Elastic compressibility modulus, 80, 124
Elastic constitutive relations, 4
   bending moment studies, 106—107
   elasticity studies, 149—150, 154—155,
     171—173
   hyperelastic membrane studies, 99—101
   isothermal, 118
   thermodynamics and, 68—70, 74, 80—81,
     118—119, 121
   viscosity studies, 129—130
Elastic curvature
   changes in, 169—180
   constitutive relations, 108—109
   effects, 68—69, 178
Elastic curvature energy, 169—180
Elastic curvature modulus, 109—111, 115
   radii, 116
Elastic energy, 78, 117
   changes in, 69—70
   shear, see Shear elastic energy
Elastic energy density, 106, 108
Elastic extensional deformation, 157—169
Elastic extension ratio, maximum, 218—219
Elastic force resultant matrix, 133—134
Elastic forces, 68
Elastic free energy
   changes in, 104
   differential, 112—113
   functional, 228—231

Elastic free energy density, 78, 84—85, 173
Elasticity
  Gibbs, 80
  hyperelasticity, see Hyperelasticity
  red blood cell membrane, 3—5
  shear, see Shear elasticity
  studies, 141—220
  theory, linear and nonlinear, 79—80
  thermodynamics of, 68—71
  thermoelasticity, see Thermoelasticity
  viscoelasticity, see Viscoelasticity
  viscoplasticity, seeViscoplasticity
Elastic modulus
  area, 140
  curvature, see Elastic curvature modulus
  red blood cell membrane, 3—4
  shear, see Elastic share modulus
Young's, 3
Elastic potential, see also Helmholtz free energy
          density, 74—77, 81—82
  changes in, 75—77
Elastic potential energy density, 74, 76—77,
          79—80, 92
Elastic potential energy function, 72—73
Elastic shear modulus
  elasticity studies, 157—160, 166—168,
          179—180
  elastomer studies, 91—92, 96—97
  micropipet aspiration studies, 166—168,
          236—237
  thermoelasticity studies, 121—122, 187—188
  viscosity studies, 140—141
Elastomers
  end-to-end distance, 92—95
  entropic quality of, 126, 129
  hyperelasticity of, 79, 91—97
  networks, 79, 92—97, 126—129
  shear hyperelasticity of, 91—97
  structure, 92
  thermoelasticity and, 187
Elastomer shear modulus, 79, 96—97, 126, 129,
          167—170
  nucleated cells, 167—170
Elongation, square element, 99
Encapsulating membranes, 133—134
End-to-end distance, elastomers, 92—95
Energy
  conservation, see Conservation of energy
  elastic, see Elastic energy
  internal, see Internal energy
  minimum deformation, 225, 234—237
  strain and, 74
  thermal, transfer of, 71—72
Ensemble average, networks, elastomer, 96
Entropic elastomers, see Elastomers, entropic
          quality of
Entropy, see also Second law of thermodynamics
  area specific function, 73
  configurational, 94—95
  density, 73, 123, 129
  elastomer studies, 91—97

thermodynamics of, 67, 72, 180—181
thermoelasticity and, 117—118, 122—123, 125
viscosity studies, 130—143
Equations of state
  internal, see Internal equations of state
  mechanochemical, see Mechanochemical
          equations of state
  surface, see Surface equations of state
  thermoelastic, 118—129
Equilibrium
  chemical, changes in, 112, 117
  curvature, 112—117
  mechanical, 46, 50
    axysymmetric membrane, 65—67
    axysymmetric surface layer, 52—62
    differential equations, 47, 61—62
    equations, 50, 143—147, 162
    flat membrane, 62—65
    force-free, 85
    force resultants, see Force resultants
    plane surface layer, 50—52
    red blood cell membrane, 3
    virtual work of traction and, 226—227
  static, defined, 50
  thermodynamic, 50, 67—75, 81
Eularian strain components
  deformation studies, 11—13, 16, 21, 25, 28,
          30—31, 33—34, 37—39, 43
  equalibrium studies, 46
  force resultant studies, 48
  isothermal constitutive equation studies, 78, 81
  maximum Eularian shear strain resultant, 78
  thermodynamics and, 70
  viscosity studies, 131, 139—140, 211, 216,
          219—220
Exact differential, 126, 131
Expansion
  heat of, see Heat of expansion; Reversible heat
          of expansion
  radial, 15—17
Expansivity, thermal area, see Thermal area
          expansivitiy
Extensibility, thermal, 121
Extension
  area, resistance to, hyperelastic membrane,
          97—99
  circular disc, 157—160
  coordinate axis, 13—15
  detformation and, 10, 13—15
  elastic, see Elastic extension
  elastomic studies, 92, 96—97
  heat of, see Heat of extension
  isothermal constitutive equation studies,
          76—78, 81
  square elements, 11—13
  thermoelasticity studies, 123,
          188—189
  uniaxial, red blood cell, 157—160
  viscoelastic recovery and response to, 203—210
Extensional deformation, 74, 77—78, 95—96, 157
  elastic, 157—169

recovery, 134
Extensional recovery, red blood cell membrane, 204—208
Extension ratios
  bending moment studies, 102—103, 105—107
  deformation and, 16—17, 25—27, 36, 42—45
    scaled, 27, 36
  free changes in, 129
  hyperelastic membrane studies, 98
  isothermal constitutive equation studies, 78—79
  plastic, 138
  viscosity studies, 137—138
Extensive variables, 70
External forces, 46—47, 50, 57, 68, 71
Extracellular fluid, Brownian motion in, 154

F

Failure, 212—214, 218
Finger's strain tensor, 83—84
Finite deformation, 78—79
First law of thermodynamics, see also
    Conservation of energy; Second law of
    thermodynamics, 67, 70—73, 75, 118, 123,
    130—131
Fixed mass, 80
Flaccid cells, red blood
  bending vs. shear rigidity in, 169—180
  elasticity studies, 157, 159
  extensional deformation of, 157
  micropipet aspiration experiments, 112,
    165—166, 225, 234—237
  stretching of, 4
Flat annular disc, 60—61
Flat membrane equilibrium, 62—65
Flat surfaces, compression between, 143—145,
  148—155
Flicker, red blood cell membrane, 111
Flow
  irrecoverable, membrane, 212—214
  liquid, membrane, 70
  plastic, see Plastic flow
  surface, incompressible, 43, 45
  viscoplastic, relaxation and, 210—220
Flow channel experiment, 157—160
Fluid mosaic, cell membrane as, 1, 7
Force couples, see also Couples, 67—68
Force equilibrium equations, 63—64
Force-free equilibrium, 85
Force moments, intrinsic, 47
Force resultant couples, 62
Force resultant matrix, 50, 133, 135
  elastic, 133—134
Force resultants, 4—5
  bending moment studies, 101—106, 108—109
  chemically induced curvature studies, 115
  equilibrium and, 46—53, 61—62
  hyperelastic membrane studies, 98—101
  intensive, see Intensive force resultants
  intrinsic, 47

  isothermal constitutive equation studies, 74,
    77—78, 80—85
  isotropic, 98, 114, 215—216
  micropipet experiments, 161
  principal, see Principal force resultants
  shear, see Shear force resultants
  temperature, dependent, 118
  thermoelasticity studies, 118, 193—195
  viscosity studies, 132—133
Forces
  balance, see Balance of forces; Equilibrium
  body, 63
  compression, 152—153
  conservative, 67—68
  deformation and, 6—45
  elastic, 68
  elasticity and, 141—220
  equilibrium and, 46—67
  external, see External forces
  principal, 75
  shear, deformation and, 157—169
  symbols, 222—223
  thermodynamics and, 67—141
Fractional change of area
  deformation and, 26—28, 36, 74
  hyperelastic membrane studies, 98—99
  isothermal constitutive equation studies, 76,
    79—80
  isotropic tension and, 148
  pressure- and tension-free state, 91
  thermoelasticity studies, 120, 181—183
Free-body layer, diagram, 47
Free energy, 124—125, 195
  bending, see Bending free energy
  changes in, 69, 77, 84—85, 113, 115—116, 227
Free energy density, 69
  bending moment studies, 104—105, 107—108,
    111—112
  chemically-induced curvature studies, 113, 115,
    117
  elastic, 78, 84—85, 173
  elastomer studies, 92, 96
  Helmholtz, see Helmholtz, free energy density
  hyperelastic membrane studies, 97—98
  induced, 116
  interfacial, see Interfacial free energy density
  isothermal constitutive equation studies, 77,
    79—80, 82
  red cell membrane, 225
  thermoelasticity studies, 129
  unitary, of transfer, 88—90

G

Gaussian probability density functions, 93—94
Geometry
  axisymmetric, 225—237
  symbols, 221-222
Gibb's principles, 80, 195
Green's strain tensor, 10

H

Heat, see also Temperature, 67
  changes in, 117—118
  dissipation, 71
  generation, internal, 69
  loss, reversible, 131
  specific, 125—126, 128—129
Heat exchange, 67, 70—72, 131—132
  incremental, 70—72
  reversible, 118, 125—126
Heat of deformation, 118, 123—124
Heat of expansion, 124—125, 185—186
  Kelvin's, 125
  natural, 124—125
  reversible, see Reversible heat of expansion
heat of extension, 96—97, 126, 187—190
Helmholtz free energy density, see also Elastic
     potential, 73—75, 79—82, 86, 92, 118,
     171—173
Hemisphere, 111—112, 116
Hole of radius, 41—43
Hydrocarbon
  model, phospholipid bilayers, 90, 140
  water interface with, 127—128, 195—197
Hydrophobic effect, 89, 181, 190
Hydrophobic interaction, 69, 89, 127—129
  interfacial free energy density of, 86, 90
Hyperelastic constitutive relations, 99
Hyperelasticity, 79, 91—97, 166, 168, 186—187
  shear, 91—97
Hyperelastic membrane materials, 97—101, 122,
     136—137
  small area compressibility, 97—101
Hysteresis, 131

I

Identity matrix, 10
Immiscible drops, 152—153
Incompressibility of surface, 36—37, 44
Incompressible flow of surface, 43, 45
Incremental deformation, 74—75
Incremental distances along meridian, 17
Incremental heat exchange, 70—72
Incremental work, 70—71
Increments, differential, 56—57
Independent deformation variables, 24—30
Independent time derivatives of strain, see also
     Time-dependent deformation, 82
Induced free energy density, 116
Induced moment coefficient, 114—117
Induced moments, 117
Inelastic constitutive equations, 69
Infinite flat surface, 41—42
Infinitesimal cartesian coordinates, 8—9
Infinitesimal displacement field, 31—33
Infinitisimal increments, first law of
     thermodynamics, 70

Infinitesimal scale, 47
Infinitesimal shear components, 13
Infinitesimal theory, 78—79
Initial coordinates, 9—10, 17, 30—31, 101
Initial matrix, 9—10
Initial state or area, 8—9, 11, 70—71, 74, 77—78,
     80—84, 95
Instantaneous axes, 46, 49
Instantaneous coordinates, 9—10, 17, 30—31, 33,
     36—39, 43, 48, 79—81, 83—84
Instantaneous matrix, 9—10
Instantaneous mechanical power density, 81—82
Instantaneous state or area, 9, 11, 42, 71, 74, 77,
     79—84, 95
Integral relations and equations, 61, 211, 220
Integrand, 230
Intensive deformation, 6—8, 68—69, 75—76
Intensive force resultants, see also Principal
     membrane tensions, 57, 75—77, 81
Intensive variables, 70
Interactions
  hydrophobic, see Hydrophobic interactions
  interface, 85—91, 119—121, 125—129,
     195—197
Enterface
  chemically induced curvature, 112
  closed, 80
  hydrocarbon and water, interaction, 127—128,
     195—197
  interactions at, see Interactions, interface
  liquid, 80
  surface tension, 80, 85—86
Interfacial free energy density, 87—90, 112,
     184—185, 195—199
Intermediate state, 70—71
Internal dissipation, 129—141
Internal energy, 67, 70, 117—118, 122—124, 126,
     180—181
  area specific function, 73
  density function, 73
Internal equations of state
  conceptual isotherm of, 86—87
  mechanochemical equation of state and, 90
  thermoelastic, 111, 121, 126—129
Internal heat generation, 69
Intrinsic force moments, 47
Intrinsic force resultants, 47
Invariants of strain, 19, 24—30, 34—37, 83—84
Irrecoverable flow and deformation of material,
     see also Permanent deformation, 6, 130,
     212—214
Irrecoverable heat loss, 130—131
Irreversible processes, 67, 68—70, 130—143
Isotherm, conceptual, 86—87
Isothermal area compressibility modules, 80—81,
     127, 181—183
Isothermal constitutive relations and equations,
     74—85, 118
Isothermal medulus of compression, 119
Isothermal process and response, 71, 119
Isothermal work, 74—75, 77—78, 81—82

Isotropic bending moments, 109—110
Isotropic dilation and condensation, 74, 79—80
Isotropic equations, 119, 124
Isotropic force resultants, 98, 114, 215—216
Isotropic matrix, 22—23
Isotropic moments, 114
Isotropic tension
    elastic area dilation produced by, 142—157
    force resultant studies, 49
    fractional change of area and, 148
    hyperelastic membrane studies, 98—99, 101
    isothermal constitutive equation studies, 77,
        79—80, 84
    pressure- and tension-free state, 86, 90
    thermoelastic studies, 118—120, 126, 181—184,
        191, 195—197
    viscosity studies, 134
Isotropic tension resultants, 79—80, 133
Isotropic thermoelastic coefficients, 119—120
isotrophy
    local, 28
    surface, defined,2

K

Kelvin's principles, 125, 133—134
Kronecker delta, 10

L

Lagrangian strain components
    deformation studies, 11—13, 16, 20, 25—26,
        29—30, 34, 37, 41—43
    isothermal constitutive equation studies,
        77—78, 81—84
    shear strain, 77—78
    strain matrix, 81—84
    thermodynamics and, 70
    viscosity studies, 135—136, 139, 212, 219—220
Langmuir trough technique, 127
Layers
    amphiphilic, see Amphiphilic layer system
        studies
    coupled, 48, 101—113
    distance between, 114—115
    free-body, diagram, 47
    lecithin, see Lecithin bilayers
    lipid, see Lipid bilayers
    molecular, coupled, 48, 101—113
    phospholipid, see Phospholipid bilayers
    single, 74
    surface, see Surface layer
    trilamellar, 111—112, 116—117
    uncoupled, 115—116
Lecithin bilayers
    bending moments, 110
    compression modulus, 119
    dehydration of, 191—192

elastic area compressibility modulus, 91
    thermoelasticity studies, 119, 128, 184—185,
        191—192, 196—203
Linear elastic theory, 78—79
lipid bilayers
    bending moments, 110—111
    chemically induced curvature studies, 116—117
    elasticity studies, 157, 159
    hyperelastic membrane studies, 99
    surface viscosity studies, 208
    thermoelasticity studies, 128, 181—203
    time-derivative behavior, 140—141
Lipid hydration, work of, 194
Lipid phases, multilamellar, area compressibility
    of, 190—203
Lipid weight fraction, 191—192
Liquid drops, 152—153, 155—156
Liquid state, effect of, 60, 70, 79—80, 134—135,
    138—139
Local bending energy, 230
Local isotropy, defined, 28
Local spatial axes, 53—54

M

Magnetic particle experiment, 6
Marker particles, small, Brownian motion of, 154
Material edges, 46—47, 49, 56, 64
Material points, 30—31, 40
Material properties, membrane, see also specific
    properties by name, 67—72, 84—85
    symbols, 223—224
Material strain, 10
Maximum elastic extension ratio, 218—219
Maximum shear force resultants, 61
Maximum shear resultants, 77—80, 84, 134
    Eularian, 78
Maxwell equation, 135—137
Mean deformation, bending moment, 102
Mean square distance, 92—95
Mechanical equilibrium, see Equilibrium,
    mechanical
Mechanical power, 131—132
Mechanical power density, 81—82, 131
Mechanical power functional, 171
Mechanochemical equations of state, 69, 80—81,
    90, 118—119, 121—122, 126—129
    internal equations of state and, 90
    isothermal constitutive equations and,
        80—81
Membranes
    axisymmetric, see Azisymmetric membrane
        surface
    biological experiments with, 141—142
    cell, see Cell membrane
    encapsulating, 133—134
    flat, equilibrium, 62—65
    hyperelastic, see Hyperelastic membrane
        materials
    layers, see Layers

liquid, see Liquid state
semisolid, see Semisolid state
solid, see Solid state
systems, see Closed membrane system; Open
    membrane system
thin, see Thin membranes
Meridional tensions, 52—62
Metric, 8—11
    initial and instantaneous, 9—10
Micropipet aspiration studies, red blood cell
    membrane, 110, 141, 143—148, 157—160,
    167—170, 176, 183—184, 187—188,
    208—220, 226—227
    membrane region local to pipet, 162, 211
    suction force and pressure, see Suction force
    and pressure
    sudden increase of pressure in pipet, 208—210
Microtether, 215
Minimum energy, deformation, red cell
    membrane, 225, 227—234
Molecular fractions, 69
Molecular layers
    coupled, 48, 101—113
    uncoupled, 115—116
Moment resultants
    axisymmetric membrane equilibrium, 65—67
    bending, 106, 108
    chemically-induced curvature studies, 114—115
    equilibrium studies, 46, 50—67
    flat membrane equilibrium, 62—65
    force resultants and, 48
    principal, 62—67, 106
Moments
    bending, see Bending moments
    chemically-induced, 112—117
    energy produced by, 67—68
    force, intrinsic, 47
    induced, and coefficient, 114—117
    isotropic, 114
    symbols, 222—223
Mosaic nature, membrane surface, 1, 7
Motion, equations of, Newton's, 50
Multilamelar lipid phases
    area compressibility of, 190—203
    thermoelasticity of, 190—203

N

Necking region, 215
Network distribution function, 93
Networks
    chemically-induced curvature studies, 116—117
    elastomes, 79, 92—97, 126—129
        chains in, 79, 94—97
        end-to-end distance effects, 92—95
        thermoelastic studies, 176—179
    spectrin, 1—2, 97
Neutral surface deformation, 102—109, 113—114
Newton-Raphson technique, 234
Newton's equations of motion, 50

Nonlinear elastic theory, 78—79
Nonlocal bending energy, 230
Nonmammalian red cells, micropipet studies,
    167—170
Nonuniform deformation, see also Uniform
    deformation, 41—45
Nonzero strain, 12—13
Notation symbols, 224
Nucleated cells, mickopipet studies, 167—170

O

Onsager equation, 132
Open membrane system studies, 85
Orthoganal axes, 8
Osmotic stress, 87
Osmotic swelling, 6
    flaccid cells, 169—180
    preswollen red blood cells, 142—143, 146
Outline, thermodynamic, 70—72

P

Parabaloid envelope, infinite, 43—45
Partial derivatives, 10, 83, 119—121, 123, 126,
    193—194, 230—231, 234
Particle, marker, small Brownian motion of, 154
Particle displacement, dimensionless force vs.,
    155—156
Path-independent deformation process, 74
Perfect plastic material, see Bingham material
Peripheral protein, see Spectrin
Permanent deformation, see also Irrecoverable
    flow and deformation of material, 69—71
Phases, lipid see Multilamellar lipid phases
Phospholipid bilayers
    area compressibility modulus, 90—91, 119
    chemicaly induced curvature studies, 117
    elastic area compressibility modulus, 97
    equation of state for, 90—91
    hydrocarbon model, 90
    thermoelasticity studies, 183, 185—186
Phospholipid vesicle, see Vesicle
Pipet suction force and pressure, see Suction
    force and pressure
Planar bilayer films, 89—90
Plane surface layer equilibrium, 50—52
Plastic deformation, 69—70
Plastic extension ratio, 138
Plastic flow, 130, 141, 210—220
    relaxation and, 210—220
Plastic material, perfect, see Bingham material
Plastic state, see Liquid state
Plates, 152—154
Point-to-point distance, 8—11
Polar head groups, 89—90
Pressure
    pipet

suction, see Suctions force and presure
sudden increase in, 208—210
surface, see Surface pressure
vapor, see Vapor pressure
Primid system, strain matrix in, 20
Principal axes systm
deformation studies, 10, 18—25, 28—30, 34—36, 39
equilibrium studies, 50—52, 64
force resultant studies, 48—49
hyperelastic membrane studies, 98—100
isothermal constitutive equations, 74—85
values and invariants, symmetrical matrix, 28—30
Principal components, 100
Principal curvature changes, 232
Principal extension ratios
bending moment studies, 102—103
deformation studies, 16—17, 25—27, 26—37, 39, 44
differential changes, 75—76
elasticity studies, 159, 163—164
isothermal constitutive equation studies, 75—76, 82
micropipet experiments, 163—164
Principal force resultants, 65, 103—107, 161—162, 173
Principal forces, 75
Principal moment resultants, 62—67, 106
Principal radii of curvature, 52—53, 57
Principal Rate of deformation, 39
Principal tensions, see also Intensive force resultants, 48, 60—61, 75—77, 84—85, 98—99
Probability density factor, 92—94
Probability density functions, Gaussian, 93—94
Process, defined, 70—71
Product summation rate, 8
Projection lengths, viscoplastic flow, 208—212, 219—220
Properties
cell membrane, chemical effects on, 4—5
material, see Material properties
Protein, peripheral, see Spectrin

R

Radial expansion, axially symmetric surface, 15—17
Radial tension, 60—61
Radial velocity, 39, 43
Radius
curvature, see Curvature, radii; Elastic curvature modulus, rodii
hole of, 41—43
tether, 216—217
Rate of deformation, 6—8, 30—45, 68—69, 81—82, 132, 135—136, 220
intensive, 6—8
matrix, 135

nonuniform, 41—45
principal, 39
shear, 35—37, 41, 43, 45
total, 135
Recovery
deformation, dynamic equation for, 134
viscoelastic, 203—210
Red blood cell membrane
area compressibility modulus, 119
bending 3—5
biconcave disc shape, 4
coupling, 111
curvature, changes in, 111—112
deformation of, 3—6, 225
discocyte, shape of, 4
elastic properties, 3—5, 155—157, 166—169
equilibrium, 3
flaccid, see Flaccid cells, red blood
flicker, 111
hyperelasticity of, 92, 97—98
mechanics of, 2—3
micropipet experiment, see Micropipet aspiration studies, red blood cell membrane
minimum energy deformation of, 225, 227—234
nonmammalian, 167—170
shape, see Shape, membrane, red blood cell
shear deformation, 211—212
shear modulus, 97
sphering of, 174—178
surface shear viscosity, 5
structure, 1—4
thermoelasticity, 5, 181—190
time derivative behavior, 140—141
uniaxial extension, 157—160
viscoelasticity, 5—6, 203—210
viscoplasticity, 210—220
Relative velocity, deformation, 31—33, 38—40
Relaxation studies, 70, 129—141, 210—220
equations, 219
Residual deformation, 211—212
Responses, viscoelastic, 203—210
Resultants
force, see Force resultants
moment, see Moment resultants
symbols, 222—223
Reversibility, process
eformation, 50, 67—72, 84
thermodynamic activities, 67, 72, 74, 180—181
thermoelastic activities, 117—118, 123—124
Reversible heat exchange, 118, 125—126
Reversible heat of expansion, see also Thermoelasticity, 181—182, 202—203
Rheology, blood, 5—6
Rigidity, see also Stiffness, 112
shear, see Shear rigidity
Rigid rotation, 32—33
Rigid spherical particle, deflection by, 143—145, 155
Rolation angle, 19—24, 48—49

Rotation angle coordinate, 49
Rubber cells, 174—175, 177
Ruler, 8

S

Scalar distance, 8
Scaled extension ratios, 27, 36, 77
Scale on surface, minimum, 67—68
Sea urchin egg studies, 1, 6, 97, 142—145,
    148—155, 157
Second law of thermodynamics, see also Entropy;
    First law of thermodynamics, 67, 70—73,
    75, 96, 123
Semisolid state studies, 134—135, 138—139, 141,
    210
Shape, membrane
    red blood cell, 4—5, 117, 176—180, 225,
        232—234
        unstressed, 170—172, 174—180
    sea urchin egg, 150—151, 157
Shear
    elastic extensional deformation produced by,
        157—169
    infinitesimal, 13
    process, 101
    resistance to, hyperelastic membrane, 97—98
Shear components, 100—101
Shear deformation
    angle, 100
    bending moment studies, 110—111
    hyperelastic membrane studies, 99—101
    red blood cell, 211—212
    shear force resultants and, 157—169
    square elements, 13—15, 37—39, 100-101
    twist of axisymmetric surface, 17—18
Shear deformation rate, 35—37, 41, 43, 45
Shear elastic energy, 116
Shear elasticity, 114, 116, 178
Shear force differences, 63
Shear force resultants, 61, 157—169, 186—187
    maximum, 61
    shear deformation and, 157—169
Shear forces, 157—169
Shear hyperelasticity, 91—97
Shear modulus
    bending moment studies, 108
    elastic, see Elastic shear modulus
    elastomer studies, 79, 96—97, 126, 129
    isothermal constitutive equation studies, 78—79
    red blood cell membrane, 97
    thermoelasticity studies, 129, 187—188
Shear resultant matrix, 49
Shear resultants
    equilibrium studies, 52, 58
Strain tensors, see Tensors
Stress
    osmotic, 87
    shear, see Shear stress
    symbols, 222—223

tractions, see Tractions
Stress-strain relations, see Isothermal constitutive
    equations
Structure
    changes in, see Deformation
    equilibrium and, 46
    red blood cell membranes, 1—4
    ultra-, cell membrane, 1—3
Sucrose solution experiments, 197—199, 202
Suction force and pressure, micropipet
    experiments, 143—147, 161—163,
    165—166, 183, 226—227, 236—237
Sudden pressure increase, projection response to,
    208—210
Summation convention, 8
Sum of moments about axis through centroid,
    63—64
Surface density, constant, 41—45
Surface density flow, 40—41
Surface equation of state
    phospholipid bilayer, 90—91
    thermoelastic studies, 181—183, 186
Surface isotropy, 2
Surface layer
    axisymmetric, see Axisymmetric surface layer
    mosaic nature of, 1, 7
    plane, 50—52
Surface pressure
    chemically-induced curvature studies, 112—113
    equilibrium studies, 50—51, 57, 59—61
    hyperelastic membrane studies, 99
    tension-free state and, 85—91
    Eularian, 78
    force resultant stidies, 46—49
    isothermal constitutive equation studies,
        77—80, 84
    maximum, 77—80, 84, 134
    thermoelasticity studies, 118—119, 121, 126
    transverse, 46—47, 65—66
    viscosity studies, 133—134, 136—141
Shear ridigity, 78—80, 117, 169—180
Shear strain components
    deformation and, 23, 25
    Eularian, see also Eularian strain components,
        78
    Lagrangian, see also Lagrangian strain
        components, 77—78
    square of, 78—79
Shear stress, 50—51, 54, 58—59
Shear viscosity, surface, red blood cell
    membrane, 5
Shell theory, 3, 5
Single layers, 68,,74
Small area compressibility, 97—101, 109, 134
Small area deformation, 73, 75, 78—80, 142
Small differential element, 53—55
Solid state studies, 70, 79—81, 126, 134—135,
    138—139
Spatial axes, 53—54
Spantial coordinates, 8—9
Spatial distribution of velocity, 32

Spatial strain tensor, 10
Specific heat, 125—126, 128—129
Spectrin, studies of, 117, 181, 183, 186, 188—190, 210
    network, 1—2
        hyperelasticity of, 97
Sphere, 58
Spherical particle, rigid, deflection by, 143—145, 155
Spherical shape, red blood cell membrane, 176—180
Sphering, of red blood cell, 174—178
Spheroid, biconcave disc swelling to, 169—180
Spheroidal cap, 164—165
Spontaneous curvature, 5
Square elements, deformation of, 13—15, 23—25, 36—39, 99—101
Square of shear strain components, 78—69
State
    equations, see Equations, of state
    functions, thermoelastic studies, 117—129
Static equilibrium, defined, 50
Stiffness, see also Rigidity, 3
Strain
    deformation and, 10—24, 33—34
    Eularian, see Eularian strain components
    invariants of, see Invariants of strain
    isothermal constitutive equation studies, 83—84
    Lugrangian strain components, see Lagrangian stress components
    nonzero, 12—13
    shear, see Shear strain components
    tensor, 10, 82—84, 134
Strain energy density, 171
Strain-energy function, 74
Strain matrix, 10
    thermoelastic studies, 119—121, 127, 181—185, 196—197, 200—202
    tractions, see Tractions
Surface tension, see also Tension, 80, 127—128
Surface viscosity, see also Viscosity, 133
Symbols
    deformation, 221—222
    forces, 222—223
    geometry, 221—222
    material properties, membrane, 229—224
    moments, 222—223
    notation, 224
    resultants, 222—223
    stresses, 222—223
    variables
        miscellaneous, 224
        thermodynamic, 223
Symmetric matrices, 48
    deformation, 32
    principal values and variants for, 28—30
    two-dimensional, 28—30
Symmetry, axis of, 40—41, 44

T

Tangent plane coordinates, 8—9

Taylor series, 79—80, 147, 227
Temperature, see also Heat
    elastomer studies, 79, 96
    equilibrium thermodynamics-deformation relations, 72—74
    isothermal constitutive equation studies, 79—81
    pressure- and tension-free state, 85—86, 90
    thermodynamics and, 67, 69, 72
    thermoelastic studies, 118—124, 127—129, 187—188, 199—200
Tension
    bending moment studies, 105
    boundary, 89
    distribution of, 155—156
    equilibrium studies, 54, 56, 59—60, 65
    force resultant studies, see also Force resultants, 48
    freedom from, see Tension-free state
    interfacial, 80, 85—86
    isotropic, see Isotropic tension
    meridional, 52—62
    principal, see Principal tensions
    radial, 60—61
    surface, see Surface tension
    thermoelastic studies, 120, 128
    uniaxial, 99
Tension-free state, 85—91, 97—98, 124
Tension resultants, see also Force resultants, 46, 49, 57
Tensors, 10, 82—84, 134
Tethers, 212—213, 215
    diameters, 218
    growth rate, 217
        dimensionless, 217—218
    radii, 216—217
Thermal area expansivity, 120, 124—125, 127—128, 183, 185, 190—191, 199—200
Thermal energy, transfer of, 71—72
Thermal extensibility, 121
Thermodynamic relations, 67—141
    constitutive, 67—70
    equilibrium, 50, 67—75, 81
    laws of, see First law of thermodynamics; Second law of thermodynamics
    outline, 70—72
    studies, 67—141
    variables, symbols, 223
Thermoelasticity, 69, 73, 117—129
    cell membranes, 5, 180—190
    coefficients, isotropic, 119—120
    multilamellar lipid phase, 190—203
    pressure- and tension-free state, 90
    red blood cell membrane, 5
Thickness, membrane and bilayer, 3, 67—68, 140—141, 191, 197—199, 201—202
Thin membranes, 46, 74, 81, 103—104, 108
Thin shell theory, 110, 148—149
Three-dimensionally isotropic (bulk) material, 149—150
Time-dependent deformation see also
        Independent time derivatives of strain, 5—6, 69—70

Time-dependent length of membrane projection, 211—212
Time-dependent recovery equation, 204—205
Time-dependent response of membrane projection, 208—210
Time-derivative, material behavior, 135—137, 139—141
Time-independent derivative of strain, 82
Time rate of change, 30—31, 33—34, 36, 69, 81—82
Tractions, 50—52, 54, 56—58, 63, 65
    virtual work, 226—227
Transverse shear resultants, 46—47, 65—66
Trapezoidal shape, 54—56, 65
Trilamellar lyaer, 111—112, 116—117
Trumcated cone, 59—60
Twist of axisymmetric surface, 17—18
Two-dimensional continuity equation, 37
Two-dimensional continuum cell at ructure, 2—3, 9, 67—68
Two-dimensional symmetric matrices, 28—30

U

Ultrastructure, cell membrane, 1—3
Uncoupled layers, see also Coupled layers, 115—116
Uniaxial extension, red cell, 157—160
Uniaxial tension, 99, 157
Uniform deformation, see also Nonuniform deformation, 15
Unitary free energy density of transfer
    hydrocarbon to aqueous solution, 90
    water to interface, 88—89
Unit vectors, principal axis, 29
Unprimed system, strain matrix in, 20
Unstressed shape of cell, 170—172, 174—180

V

Vander Waal's equations, 90—91, 181—183
Vapor pressure, 191, 200—201
Variables
    deformation
        independent, 24—30
        isothermal constitutive equation studies, 74
    extensive, 70
    intensive, 70
    symbols, 223
Vector projection along principal axis, 29
Velocity
    deformation, 31—34, 37—41, 43, 45
    radial, 39, 43
    relative, 31—33, 38—40
    spatial distribution of, 32
Viscosity studies, 139—146, 211, 215—217
Vesicles, studies of, 143, 152—153, 155—156, 159, 167, 181, 183, 190
Virtual work, tractions, 226—227

Viscoelastic deformation, 70
Viscoelasticity
    cell membrane, 5—6
    red blood cell membrane, 5—6
    thermodynamics, of, 129—141
Viscoelastic recovery, 203—210
Viscoelastic response, 203—210
Viscoplasticity
    cell membrane, 5—6
    relaxation studies, 210—220
    thermodynamics of, 70, 129—141
Viscosity, 129—141
    surface, 133
        coefficient, 130—143, 213, 217—218
    surface shear, red blood cell membrane, 5
Volume conservation equations, 194

W

Wall friction, micropipet, 161
Water
    area compressibility of, 190—203
    chemical equilibrium with bulk aqueous phase, 87—89
    chemical potential for in interface, 87—88
    gap distance between bilayers, 191—193, 199—202
    hydrocarbon interfaces with, 127—128, 195—197
    lipid affinity for, 191
    thermoelasticity of, 190—203
    work, see Work, water exchange
Work
    changes in, 105
    deformation, 193—195
    differential, 72—74
    elastomer studies, 92, 96
    incremental, 70—71
    irrecoverable, 131
    irrecoverable heat-loss and, 130—131
    isothermal, 74—75, 77-78, 81—82
    lipid hydration, 194
    mechanical, 117—118, 123
    pressure- and tension-free state, 86
    virtual, tractions, 226—227
    water exchange, 191—195, 200—202

X

X-ray diffraction measurements, 191—192

Y

Yield shear, 141, 212—213, 218—219
Young's elastic modulus, 3

Z

Zero acceleration, 50